Flux Modulation Machine

Ronghai Qu · Dawei Li · Xiang Ren · Yuting Gao

Flux Modulation Machine

Theory and Topologies

Springer

Ronghai Qu
Huazhong University of Science
and Technology
Wuhan, Hubei, China

Dawei Li
Huazhong University of Science
and Technology
Wuhan, Hubei, China

Xiang Ren
Huazhong University of Science
and Technology
Wuhan, Hubei, China

Yuting Gao
Wuhan University
Wuhan, Hubei, China

ISBN 978-981-95-4745-6 ISBN 978-981-95-4746-3 (eBook)
https://doi.org/10.1007/978-981-95-4746-3

Translation from the Chinese Simplified language edition: “磁场调制电机” by Ronghai Qu et al.,

Previously published by China Science Publishing & Media Ltd.

This Springer imprint is published by the registered company Springer Nature Singapore Pte Ltd.
The registered company address is: 152 Beach Road, #21-01/04 Gateway East, Singapore 189721, Singapore

Foreword

It has been almost two centuries since the invention of the first electrical machine. While technologies from the same era, such as the steam engine, have gradually receded from the forefront, electrical machines have continually evolved. Over the past two decades, in particular, electrical machines have undergone rapid development and found widespread applications in electric vehicles, railway transportation, robotics, digital equipment, and more, establishing themselves as the preferred driving force in the era of human intelligence. Furthermore, they are anticipated to play a pivotal role in the "electrification" reform across aviation, aerospace, defense equipment, and other fields.

The demand for advancements drives scientific progress, with theory laying the foundation for technological innovation. The increasing performance requirements of electrical machines have spurred academia and industry to continuously explore the "new boundaries" of electrical machine theory and technology. In this context, a series of innovative high-quality machines has emerged, including the vernier permanent magnet machine, flux switching machine, magnetic geared machine, and others. However, these machines present challenges for study and analysis using traditional electrical machine theory, often necessitating separate analysis as special cases. Each of these machines tends to have its own isolated theoretical system, topological research, and design method, resulting in incompatible and fragmented research outcomes. This fragmentation impedes the theoretical development, technological innovation, and practical application of new high-quality machines.

Dr. Ronghai Qu, the primary author of this book, brings extensive research and development experience in electrical machines from his tenure at the internationally renowned General Electric in the USA. Subsequently, he joined Huazhong University of Science and Technology (HUST) in China, effectively bridging academia and industry. This unique position provided him with a comprehensive and pragmatic perspective on the theories and technologies of electrical machines. Dr. Ronghai Qu and his dedicated team have spent years studying high-quality electrical machines. Their efforts culminated in the identification of a unified principle, the establishment of flux modulation theory, and the subsequent naming of these high-quality machines as flux modulation machines (FMMs). This book represents the culmination of their research achievements in FMMs. It elucidates the phenomenon of the flux modulation principle from a physical concept, clarifies the differences and connections among various FMMs via the topological evolution process, compares the commonalities and characteristics of FMMs within the same theoretical framework, and ultimately proposes development directions and ideas for high-performance FMM topologies from the perspective of flux modulation theory. This not only consolidates their findings but also offers valuable predictions for the future academic frontiers in this field.

In summary, this book introduces a unified analysis method of "flux modulation" for various high-quality electrical machines. It emphasizes both the similarities and differences in the structure and performance of these machines, forming a self-contained theoretical system. Starting from basic physical concepts, the book gradually progresses to the academic frontier, providing an in-depth yet accessible understanding. I highly recommend it as a significant reference for the development of electrical machine theory, as well as for researchers and practitioners engaged in the exploration and application of new high-quality machine topologies.

Weiming Ma
Academician of the Chinese Academy
of Engineering
Beijing, China

Preface

Electrical machines, functioning as energy conversion devices between electrical and mechanical energy, have been indispensable power sources for industrial equipment for over a century. In recent years, driven by global economic development, advancements in engineering science, and improved living standards, there has been an increasing demand for enhanced performance in electrical machines across various sectors. High torque density has become a primary focus, particularly in areas such as electric vehicles, high-end computer numerical control machine tools, naval, and aerospace applications. Essentially, the torque density of electrical machines is a critical factor in determining the success or failure of high-end equipment.

Traditionally, torque density improvement relies on increasing electric and magnetic loading. However, this conventional method is constrained by current material properties and processing technology, approaching its upper limit and proving difficult to enhance further. Consequently, the pursuit of significantly improving electromechanical energy conversion capacity to develop ultra-high torque density machines has become a paramount challenge for researchers worldwide. In response to these pressing circumstances, flux modulation machines (FMMs) have emerged.

In 2005, during my tenure at the General Electric Global Research and Development Center (GE GRC), I observed the vernier machine, a type of FMM, exhibiting exceptional torque amplification capability. This discovery led to a comprehensive study into the differences between vernier machines and conventional machines. It was revealed that the stator and rotor pole numbers of the vernier machine varied, resulting in torque amplification. In collaboration with Dr. Manoj R. Shah at GE GRC and German researchers, we applied for the first patent on the vernier machine in 2006.

In 2010, I joined Huazhong University of Science and Technology (HUST) in China and established the Center for Advanced Electrical Machines and Drives (CAEMD) to systematically explore FMMs. The primary objective was to identify a third effective method to enhance the torque density of electrical machines, supplementing the conventional approaches of increasing electric loading and magnetic loading. In 2011, our paper presented at the International Conference on Electrical Machines and Systems (ICEMS) in Beijing elucidated that magnetic gears and vernier machines operate on the same flux modulation principle. This paper attracted the attention of Prof. Thomas A. Lipo, an IEEE Fellow and member of the US National Academy of Engineering, who contacted us seeking insights into the advantages and disadvantages of vernier machines. Following several email exchanges, Prof. Lipo decided to collaborate with us to address a key issue of vernier machines—their low power factor—which was not yet widely recognized in academia. I led the team at HUST in investigating radial-flux machine solutions, while Prof. Lipo guided his team in South Korea in exploring axial-flux machine solutions.

Our subsequent research revealed that the flux modulation principle applies not only to vernier machines and magnetic gears but also to other types such as transverse-flux machines, flux reversal machines, and flux switching machines. These machines share common characteristics, including high torque density and low power factor. As a result, I proposed unifying the terminology for these machines under the name FMM. To validate this proposal, I consulted Dr. Thomas M. Jahns, another renowned professor from the Wisconsin Electric Machines and Power Electronics Consortium (WEMPEC), and also an IEEE Fellow and member of the US National Academy of Engineering. After receiving his positive feedback, we began to consistently refer to these machines as FMMs in our publications.

Over the years, my team systematically explored the working principles, unified theory, analysis and design methods, driving topology, and control strategies of FMMs. This comprehensive effort resulted in the gradual development of a more complete flux modulation theory, uncovering the profound effects of flux modulation on machine topology and performance. Importantly, this theory not only provided theoretical guidance but also opened new research avenues for discovering high-performance machines.

Through our in-depth study of FMMs, a continuous stream of high-quality machine topologies has been invented, revealing more underlying principles and significantly increasing the volume of related publications. FMMs have emerged as a prominent research focus in the field of electrical machines globally, and the theoretical framework surrounding FMMs has matured. For example, Prof. Z. Q. Zhu, an IEEE Fellow and member of the Royal Academy of Engineering, conducted extensive research on FMMs. Prof. Fengge Zhang from Shenyang University of Technology, China, systematically researched flux modulation brushless doubly-fed machines. Additionally, Prof. Ming Cheng from Southeast University, China, has diligently extended flux modulation theory to various machine types. Scholars from the Harbin Institute of Technology, Jiangsu University, and several other universities have also made significant contributions to FMM research. Their distinguished studies are crucial to the ongoing development of FMMs.

This book systematically summarizes the scientific achievements by the CAEMD at HUST in the theoretical research and application technology of FMMs over the past decade. We anticipate that this compilation will play a pivotal role in promoting the development of electrical machine theory, enhancing machine performance, advancing design methodology, and facilitating the application of FMMs in industry. It is our hope that this work will have a significant impact not only on classical machine theory but also on the broader landscape of the machine industry. Throughout the process of writing this book, numerous scholars and students associated with HUST CAEMD have made substantial contributions. We express our heartfelt thanks to Shaofeng Jia, Tianjie Zou, Kangfu Xie, Chaojie Shi, Mengxuan Lin, Li Fang, Ziyi Liang, Yu Zhao, Peng Yan, and others, for their invaluable help and support. Additionally, we have consulted a vast array of literature, duly credited in the references at the end of each chapter. We extend our sincere gratitude to the authors of these works as well.

The research work for this book was accomplished under the financial support of the National Natural Science Foundation of China, under Project No. 51991382, 51337004, 51520105010, 51607079, and 51977094. We are deeply grateful for this support.

Considering the rapid development of FMMs, and the evolving nature of the research, and our limited English proficiency levels, this book may not be perfect. There may be inevitable mistakes or inaccuracies in the content, and we invite your valuable comments and active discussions. With heartfelt thanks!

Wuhan, China — Ronghai Qu

Introduction

Flux modulation machines (FMMs) represent a revolutionary class of electrical machines that have emerged in recent years. Unlike conventional electrical machines, FMMs integrate a unique component—the modulation unit—into their design, which results in a more diverse topology and endows these machines with distinct advantages, such as high torque density, reduced pulsating torque, and brushless excitation. These characteristics make FMMs highly versatile, with potential applications spanning aerospace, renewable energy generation, servo processing, and large-scale industrial production. This book comprehensively introduces the concept, principles, structure, and performance characteristics of FMMs, presenting the latest research advancements in this field. The content is organized into six detailed chapters: Chap. 1 defines FMMs and describes their fundamental types; Chap. 2 systematically presents flux modulation theory and outlines methodologies for calculating air gap magnetic fields across various FMM topologies; Chap. 3 categorizes FMMs based on their topological evolution; Chap. 4 provides a theoretical analysis of the primary performance characteristics of different FMM types; Chap. 5 introduces innovative new FMM topologies and highlights their specific advantages; and Chap. 6 summarizes the current state of FMM research and forecasts future development trends.

This book serves as an invaluable reference for researchers in electric machinery, offering both foundational knowledge and cutting-edge insights. It is also designed as a comprehensive textbook for undergraduate and postgraduate students specializing in electric machinery. Furthermore, it serves as a crucial resource for engineers and technicians involved in the design, manufacture, and testing of electric machinery or related fields.

Contents

About the Authors

Dr. Ronghai Qu (Fellow, IEEE) received his B.E. and M.S. degrees from Tsinghua University, Beijing, China, in 1993 and 1996, respectively, and his Ph.D. degree from the University of Wisconsin-Madison, Madison, WI, USA, in 2002, all in electrical engineering. He is currently the Director of the State and Province Joint Engineering Research Center of Novel Electrical Machines and the Director of the Center for Advanced Electrical Machines and Drives (CAEMD) at Huazhong University of Science and Technology (HUST).

From 2003 to 2010, Dr. Qu worked at the General Electric (GE) Global Research Center in Niskayuna, NY, USA, as a Senior Electrical Engineer in the Electrical Machines and Drives Laboratory. In 2010, he joined HUST in Wuhan, China. From 2012 to 2016, he served as the Deputy Dean of the School of Electrical and Electronic Engineering at HUST. Dr. Qu has authored or coauthored over 400 technical papers, including 14 IEEE award-winning papers, and holds more than 170 patents. His research interests focus on the design and drive of electrical machines. Dr. Qu has received numerous accolades throughout his career. He was the recipient of 11 GE GRC awards, including the EPST Technical Achievement Award and the Outstanding Teamwork and Management Award. He has also won several other prestigious awards, such as the 2020 Science and Technology Invention Award (1st prize) from the China Electrotechnical Society, the 7th Nagamori Award from the Nagamori Foundation in Japan (2021), two gold medals from the 47th Exhibition of Inventions of Geneva (2019), and the 12th Innovation and Entrepreneurial Figure Award from the China Association of Inventions (2022). He is an active member of the international electrical engineering community. He currently serves on the AdCom of the ICEM NPO and is the Chair of the IEEE Industry Application Society (IAS) Wuhan Chapter.

Dr. Dawei Li (Senior Member, IEEE) was born in China. He received his B.S. degree in electrical engineering from the Harbin Institute of Technology, Harbin, China, in 2010, and his Ph.D. degree in electrical engineering from HUST, Wuhan, China, in 2015. In 2015, he joined HUST, where he now serves as a Professor. He has authored over 200 technical papers and holds more than 50 patents and patent applications. His research interests focus on the design and analysis of flux-modulation permanent-magnet brushless machines.

Dr. Xiang Ren (Member, IEEE) was born in Hunan, China. He received his B.Eng. and Ph.D. degrees in electrical engineering from HUST, Wuhan, China, in 2014 and 2019, respectively. He is currently a Postdoctoral Researcher with the Power Electronics Machines and Control Group (PEMC) at the University of Nottingham, Nottingham, UK. His research interests include the design and analysis of flux modulation machines.

Dr. Yuting Gao (Senior Member, IEEE) received her B.S. and Ph.D. degrees in electrical engineering from HUST, Wuhan, China, in 2012 and 2017, respectively. Since 2023, she has been a Professor at Wuhan University, Wuhan, China. Her research interests include the design and analysis of flux modulation permanent-magnet machines.

Chapter 1
Introduction

Flux modulation machines (FMMs) represent a groundbreaking class of electrical machines, distinct from conventional designs in their principles, structure, and characteristics. This chapter offers a systematic definition of FMMs, starting with an in-depth exploration of the term "modulation" and gradually differentiating FMMs from traditional machines. Building on this foundation, the chapter introduces various types of FMMs extensively studied in academia, providing concise descriptions of their performance characteristics and appropriate application areas.

1.1 Basic Concept of Modulation

The term "modulation" originates from the field of signal transmission, describing the mathematical transformation of an original signal, its integration onto a carrier wave, and conversion into a form suitable for transmission. In wireless communication, for example, the original signal is often a sound signal. Human audible sound signals range from 20 Hz to 20 kHz, corresponding to ultra-long electromagnetic waves, requiring large antennas for transmission, which are costly and impractical. Additionally, transmitting multiple sound signals simultaneously in the same frequency band can cause mutual interference. To overcome these challenges, sound signals are modulated onto high-frequency carrier waves, converting low-frequency sound signals into high-frequency transmission signals. This method reduces the need for large antennas and allows loading of sound signals onto different frequency carrier waves, preventing interference.

In modulation, the carrier wave is typically a sine wave characterized by three key parameters: amplitude, frequency, and phase. To transmit the original signal's information, one of these parameters must be adjusted according to the original signal. Amplitude modulation (AM) and frequency modulation (FM) are common

R. Qu et al., *Flux Modulation Machine*, https://doi.org/10.1007/978-981-95-4746-3_1

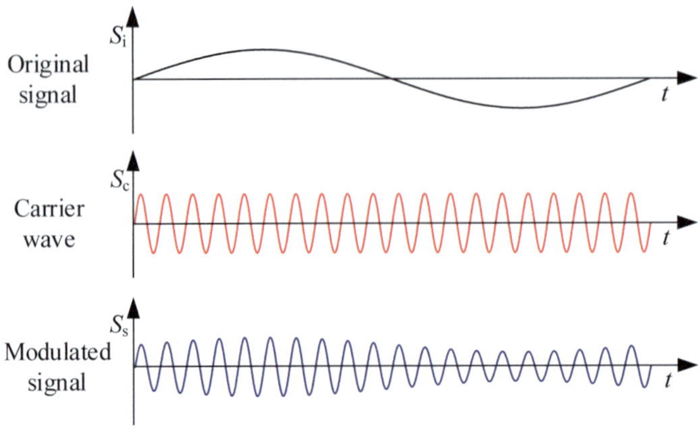

Fig. 1.1 Basic process of amplitude modulation

techniques, involving changes to the carrier wave's amplitude and frequency, respectively.

The basic process of amplitude modulation is shown in Fig. 1.1. The carrier wave can be represented mathematically as:

$$S_c(t) = S_m \sin(\omega_c t) \quad (1.1)$$

where S_m represents the amplitude of the carrier wave, ω_c denotes the frequency of the carrier wave, and t indicates the time.

Assuming the original signal is $S_i(t)$, then the modulated signal can be described as:

$$S_s(t) = (S_m + k_S S_i(t)) \sin(\omega_c t) \quad (1.2)$$

where k_S denotes the scaling factor.

As delineated in Eq. (1.2), the frequency of the modulated signal matches that of the carrier wave precisely, while its amplitude varies in synchrony with the original signal. The greater the amplitude of the original signal, the lesser the amplitude of the modulated signal, and vice versa. In more straightforward terms, the envelope shape of the modulated signal closely emulates the original signal, thereby preserving all pertinent information.

If the information from the original signal is encoded in the frequency of the carrier wave, the resulting modulated signal S_s can be described as follows:

$$S_s(t) = \sin(\omega_c + k_S S_i(t))t \quad (1.3)$$

This method of modulation is known as frequency modulation, and its fundamental mechanism is depicted in Fig. 1.2. In this modulation type, the amplitude of the

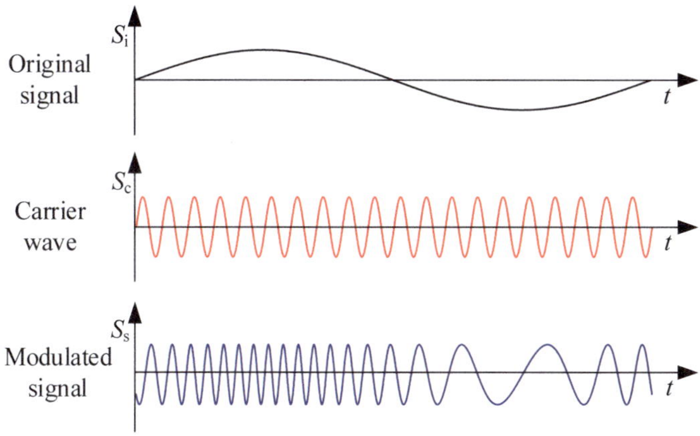

Fig. 1.2 Basic process of frequency modulation

modulated signal remains constant, whereas its frequency varies according to the instantaneous value of the original signal.

With advancements in power electronics, the term "modulation" has acquired broader significance. A notable technology in power electronics is pulse width modulation (PWM). PWM employs power switches such as insulated gate bipolar transistors (IGBTs) in inverters to enable flexible control over the output voltage. The single-phase bridge inverter circuit is shown in Fig. 1.3. In this configuration, the voltage applied to the load is U_{dc} when switches T_1 and T_4 are activated, and T_2 and T_3 are deactivated. Conversely, the voltage reverses to $-U_{dc}$ when switches T_2 and T_3 are activated, and T_1 and T_4 are deactivated. By manipulating these four power switches, any desired voltage signal can be achieved, provided that the supply voltage U_{dc} is adequately high.

The PWM process is illustrated in Fig. 1.4. Contrary to the sine carrier wave typically used in traditional signal transmission, the carrier wave in PWM is a high-frequency triangular wave. The modulated signal, or output voltage, is defined by the following expression:

Fig. 1.3 Single-phase bridge inverter circuit

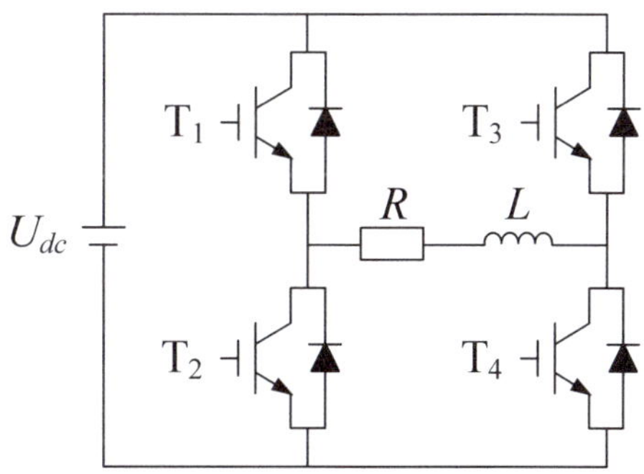

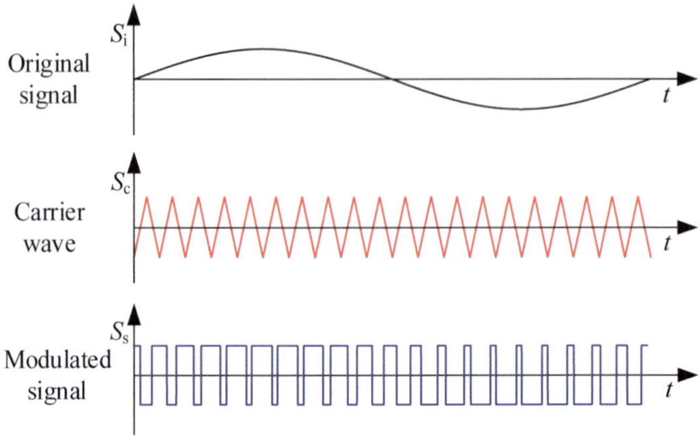

Fig. 1.4 Pulse width modulation (PWM) process

$$S_s(t) = U_{dc} \times \text{sgn}(S_i(t) - S_c(t)) \tag{1.4}$$

where sgn is a signum function. The value of sgn is + 1 when the independent variable is positive, and − 1 when it is negative.

The interpretation of Eq. 1.4 is elucidated as follows: When the original signal exceeds the carrier wave, switches T_1 and T_4 are activated, yielding an output of U_{dc}. Conversely, when the original signal drops below the carrier wave, switches T_2 and T_3 are activated, resulting in an output of $-U_{dc}$. A comparison of the original and modulated signals, as shown in Fig. 1.4, reveals significant waveform differences. However, applying Fourier decomposition to these waveforms reveals that they are nearly identical in the low-frequency range, although the modulated signal exhibits numerous high-frequency harmonics. Furthermore, given that the load in Fig. 1.3 is inductive, the currents induced by the high-frequency voltage harmonics are negligible and can be disregarded. Consequently, the PWM process described facilitates the generation of a sinusoidal current, aligning closely with practical requirements.

In summary, modulation is fundamentally a mathematical or physical process wherein the carrier wave is transformed under the influence of the original signal according to specific rules. Moreover, modulation exhibits a critical characteristic, i.e. while the original signal may vary arbitrarily, the carrier wave remains constant once the modulation method is established. Thus, the original signal acts as an active and variable quantity, whereas the carrier wave serves as a passive and constant quantity. Despite the carrier wave forming the basis of the modulated signal, it encapsulates all the information from the original signal. This modulation process in power electronics parallels the flux modulation process in FMM, which are discussed in the subsequent Sections.

1.2 Basic Concept of Flux Modulation Machine

During the initial phase of research, high-torque-density machines such as the vernier permanent magnet machine, flux switching machine, flux reversal machine, and magnetic geared machine were studied independently. At this stage, the concept of "flux modulation" had not yet been established. It was not until 2015 that our team identified a common working principle among these high-torque-density machines: the flux modulation principle. These machines can be structurally divided into three functional units, referred to as the "three-unit structure." The introduction of the FMM family concept marked a significant milestone in the international academic community. Consequently, all these machine types were classified as FMMs, differing only in the physical configurations of their three functional units. Their operational principles can be unified under the flux modulation theory [1].

This novel breakthrough attracted extensive attention from scholars worldwide, spurring further research on FMMs. For example, Prof. Zhu from The University of Sheffield, UK, analyzed the operating principle of stator permanent-magnet machines using flux modulation theory [2]. Prof. Ming Cheng from Southeast University, China, translated the flux modulation concept into the mathematical form of "three elements" and investigated the electromagnetic performance of various machines based on this foundation [3]. Prof. Refaie from Marquette University, USA, elucidated the high torque density feature of the vernier machine by employing the "three-unit structure" and flux modulation principle in his review paper [4]. The "three-unit" structure of FMM has gained considerable recognition within the academic community and will serve as the foundational structure in the subsequent content of this book.

The basic structure of FMM is shown in Fig. 1.5b. Unlike the conventional surface-mounted permanent magnet machine depicted in Fig. 1.5a, the FMM includes a flux modulation unit (modulation unit) consisting of several ferromagnetic blocks placed between the armature winding and the permanent magnet rotor (excitation unit). This modulation unit distorts the magnetic field produced by the permanent magnets, as illustrated in Fig. 1.6. From a spatial harmonics perspective, the modulation unit introduces new flux density harmonics in the air gap. Without the modulation unit, the primary harmonic component of the air gap flux density aligns with the 22 pole pairs of the permanent magnet excitation's magneto-motive force (MMF). However, with the modulation unit, a new 2-pole-pair flux density harmonic is generated in the air gap. Therefore, the modulation unit functions similarly to a carrier wave in power electronics, influencing the original signal (permanent magnet MMF) as introduced in Sect. 1.1. As the modulation effect pertains to flux density, these machines are collectively referred to as FMMs, as shown in Figs. 1.5 and 1.6.

According to the theory of electromechanical energy conversion, an electrical machine requires that the pole pair number of its armature winding matches the no-load air gap flux density. Consequently, the FMM should have a pole pair number that corresponds to the new air gap flux density generated by the flux modulation effect, i.e., 2 pole pairs, as depicted in Fig. 1.6. However, a non-modulated flux density

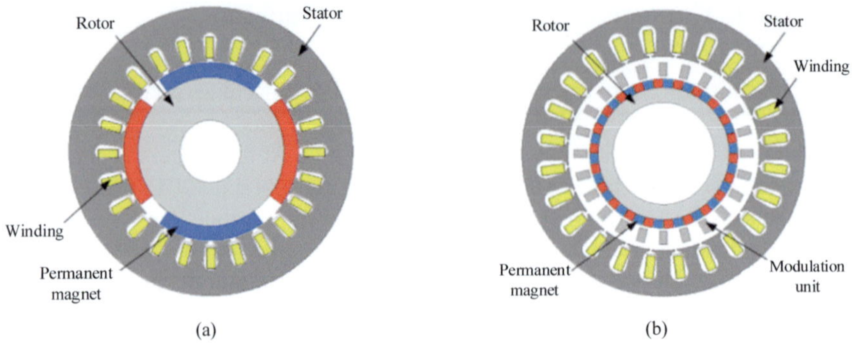

Fig. 1.5 Schematic diagrams of: **a** conventional surface-mounted permanent magnet machine, **b** FMM

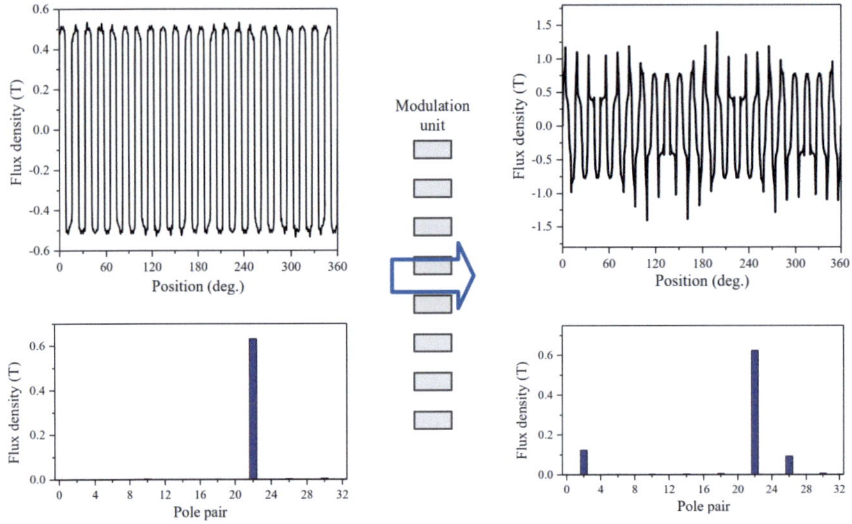

Fig. 1.6 Flux modulation process in FMM

harmonic of 22 pole pairs also exists. If the armature winding's pole pair number is chosen as 22, the machine fails to utilize the flux modulation effect and cannot be classified as an FMM. Thus, a key characteristic of an FMM is that the pole pair numbers of the excitation unit and the armature winding are not identical.

1.3 Research Status of Flux Modulation Machine

In terms of electromagnetic structure, a key distinction of FMMs compared to conventional permanent magnet machines is the unequal pole pair numbers of the excitation unit and the armature winding. This unique feature has led to the development of several new machine topologies in recent years. These include the vernier permanent magnet machine, flux switching permanent magnet machine, flux reversal permanent magnet machine, and electrically excited doubly-salient machine, all of which are classified under the FMM family.

1.3.1 Vernier Permanent Magnet Machine

The vernier reluctance machine, originating in 1963 [5], evolved from the conventional reluctance machine, as illustrated in Fig. 1.7. Unlike the conventional reluctance machine, the stator tooth pitch in the vernier reluctance machine differs from that of the rotor. This discrepancy causes the small displacement of the rotor to be amplified by the stator teeth, resulting in significant movement of the air gap permeance axis. This amplification effect, analogous to the vernier caliper, gives the machine its name "vernier reluctance machine." In 1995, Prof. Ishizaki introduced a vernier permanent magnet machine topology featuring permanent magnets on both the stator and rotor [6], as shown in Fig. 1.8. He explored the working principle and optimization design method of this topology through finite element simulations and prototype experiments. The results indicated that, compared to the regular vernier reluctance machine, his proposed machine demonstrated higher torque density, efficiency, and power factor.

In 1999, Prof. Lipo and Dr. Toba collaborated to propose a surface-mounted vernier permanent magnet machine [7], depicted in Fig. 1.9. This topology represents the typical structure of a vernier permanent magnet machine, comprising a permanent magnet rotor and a wound stator. While its electromagnetic structure is similar to that of the conventional surface-mounted permanent magnet machine, it is distinguished by the unequal pole pair numbers of the rotor permanent magnets and the stator winding. Specifically, the stator winding pole pair number equals the difference between the number of stator teeth and the rotor permanent magnet pole pairs. Notably, the stator teeth serve a dual purpose: facilitating magnetic conduction and playing a role in flux modulation. Analytical analysis and finite element simulations have confirmed the high torque density characteristic of the surface-mounted vernier permanent magnet machine.

In the vernier permanent magnet machine, the rotor pole pair, winding pole pair, and the number of stator teeth responsible for flux modulation must satisfy a specific mathematical relationship. This often results in a significantly larger number of stator teeth compared to the pole pair number of the stator winding. Consequently, this leads to a slot per pole per phase (≥ 1), necessitating an integer-slot overlapping-winding.

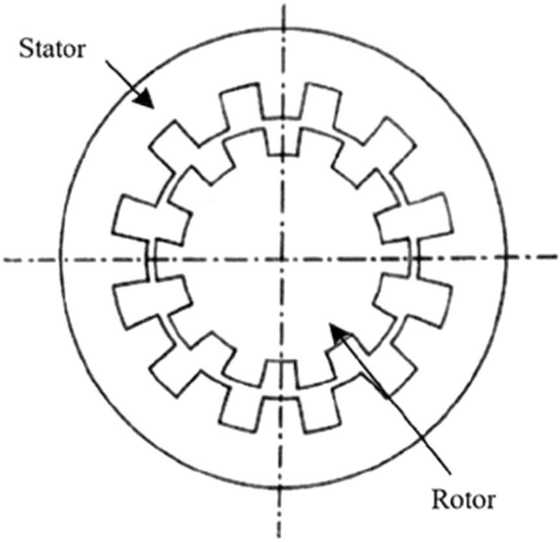

Fig. 1.7 Vernier reluctance machine [5]

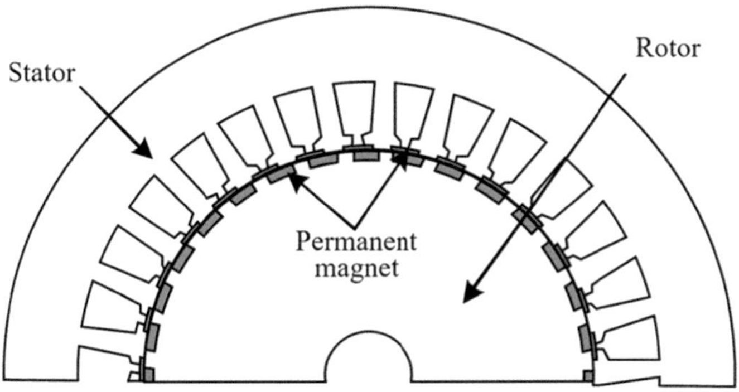

Fig. 1.8 Vernier permanent magnet machine [6]

Unfortunately, this winding type has a long end-winding length, occupying excessive space and diminishing the advantage of high torque density. To address this issue, a split-tooth vernier permanent magnet machine was proposed [8], in which each primary tooth is connected with several auxiliary teeth, forming a split-tooth structure (Fig. 1.10). The auxiliary teeth modulate the rotor permanent magnet MMF, while the primary teeth embed the stator winding. This structure enables the separation of the teeth for the embedded winding and the teeth for flux modulation, allowing for a non-overlapping stator winding, shorter end-winding length, and more flexible slot-pole combination selection [9, 10]. Building on this, a vernier permanent magnet machine with a hybrid-tooth structure was proposed [11] (Fig. 1.11). This stator teeth

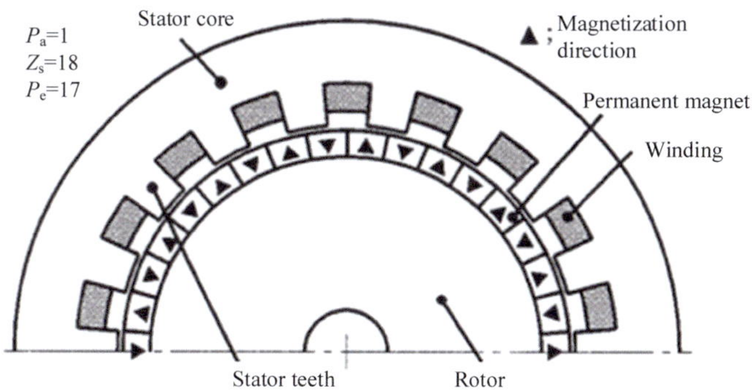

Fig. 1.9 Surface-mounted vernier permanent magnet machine [7]

structure integrates parallel-tooth and split-tooth designs to enhance fault-tolerance performance and reduce silicon steel sheet consumption [12].

Apart from vernier permanent magnet machines, researchers have explored vernier superconducting machines [13, 14]. The principal line of inquiry involves adapting the topologies of existing vernier permanent magnet machines by substituting permanent magnets with superconducting wires or replacing non-magnetic components with superconducting bulks to enhance the torque output capabilities of these machines.

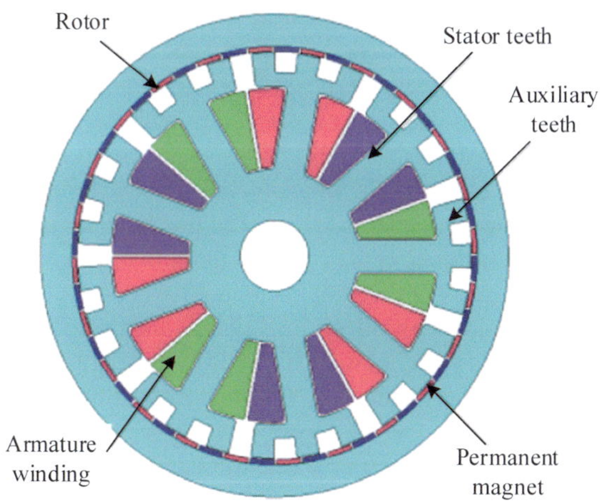

Fig. 1.10 Split-tooth vernier permanent magnet machine [8]

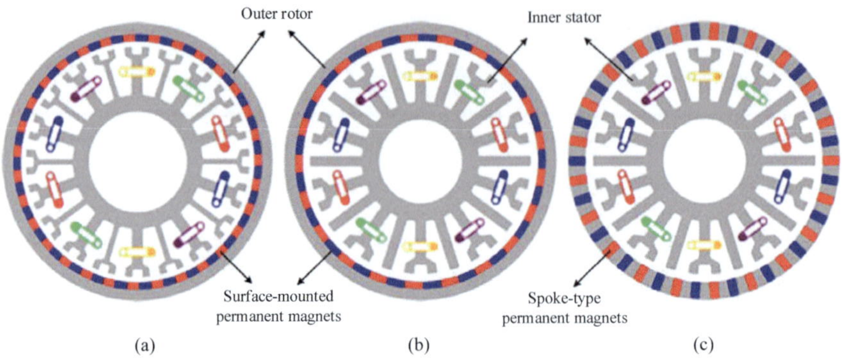

Fig. 1.11 Hybrid-tooth vernier permanent magnet machine [11]: **a** auxiliary-tooth vernier permanent magnet machine, **b** hybrid-tooth surface-mounted vernier permanent magnet machine, **c** hybrid-tooth spoke-type vernier permanent magnet machine

1.3.2 Flux Switching Permanent Magnet Machine

The initial proposal for the first flux switching permanent magnet machine was made in 1955 by Rauch and Johnson [15], presented as a single-phase machine in Fig. 1.12. The principal advantages of flux switching machines derive from the arrangement of both the windings and permanent magnets in the stator, while the rotor is equipped with a reluctance structure to enhance robustness. However, due to constraints in material technology and machine control at that time, the machine did not demonstrate superior electromagnetic performance, such as high torque density, and consequently, it did not attract significant attention. Research into flux switching permanent magnet machines was revived years later with the development of high-performance permanent magnet materials and advanced control technology.

The three-phase flux switching permanent magnet machine, proposed by Dr. Hoang in 1997 [16], is illustrated in Fig. 1.13. In this machine, the permanent magnets are spoke-magnetized, positioned at the center of the stator teeth, and extend through the entire stator core. Adjacent permanent magnets have opposite polarities. Both the stator and rotor are laminated with silicon steel sheets, and the rotor features a simple reluctance structure. The stator winding, which is wound around the stator teeth, can be either concentrated or distributed. The flux line distribution in a flux switching machine is shown in Fig. 1.14. Due to the opposite polarities of adjacent permanent magnets, the flux lines are concentrated within the stator teeth. These flux lines then cross the air gap to the rotor and return through another stator tooth, forming a closed loop. As the rotor teeth move relative to the permanent magnets, the flux direction in the winding changes. In Fig. 1.14a, the winding flux linkage reaches its positive maximum, while in Fig. 1.14b, it reaches its negative maximum. This bipolar flux linkage enables the flux switching permanent magnet machine to generate high back-electromotive force (EMF) and torque. As a result, the flux switching permanent

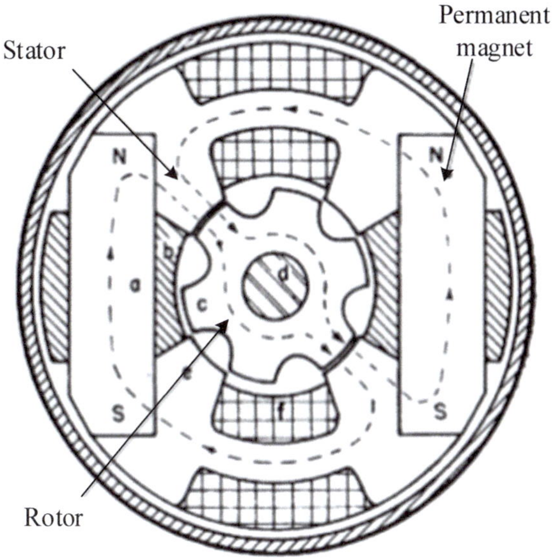

Fig. 1.12 Single-phase flux switching permanent magnet machine [15]

magnet machine has become a significant research area, leading to the development of various new topologies.

The conventional flux switching permanent magnet machine uses a U-shaped stator core. To reduce the use of permanent magnets and lower costs, several new

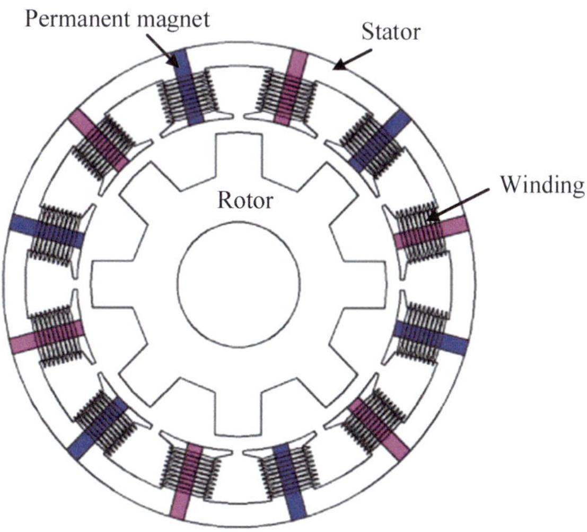

Fig. 1.13 Three-phase flux switching permanent magnet machine

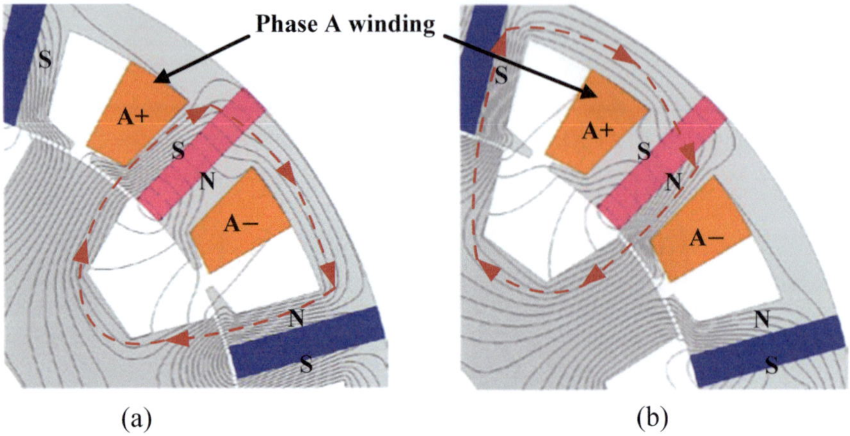

Fig. 1.14 Flux line distribution in a flux switching permanent magnet machine at different rotor positions: **a** position A, **b** position B

stator core structures have been proposed [17], including C-core, E-core, split-tooth, and hybrid-tooth designs, as shown in Fig. 1.15.

The rotor of the flux switching permanent magnet machine features a reluctance structure without permanent magnets and windings, making it highly suitable for high-speed applications. To further enhance the flux-weakening performance, several hybrid excitation flux switching permanent magnet machines have been proposed [18], as illustrated in Fig. 1.16. The key idea behind hybrid excitation topologies is to leverage the magnetic field regulation capability of field winding excitation while maintaining the performance of permanent magnet excitation. Hybrid excitation flux switching machines can be classified into two categories based on the flux paths of the field winding: series type and parallel type. In a series hybrid excitation flux switching machine, the flux excited by the field winding passes through the permanent magnets. Conversely, in a parallel hybrid excitation flux switching machine, the flux bypasses the permanent magnets. Series hybrid excitation offers advantages in construction simplicity and overall flux density reduction. However, because the flux from the field winding traverses the magnets, there is a risk of demagnetization during flux weakening operations. Additionally, the flux regulation effect of the excitation coils may be limited due to the relatively high magnetic reluctance of the permanent magnets. In parallel hybrid excitation, since the flux excited by the field winding bypasses the permanent magnets, it reduces the risk of demagnetization and exhibits a stronger flux regulation capability compared to series hybrid excitation.

Since the permanent magnets penetrate the stator teeth and the armature winding is wound around the stator teeth, the flux switching machine is prone to saturation. To address this, Prof. Zhu from The University of Sheffield, UK, introduced a partitioned-stator flux switching permanent magnet machine [19], as depicted in Fig. 1.17. This topology separates the permanent magnets from the armature winding, placing the armature winding in the outer stator and the permanent magnets in the

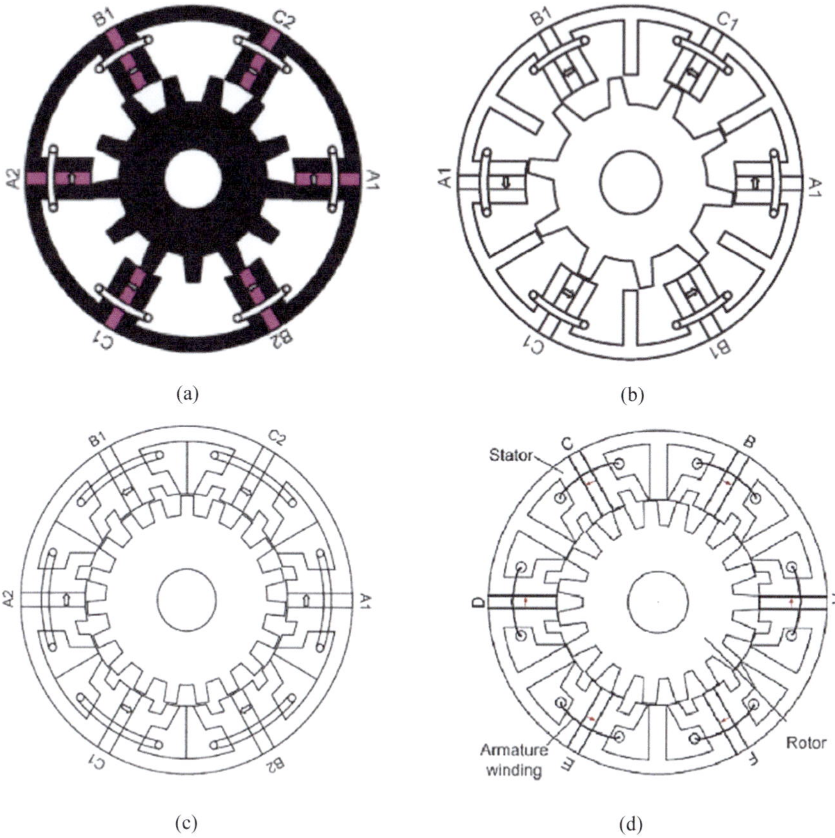

Fig. 1.15 Flux switching permanent magnet machines with different stator core structures [17]: **a** C-core, **b** E-core, **c** split-tooth, **d** hybrid-tooth

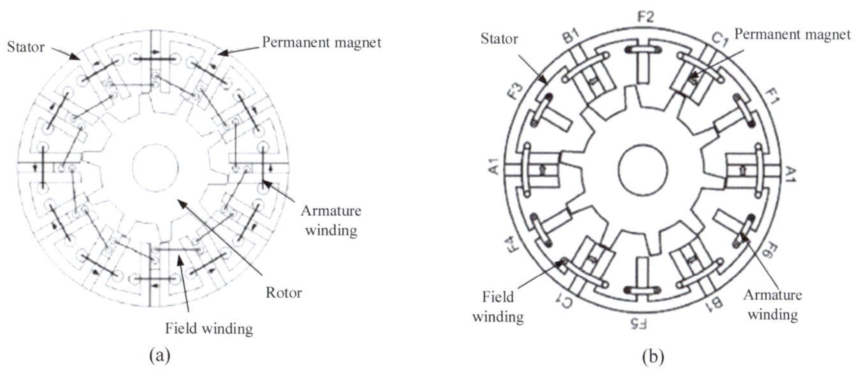

Fig. 1.16 Hybrid excitation flux switching machines [18]: **a** series hybrid excitation, **b** parallel hybrid excitation

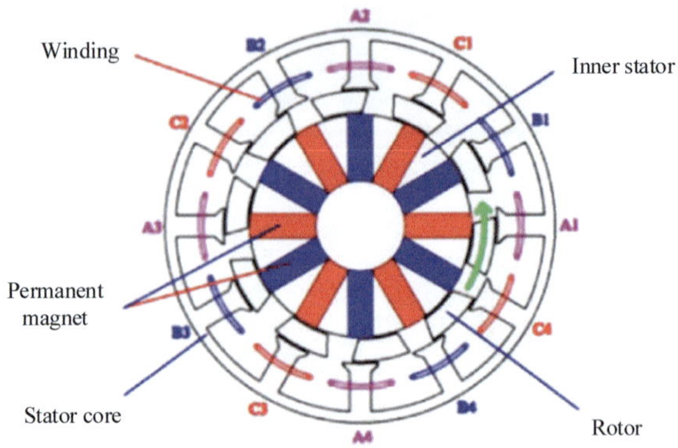

Fig. 1.17 Partitioned-stator flux switching permanent magnet machine [19]

inner stator, with the rotor sandwiched between these two stators. Due to the efficient utilization of inner space, the partitioned-stator flux switching permanent magnet machine exhibits much higher torque density than the conventional flux switching permanent magnet machine.

1.3.3 Flux Reversal Permanent Magnet Machine

Similar to the flux switching permanent magnet machine, the earliest flux reversal permanent magnet machine was also single-phase [20], as shown in Fig. 1.18. In this design, permanent magnets are attached to the inner surface of the stator teeth near the air gap, and the rotor has a reluctance structure. In 1999, Prof. Boldea proposed a three-phase flux reversal permanent magnet machine [21], as illustrated in Fig. 1.19. This paper derived analytical expressions for key performance metrics and presented the design method for the flux reversal machine.

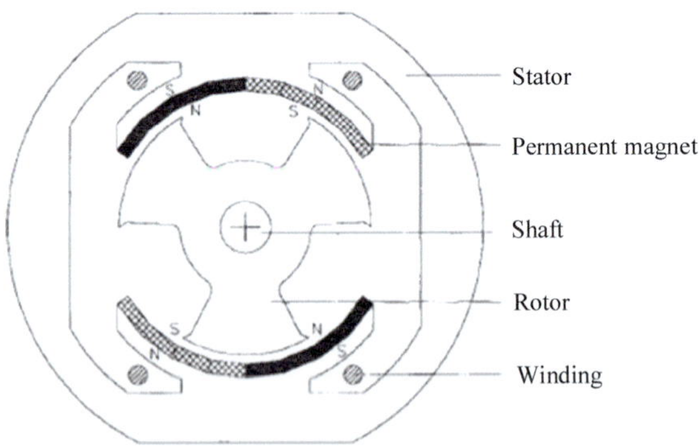

Fig. 1.18 Single-phase flux reversal permanent magnet machine [20]

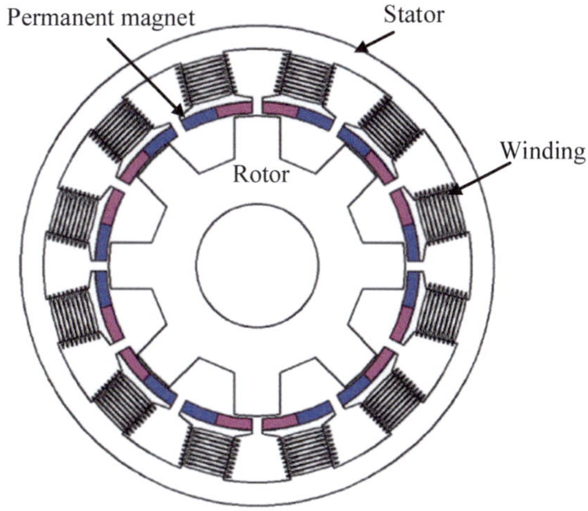

Fig. 1.19 Three-phase flux reversal permanent magnet machine

The working principle of the flux reversal permanent magnet machine is as follows: Each stator tooth has at least one pole pair of radially magnetized permanent magnets. As the rotor rotates, the alignment of the rotor tooth with the N-pole permanent magnet results in the winding flux linkage reaching its positive maximum. Conversely, alignment with the S-pole permanent magnet causes the winding flux linkage to reverse direction and reach its negative maximum. Therefore, similar to the flux switching permanent magnet machine, the flux reversal permanent magnet machine also has a bipolar winding flux linkage, allowing it to generate high back-EMF and output torque [22, 23].

Flux reversal permanent magnet (FRPM) machines predominantly utilize fractional-slot concentrated windings. However, for specific slot-pole combinations, the winding factor of fractional-slot concentrated windings can be relatively low. In such instances, adopting distributed winding can enhance the winding factor, thereby increasing torque density [24, 25]. Additionally, Prof. Zhu introduced various partitioned-stator flux reversal permanent magnet machines [26], as illustrated in Fig. 1.20. These machines feature an outer stator with armature windings, an inner stator with permanent magnets, and a sandwiched rotor serving as the flux modulator. The original topology, the surface-mounted partitioned-stator flux reversal machine (Fig. 1.20a), positions the permanent magnets of the conventional flux reversal machine on the inner stator, while other components remain unchanged. This design significantly increases torque density compared to the conventional flux reversal machine due to the efficient utilization of inner space. Subsequently, several other topologies have been derived from the surface-mounted partitioned-stator flux reversal machine, including the consequent-pole type, embedded-pole type, and V-shaped type, as shown in Fig. 1.20b–d, respectively. Furthermore, the literature [27] proposed an FRPM machine with two pairs of permanent magnets placed on each stator tooth, as depicted in Fig. 1.21, which further enhances torque density.

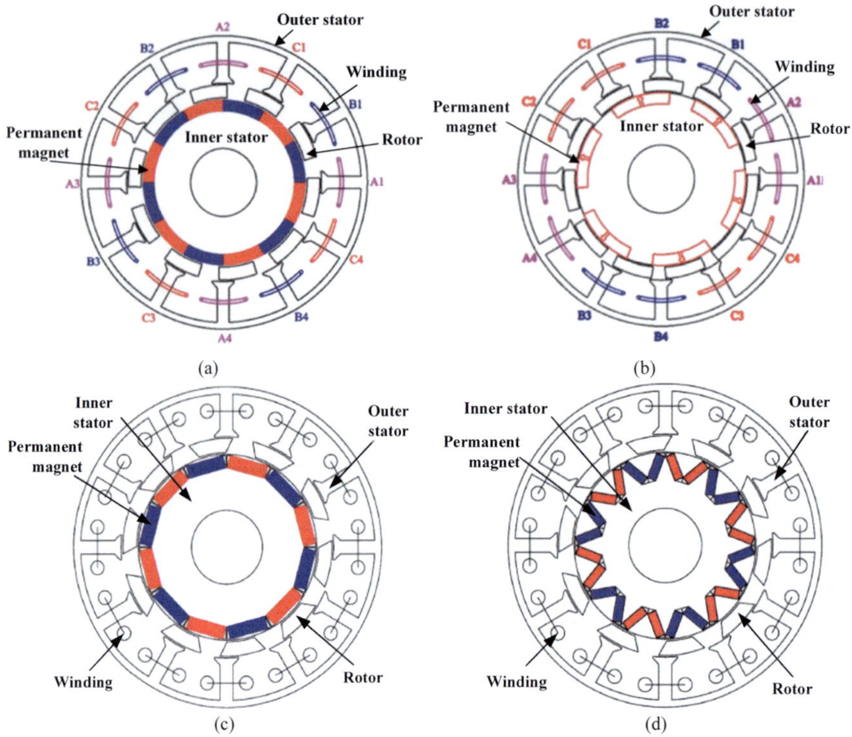

Fig. 1.20 Partitioned-stator flux reversal permanent magnet machines [26]: **a** surface-mounted, **b** consequent-pole, **c** embedded-pole, **d** V-shape

Fig. 1.21 Flux reversal permanent magnet machine with two pairs of permanent magnets on each stator tooth [27]

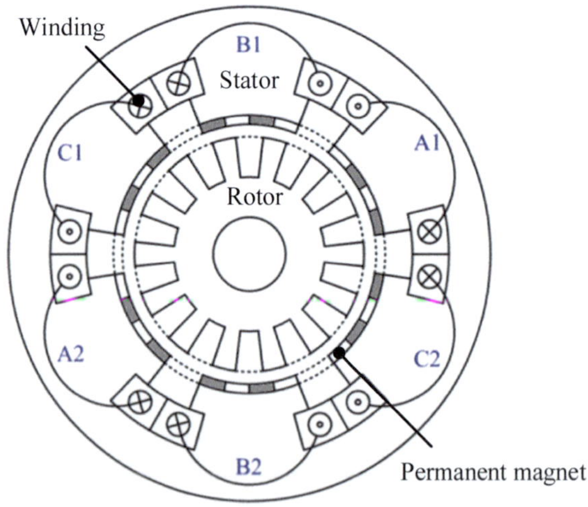

The pulsating torque of FRPM machines is typically substantial, leading to significant vibrations and noise that can affect operational reliability. Consequently, extensive research has been conducted to mitigate this pulsating torque. Traditional rotor slot skewing, applicable to FRPM machines, requires that the rotor pole pair number (i.e., rotor slot number) be unequal to the stator winding pole pair number. In this scenario, the skewing angle of the rotor slots should be calculated based on the rotor pole pair number rather than the winding pole pair number. Other effective methods to reduce pulsating torque include rotor teeth pairing and rotor dummy slots. Rotor teeth pairing utilizes the principle that pulsating torques generated by large and small rotor teeth are in different phases, offsetting each other and thereby decreasing the resultant pulsating torque. The rotor dummy slot method reduces cogging torque by increasing the least common multiple of the stator slot number and the rotor slot number. Additionally, improving the air gap structure to make the air gap flux density more sinusoidal helps reduce cogging torque. This can be achieved by employing methods such as rotor chamfering, rotor tooth asymmetry, permanent magnet chamfering, rotor tooth surface optimization, and setting gaps between permanent magnets.

1.3.4 Electrically Excited Flux Modulation Machine

Various types of electrically excited FMMs exist. This section primarily introduces three types: the brushless doubly-fed reluctance machine, the electrically excited vernier reluctance machine, and the electrically excited flux switching machine.

The brushless doubly-fed reluctance machine, proposed by Broadway et al. in the 1970s (Fig. 1.22), adopts a doubly-salient structure similar to that of a switched reluctance machine. It features two sets of three-phase windings in the stator—one control winding and one power winding. The difference between the stator slot number and the rotor slot number equals the difference between the pole pair number of the power winding and the pole pair number of the control winding.

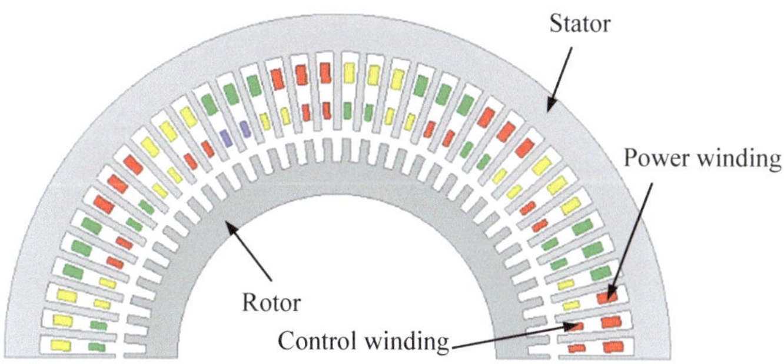

Fig. 1.22 Brushless doubly-fed reluctance machine [28]

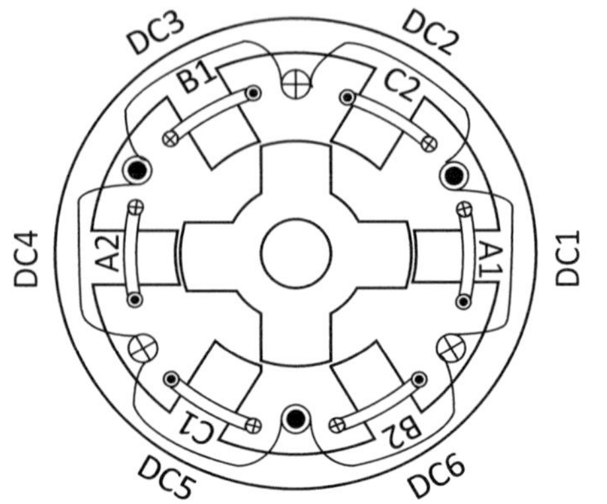

Fig. 1.23 Electrically excited vernier reluctance machine [29]

The electrically excited vernier reluctance machine (Fig. 1.23), also known as a variable reluctance machine or DC excitation switched reluctance machine, primarily employs concentrated windings in the stator. Both the field excitation winding and the armature winding are located in the stator, which eliminates the need for brushes and slip rings, thereby enhancing the machine's reliability. The rotor, characterized by a simple and robust reluctance structure, further contributes to its reliability [29, 30].

In 2005, Prof. Chau introduced a doubly-salient machine with distributed electrically excited windings (Fig. 1.24). This design features a distributed field winding and a concentrated armature winding, allowing each field coil to provide excitation flux for multiple armature coils. The machine operates with trapezoidal, rather than sinusoidal, current waveforms in the armature winding. This characteristic makes it suitable for applications such as wind power generation and electric vehicles, offering adjustable excitation and a wide range of constant power operation.

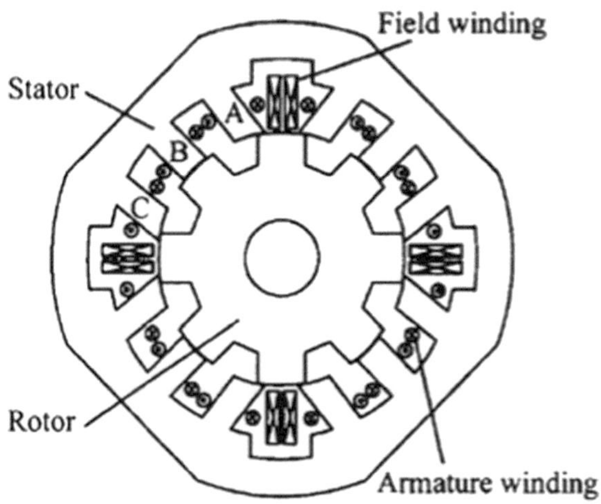

Fig. 1.24 Electrically excited doubly-salient machine [31]

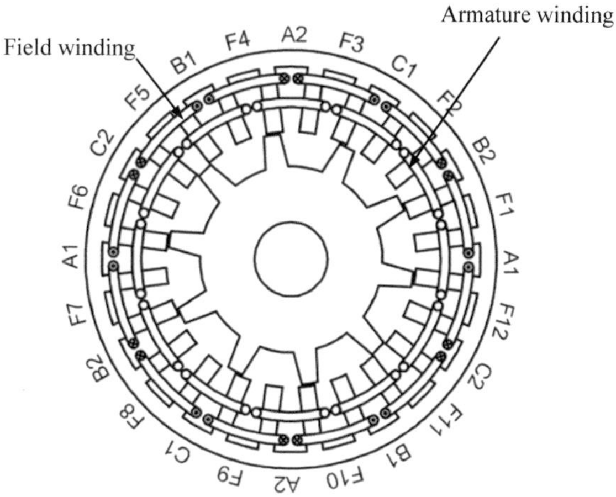

Fig. 1.25 Electrically excited flux switching machine [32]

Electrically excited flux switching machines are derived from flux switching permanent magnet machines. Researchers replaced some or all of the permanent magnets with field excitation coils to create the electrically excited flux switching machine (Fig. 1.25). The armature winding in this topology is a two-slot-pitch concentrated winding, which differs from the previous two electrically excited FMMs. This design necessitates bipolar winding flux linkage, ensuring that one stator tooth carries positive flux while another carries negative flux, resulting in a bipolar flux in the stator coil.

1.4 Application Prospects of Flux Modulation Machines

The earliest FMMs, such as the vernier reluctance machine, were used in small-power devices like money-counting machines, fax machines, and assembly robots. However, their limited efficiency and power factor hindered broader adoption. Section 1.3 introduces four emerging FMM technologies, detailing their unique advantages and disadvantages. Ongoing global research efforts have led to promising prospects for FMM applications across various fields.

Vernier permanent magnet machines offer the highest torque density among FMMs, meaning they provide higher torque per unit of armature winding current or equivalent torque with smaller winding copper loss. However, these machines often experience high iron losses due to their rich high-order harmonic contents and large pole ratio, making them more suitable for low-speed applications where iron loss is less significant. An essential advantage of vernier permanent magnet machines is their low pulsating torque, which makes them ideal for high-precision applications.

The significant drawback of vernier permanent magnet machines lies in their pronounced armature reaction. This strong armature reaction first manifests as a low power factor. Specifically, while the power factor can be maintained above 0.8

under light-load conditions with ambient cooling, it sharply drops below 0.6 under heavy-load conditions. If the output power is fixed, the low power factor necessitates a larger converter capacity, which increases the size, weight, and cost of the entire machine system. Furthermore, the strong armature reaction results in poor overload capability. Vernier permanent magnet machines are highly susceptible to saturation, which weakens their output torque capabilities compared to regular permanent magnet machines under heavy-load and overload conditions.

In summary, vernier permanent magnet machines are best suited for high-torque-density, low-speed, and light-load applications where cooling conditions are limited. They are particularly useful in low- and medium-speed servo systems, direct-drive wind generators, fans, and pumps, leveraging their high-quality torque.

Stator permanent magnet FMMs, including flux switching permanent magnet machines and flux reversal permanent magnet machines, share electromagnetic features with vernier permanent magnet machines. The critical difference is that stator permanent magnet FMMs exhibit lower torque density but higher power factor than vernier permanent magnet machines. Another essential advantage of stator permanent magnet FMMs is the simplicity of their rotor, which features a reliable reluctance structure. This makes them suitable for high-speed applications. However, to reduce iron losses at high speeds, the rotor teeth number should not be too high. Additionally, stator permanent magnet FMMs can be utilized as linear motion machines. In this configuration, the long secondary is a reluctance structure made with a silicon steel sheet, with both the permanent magnets and windings placed in the short primary, significantly reducing processing and maintenance costs.

Electrically excited FMMs share a robust rotor structure with stator permanent magnet FMMs. Without permanent magnets, they avoid self-demagnetization issues. Both the field excitation winding and armature winding are located in the stator, eliminating the need for brushes and slip rings. This design makes electrically excited FMMs suitable for applications in aviation starting and generating fields, where high reliability and self-demagnetization capabilities are crucial.

In conclusion, FMMs have become a focal point in the field of electrical machines due to their advantages such as high torque density and low pulsating torque. Their wide application prospects span aerospace, transportation, wind power generation, numerical control equipment, and more. Unlike conventional electrical machines, which have been well-developed over many years, FMMs are still in the early stages of research. Existing studies on various FMMs are scattered, and the theoretical system remains ununified. This lack of cohesion hinders continuous development and the ability to address drawbacks like high iron losses and strong armature reactions.

The authors' team recognized these issues and conducted in-depth research to establish the unified flux modulation theory. Building on this foundation, various FMMs were proposed, classified, and compared under the same framework. Guided by this unified theory, several high-performance FMM topologies were developed [33–53], with detailed research results explained in subsequent chapters.

1.5 Summary of This Chapter

This chapter provides a comprehensive overview of FMMs, beginning with the concept of "modulation." It reviews the current research status and discusses the applications and development prospects of FMMs:

1. *Fundamental structure and features*: FMMs are distinguished from conventional electrical machines by incorporating an additional modulation unit. This unit modulates the excitation MMF and generates new flux density harmonics in the air gap. This unique feature results in an unequal pole pair number between the rotor and stator armature winding, which is the defining characteristic of FMMs.
2. *Emergence of various FMM types*: Recent years have seen the emergence of various FMM types, including vernier permanent magnet machines, flux switching permanent magnet machines, flux reversal permanent magnet machines, and electrically excited doubly-salient machines. This chapter examines their main structural topologies and summarizes the current state of research.
3. *Application and suitability*:

   - Vernier permanent magnet machines: These machines offer high torque density and low pulsating torque but face challenges such as high iron losses, low power factor, and limited overload capability. They are particularly suitable for high-torque-density, low-speed, and light-load applications, such as low- and medium-speed servo systems and direct-drive wind generators.
   - Linear stator permanent magnet FMMs: These machines that do not use permanent magnets or windings in the long secondary are promising for applications such as railway transportation and cordless elevators owing to their simplified design and lower maintenance costs.
   - Electrically excited FMMs: Featuring brushless excitation and resistance to self-demagnetization, these machines are ideal for aviation starting and generating systems, where high reliability is essential.

In conclusion, FMMs have become a focal point in the field of electrical machines due to their high torque density and low pulsating torque. They have promising applications in various fields, including aerospace, transportation, and wind power generation. However, despite their potential, FMMs are still in the early stages of research, with a fragmented theoretical framework that hinders continuous development.

References

1. Li D, Qu R, Li J (2015) Topologies and analysis of flux-modulation machines. In: 2015 IEEE Energy conversion congress and exposition (ECCE), pp 2153–2160
2. Yang H, Zhu ZQ, Lin H, Li H, Lyu S (2019) IEEE Trans Ind Electron 67(3):2107
3. Cheng M, Han P, Hua W (2017) IEEE Trans Trans Ind Electron 64(8):6063
4. Wu F, El-Refaie A (2019) IET Electric Power Appl 13(2):127

5. Lee CH (1963) IEEE Trans Power Appar Syst 82(66):343
6. Ishizaki A, Tanaka T, Takasaki K et al (1995) Theory and optimum design of PM vernier machine. In: Seventh international conference on electrical machines and drives, Durham, pp 208–212
7. Toba A, Lipo T (1999) IEEE Thirty-fourth IAS annual meeting, Phoenix, pp 2539–2544
8. Li JG, Chau K, Jiang J et al (2010) IEEE Trans Magn 46(6):1475
9. Zou TJ, Li DW, Qu RH et al (2017) IEEE Trans Ind Appl 53(6):5295
10. Zou TJ, Li DW, Chen C et al (2018) IEEE Trans Magn 54(11):8109605
11. Xu L, Liu GH, Zhao WX et al (2017) IEEE Trans Trans Ind Electron 64(1):179
12. Li DW, Zou TJ, Qu RH et al (2018) IEEE Trans Ind Appl 54(2):1320
13. Jia SF, Qu RH, Li J et al (2016) IEEE Trans Appl Supercond 26(7):5207205
14. Xie KF, Li DW, Qu RH et al (2018) IEEE Trans Ind Appl 55(2):1420
15. Rauch S, Johnson L (1955) Trans Am Inst Electr Eng Part III: Power Appar Syst 74(3):1261
16. Hoang E, Ahmed H, Lucidarme J (1997) Switching flux permanent magnet polyphased synchronous machines. In: The 7th European conference on power electronics and applications, Aalborg, pp 903–908
17. Zhu ZQ (2011) Switched flux permanent magnet machines—innovation continues. In: 2011 International conference on electrical machines and systems, Beijing, pp 1–10
18. Chen JT, Zhu ZQ, Iwasaki S et al (2011) IEEE Trans Veh Technol 60(4):1365
19. Wu ZZ, Zhu ZQ (2016) IEEE Trans Energy Convers 31(4):1239
20. Deodhar R, Andersson S, Boldea I et al (1996) IEEE Trans Ind Appl 33(4):925
21. Boldea I, Wang CX, Nasar S (1999) Electr Mach Power Syst 27(8):849
22. Gao YT, Qu RH, Li DW et al (2017) IEEE Trans Ind Appl 53(3):2110
23. Xie KF, Li DW, Qu RH et al (2019) IEEE Trans Ind Appl 55(2):1250
24. Gao YT, Li DW, Qu RH et al (2018) IEEE Trans Energy Convers 33(4):2153
25. Gao YT, Qu RH, Li DW et al (2016) IEEE Trans Ind Appl 52(4):2856
26. Zhu ZQ, Hua H, Wu D et al (2016) IEEE Trans Ind Appl 52(1):199
27. More D, Kalluru H, Fernandes B (2008) Comparative analysis of flux reversal machine and fractional slot concentrated winding PMSM. In: Annual conference of IEEE Industrial Electronics, Orlando, pp 1131–1136
28. Tabi S, Tounzi A, Piriou F (2006) IEEE Trans Energy Convers 21(4):823
29. Jia SF, Qu RH, Li J et al (2016) IEEE Trans Energy Convers 31(3):935
30. Jia SF, Qu RH, Li J et al (2017) IEEE Trans Ind Appl 53(2):1082
31. Fan Y, Chau K (2008) IEEE Trans Ind Appl 44(3):727
32. Chen JT, Zhu ZQ, Iwasaki S et al (2010) Low-cost flux-switching brushless AC machines. In: IEEE Vehicle power and propulsion conference, Lille. IEEE, pp 1–6
33. Zhou Y, Shi C, Qu R, Li D, Gao Y, Li R (2021) IEEE Trans Ind Appl 57(6):5841
34. Cheng Y (2021) Research on key technologies for large double-stator superconducting flux modulation wind turbines (in Chinese). Huazhong University of Science and Technology, Wuhan
35. Yu Z (2021) Research on the control technology of DC biased vernier reluctance machine system (in Chinese). Huazhong University of Science and Technology, Wuhan
36. Li R (2021) Theory and design of fractional pole permanent magnet machine (in Chinese). Huazhong University of Science and Technology, Wuhan
37. Liu X (2021) Research on the control strategy of integrated winding dual electromechanical port machine (in Chinese). Huazhong University of Science and Technology, Wuhan
38. Yu Z (2021) Research on flux modulation switched reluctance machine drive system (in Chinese). Huazhong University of Science and Technology, Wuhan
39. Xie K (2020) Theory and topology study of electromagnetic composite and permanent magnet based machines (in Chinese). Huazhong University of Science and Technology, Wuhan
40. Lin M (2020) Study on topology and starting performance of asynchronous starting vernier permanent magnet machine (in Chinese). Huazhong University of Science and Technology, Wuhan

41. Shi C (2020) Study on electromagnetic characteristics and topology of vernier permanent magnet linear machine (in Chinese). Huazhong University of Science and Technology, Wuhan
42. Han X (2020) Research on the control strategy of brushless dual electromechanical port machine (in Chinese). Huazhong University of Science and Technology, Wuhan
43. Ren X (2019) Research on multi-electro-mechanical port machines (in Chinese). Huazhong University of Science and Technology, Wuhan
44. Xiong G (2019) Control of high performance permanent magnet vernier linear machine (in Chinese). Huazhong University of Science and Technology, Wuhan
45. Yan P (2019) Design and topology study of claw-pole vernier permanent magnet machine (in Chinese). Huazhong University of Science and Technology, Wuhan
46. Lin F (2018) Electromagnetic study of fully superconducting magneto-gear machine (in Chinese). Huazhong University of Science and Technology, Wuhan
47. Yu Z (2018) Research on harmonic current suppression technology for DC biased vernier reluctance machines (in Chinese). Huazhong University of Science and Technology, Wuhan
48. Zou T (2018) Theory and topology of high performance flux modulation machine (in Chinese). Huazhong University of Science and Technology, Wuhan
49. Zhang R (2017) Design and analysis of axial flux modulation permanent magnet machine (in Chinese). Huazhong University of Science and Technology, Wuhan
50. Huo Y (2017) Design of high thrust vernier permanent magnet linear machine (in Chinese). Huazhong University of Science and Technology, Wuhan
51. Jia S (2017) Research on electrically excited vernier magneto-resistive machine (in Chinese). Huazhong University of Science and Technology, Wuhan
52. Gao Y (2017) Theoretical analysis and topological study of flux reversal machine (in Chinese). Huazhong University of Science and Technology, Wuhan
53. Li D (2015) Research on flux modulation permanent magnet machine (in Chinese). Huazhong University of Science and Technology, Wuhan

Chapter 2
Basic Theory of Flux Modulation

In Chap. 1, the pole-changing effect of FMMs was explained as being achieved through the utilization of the modulation unit. However, the physical principle underlying the modulation unit's influence on altering the pole number of the magnetic field requires further elucidation. Additionally, the specific magnetic field components generated under the flux modulation effect have not yet been determined. Consequently, this chapter will initially analyze the variation pattern of the magnetic field under a non-uniform air gap structure. Subsequently, the MMF and specific permeance model will be established. Building on this foundation, the flux modulation processes of different structures—namely, the surface-mounted permanent magnet structure, the doubly-salient structure, and the embedded permanent magnet structure—will be examined, culminating in the development of the fundamental theory of flux modulation.

2.1 Physical Process of Flux Modulation

An FMM functions as an electrical machine based on the principles of electromagnetic induction. Fundamentally, it involves one armature/excitation field driving another excitation/armature field in rotational or linear motion to generate torque or force. Notably, the armature and excitation fields in FMMs do not directly interact; rather, their interaction occurs through flux modulation induced by the modulation unit. Therefore, understanding the operational principle of FMMs hinges on analyzing the flux modulation effect exerted by the modulation unit on the armature and excitation fields.

The basic model depicted in Fig. 1.5 is utilized to qualitatively analyze the physical process of flux modulation. To streamline the analysis, the following assumptions are made:

R. Qu et al., *Flux Modulation Machine*, https://doi.org/10.1007/978-981-95-4746-3_2

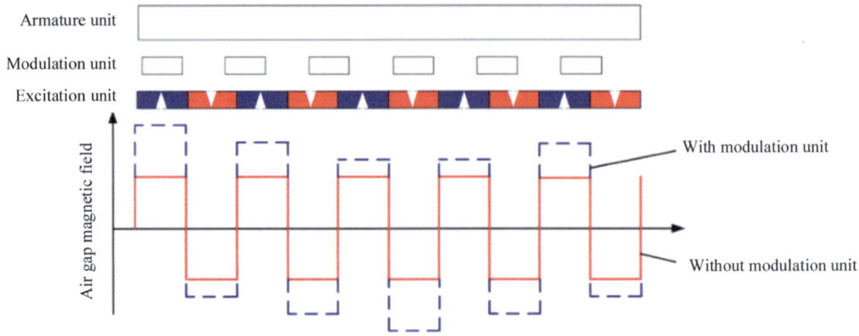

Fig. 2.1 Air gap magnetic field with and without the modulation unit

1. The excitation unit rotates.
2. The modulation unit remains stationary.
3. The number of ferromagnetic blocks (P_f) in the modulation unit exceeds the pole pair number (P_e) of the excitation unit.
4. At time = 0, the centerline of the N-pole permanent magnet aligns with the centerline of the ferromagnetic block.

Initially, in the absence of a modulation unit in the air gap, the air gap magnetic field generated by the permanent magnets exhibits a symmetrical square wave, as illustrated by the solid line in Fig. 2.1.

However, with the addition of the modulation unit to the air gap, the high magnetic permeability of the ferromagnetic blocks causes almost all of the magnetic field to pass through them. This increases the air gap flux density beneath the ferromagnetic blocks, as depicted by the dashed line in Fig. 2.1. The degree of this increase correlates with the overlap between the ferromagnetic block and the permanent magnet. Consequently, the presence of the modulation unit introduces irregularities in the air gap magnetic field, leading to the emergence of new harmonic components.

The analysis of these newly introduced spatial magnetic field harmonics is detailed below. Given the two polarities of permanent magnets, the investigation into the new magnetic field harmonics produced by the N and S poles is conducted separately. The modulated magnetic field generated by the N-pole permanent magnets is illustrated in Fig. 2.2. Notably, the alignment of the first permanent magnet with the centerline of the first ferromagnetic block results in the modulated magnetic field reaching its positive maximum. Since the number of ferromagnetic blocks (P_f) is larger than the number of pole pairs (P_e) in the excitation unit, the distance between the second N-pole permanent magnet and the centerline of the second ferromagnetic block, denoted as θ_{ef}, is calculated as:

$$\theta_{ef} = \frac{2\pi}{P_e} - \frac{2\pi}{P_f} \tag{2.1}$$

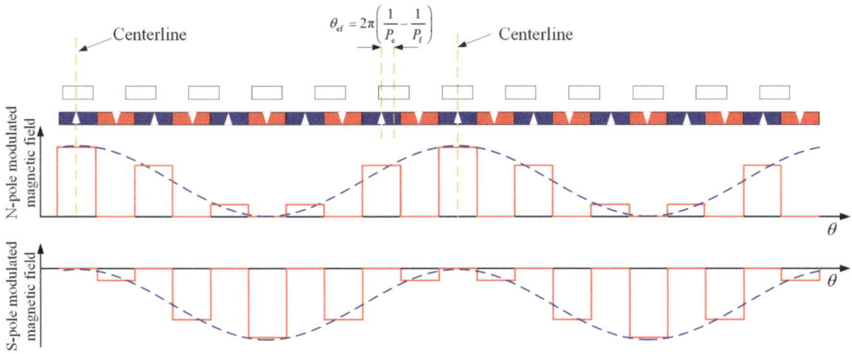

Fig. 2.2 Modulated magnetic field in the air gap

Due to the smaller overlapping area of the second N-pole permanent magnet and the ferromagnetic block, the modulated magnetic field produced by the second N-pole permanent magnet is lower compared to that produced by the first N-pole permanent magnet. Similarly, the modulated magnetic field from subsequent N-pole permanent magnets decreases sequentially until the permanent magnet and the ferromagnetic block begin to overlap once again. Assuming that the n_{ef}th N-pole permanent magnet realigns with the centerline of the ferromagnetic block, it must satisfy the following equation:

$$n_{ef}\theta_{ef} = \frac{2\pi}{P_f} \tag{2.2}$$

By combining Eqs. (2.1) and (2.2):

$$n_{ef} = \frac{P_e}{P_f - P_e} \tag{2.3}$$

The modulated air gap magnetic field completes a full cycle within the range of n_{ef} N-poles. Meanwhile, there are P_e N-poles in a complete circumference. Consequently, the pole pair number of the fundamental modulated air gap magnetic field, denoted as P, is determined as follows:

$$P = \frac{P_e}{n_{ef}} = P_f - P_e \tag{2.4}$$

Subsequently, the magnetic field generated by the S-pole permanent magnets is analyzed. When an N-pole permanent magnet aligns with the centerline of the ferromagnetic block, its adjacent S-pole permanent magnets directly face the slot. This configuration causes the magnetic field generated by the two S-pole permanent magnets to reach its minimum value. Additionally, since the magnetic field direction induced by the S-pole is opposite to that of the N-pole, the fundamental waves of the

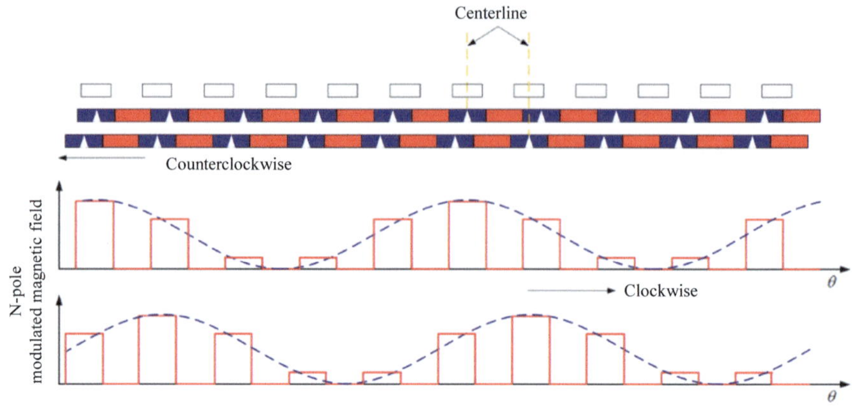

Fig. 2.3 N-pole modulated magnetic field at different positions of the excitation unit

modulated magnetic field from the N and S poles are in phase, with equal magnitudes. Consequently, the amplitude of the final $(P_\mathrm{f} - P_\mathrm{e})$ pole pair magnetic field is twice that of the N-pole modulated magnetic field.

Subsequently, the motion characteristics of the modulated magnetic field harmonics are examined. Initially, at the starting position, an N-pole permanent magnet aligns with the centerline of the ferromagnetic block, as shown in Fig. 2.3. Upon counterclockwise rotation of the excitation unit by θ_ef degrees, the adjacent N-pole permanent magnet aligns with the centerline of the subsequent ferromagnetic block. As a result, the modulated magnetic field rotates clockwise by the spatial angle θ_P degrees, defined as follows:

$$\theta_P = -\frac{2\pi}{P_\mathrm{f}} \tag{2.5}$$

Assuming the rotation speed of the excitation unit is Ω_e, the rotation speed of the fundamental wave of the modulated magnetic field can be derived as follows:

$$\Omega_P = \frac{\theta_P}{\theta_\mathrm{ef}}\Omega_\mathrm{e} = -\frac{P_\mathrm{e}}{P_\mathrm{f} - P_\mathrm{e}}\Omega_\mathrm{e} \tag{2.6}$$

By combining Eqs. (2.4) and (2.6), it becomes evident that the electric angular speed of the excitation unit is equivalent to the fundamental wave of the modulated magnetic field, i.e., $P_\mathrm{e}\Omega_\mathrm{e}$. Therefore, owing to the flux modulation effect of the modulation unit, the air gap magnetic field generated by the excitation unit becomes:

$$B_\mathrm{e}(\theta, t) \approx B_{P_\mathrm{e}} \cos(P_\mathrm{e}\theta - \omega_\mathrm{e}t) + B_{|P_\mathrm{f} - P_\mathrm{e}|} \cos[(P_\mathrm{f} - P_\mathrm{e})\theta + \omega_\mathrm{e}t] \tag{2.7}$$

where B_{Pe} and $B_{|Pf-Pe|}$ represent the amplitudes of the air gap magnetic field harmonic with P_e and $|P_f - P_e|$ pole pairs, respectively; ω_e denotes the electric angular frequency of the excitation unit.

In FMMs, the values of P_f and P_e are typically close, resulting in P_e being significantly larger than $|P_f - P_e|$. This implies that the pole pair number of the fundamental wave of the modulated magnetic field is considerably lower than that of the excitation unit. If the armature winding is designed to match the pole pair number of the fundamental wave of the modulated air gap magnetic field, denoted as $|P_f - P_e|$, the armature magnetic field can interact with the modulated air gap magnetic field, ultimately delivering steady power output. Therefore, the excitation unit and armature unit have different pole pair numbers, and it is the modulation unit that transforms these pole pair numbers, giving rise to the term FMM.

Considering that P_e may actually be larger than P_f, the pole pair number of the armature winding in FMMs, denoted as P_a, can be uniformly expressed as:

$$P_a = |P_f - P_e| \tag{2.8}$$

While the earlier qualitative analysis provides an intuitive understanding of the flux modulation phenomenon, it lacks quantitative calculations. It is observed that the air gap magnetic field with the modulation unit exhibits multiple harmonics, including a $(P_f + P_e)$ pole pair, in addition to the modulated magnetic field harmonic with $|P_f - P_e|$ pole pairs. For instance, Fig. 1.6 illustrates the air gap magnetic field of an FMM with a 22-pole-pair excitation unit and a 24-pole-pair modulation unit. The rotor MMF exhibits 22 pole pairs, and without the flux modulation unit, the primary harmonic component of its air gap magnetic field is 22 pole pairs. Along with the flux modulation unit, additional air gap magnetic fields possessing 2, 6, and other numbers of pole pairs are present. To comprehensively analyze the flux modulation process and the multiple harmonics, it is essential to establish a mathematical model, facilitating accurate quantitative calculations.

2.2 Flux Modulation Theory

The air gap magnetic field acts as the medium for achieving electromechanical energy conversion in electrical machines. Accurate determination of the air gap magnetic field is essential for comprehending the performance of electrical machines. Consequently, the fundamental aspect of developing flux modulation theory involves constructing a mathematical model that accurately represents the physical process of air gap flux modulation, facilitating precise determination of the air gap magnetic field distribution. This section will initially review the magnetic field calculation process for a uniform air gap, examining its physical characteristics, and then extend the analysis to cases where the modulation unit is introduced. For simplification, the magnetoresistance of the ferromagnetic components is neglected, assuming zero magnetic field intensity within these materials.

2.2.1 Armature Magnetic Field with Uniform Air Gap

As depicted in Fig. 2.4, assuming the rotor consists of a smooth iron core, the magnetic field within the air gap has only a radial component. Furthermore, the stator slot opening is minimal and can be disregarded. Subsequently, the calculation of the armature magnetic field is detailed. Firstly, an arbitrary spatial position within the machine, denoted as *AB* in Fig. 2.4, is taken as the reference point. The air gap magnetic field at another position, denoted as *CD*, must satisfy:

$$\oint_{\mathrm{ABCD}} \vec{H}_{\mathrm{a}} \mathrm{d}s = \sum N_{\mathrm{c}} i_{\mathrm{c}} \tag{2.9}$$

where N_{c} represents the number of coil turns, i_{c} denotes the current through a single coil, and ds is the segment length. In Eq. 2.9, as the air gap magnetic field is generated by the armature, a subscript "a" is added to the magnetic field intensity H, and an arrow is placed above it to indicate that H is a spatial vector. Notably, in the subsequent text, symbols for air gap magnetic flux density and magnetic field intensity without arrows will represent only the magnitude of the physical quantity, not its direction.

The right side of the formula represents the total ampere-turns enclosed by the loop *ABCD*. As *CD* rotates along the circumferential direction, the value of the total enclosed ampere-turns changes accordingly, making it a function of the spatial position θ, denoted as $F_{\mathrm{AB}}(\theta)$. The subscript "*AB*" is included because the value of $F_{AB}(\theta)$ is influenced by the selection of the initial position *AB*. For instance, if *AB* is replaced by *A′B′* in Fig. 2.4, even if the other side of the loop remains at the same *CD*

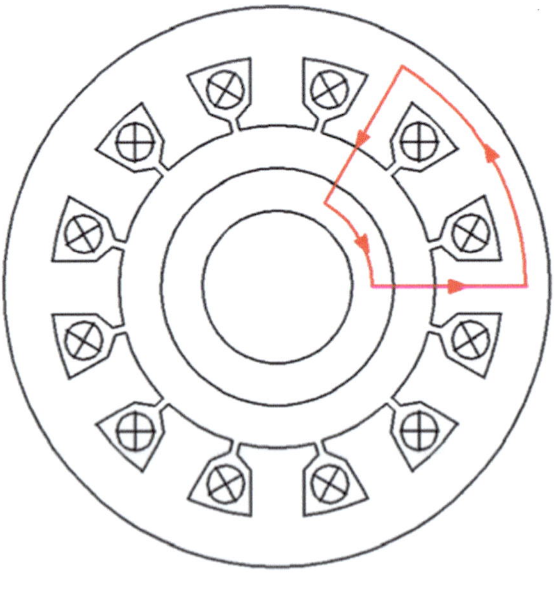

Fig. 2.4 Positions in a uniform air gap

position, the total ampere-turns enclosed by the loop $A'B'CD$ will differ from those enclosed by $ABCD$. Hence, specifying the initial position is essential in defining the function $F_{\mathrm{AB}}(\theta)$.

Given that the magnetic permeability of the iron core is infinite, the internal magnetic field intensity $H_{\mathrm{a}} = 0$. According to Eq. (2.9), the following can be deduced:

$$(H_{CD} - H_{AB})g = F_{AB}(\theta) \tag{2.10}$$

where H_{AB} and H_{CD} represent the magnetic field intensity in the air gap on lines AB and CD, respectively. As indicated in Eq. (2.10), if AB *is* fixed, H_{CD} becomes a function of θ, and is denoted as $H_{\mathrm{a}}(\theta)$. Additionally, the air gap flux density excited by the armature winding can be calculated as:

$$B_{\mathrm{a}}(\theta) = \mu_0 H_{\mathrm{a}}(\theta) = \mu_0\left[\frac{F_{AB}(\theta)}{g} + H_{AB}\right] \tag{2.11}$$

In this equation, only H_{AB} is unknown and needs to be solved. According to the principle of magnetic flux continuity:

$$\int_0^{2\pi} B_{\mathrm{a}}(\theta)\mathrm{d}\theta = \mu_0 \int_0^{2\pi} \left[\frac{F_{AB}(\theta)}{g} + H_{AB}\right]\mathrm{d}\theta = 0$$

$$H_{AB} = -\frac{1}{2\pi g}\int_0^{2\pi} F_{AB}(\theta)\mathrm{d}\theta \tag{2.12}$$

If we define:

$$F_{\mathrm{a}}(\theta) = F_{AB}(\theta) - \frac{1}{2\pi}\int_0^{2\pi} F_{AB}(\theta)\mathrm{d}\theta \tag{2.13}$$

the air gap flux density can be calculated as:

$$B_{\mathrm{a}}(\theta) = \frac{\mu_0}{g}F_{\mathrm{a}}(\theta) \tag{2.14}$$

The $F_{\mathrm{a}}(\theta)$ represents the AC winding MMF introduced in electrical machinery [1]. This can be further analyzed as follows:

(1) Subscript "a" for armature: In the MMF symbol F_{a}, the subscript "a" denotes the armature, indicating that F_{a} is generated by the armature unit. Moreover, F_{a} is independent of the initial position AB because its definition in Eq. (2.12) involves subtracting the average value, thereby eliminating any average component from

Fig. 2.5 Armature MMF waveform

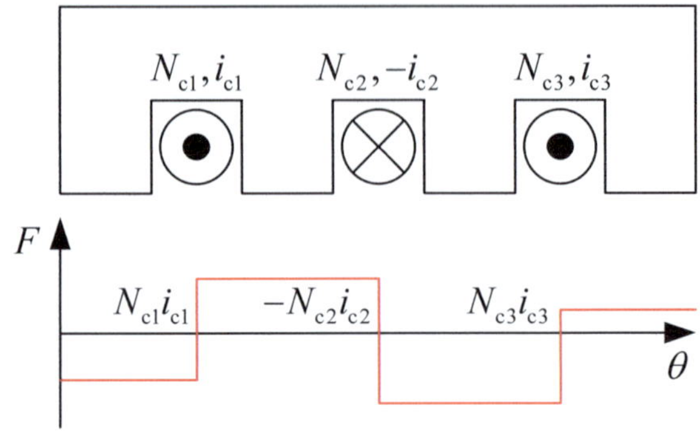

F_a. Consequently, changing the initial position does not alter the total ampere-turns of the loop, meaning the armature-generated MMF is unaffected by the choice of the initial position *AB*.

(2) Step function composition: The armature MMF consists of multiple step functions. Each time the MMF passes through a conductor, the waveform steps up positively if the current in the conductor is directed outward and steps down negatively if it is inward, as illustrated in Fig. 2.5.

(3) Dependence on winding structure and current: In a smooth air gap, the armature MMF depends solely on the winding structure and the current, and is unrelated to other dimensional parameters of the machine. When the armature is supplied with AC, F_a becomes a function of both the spatial position θ and time t.

The above analysis focuses on the mathematical characterization of the MMF. However, an explanation of its physical essence is also necessary for the subsequent introduction of flux modulation theory. In electromagnetism, the scalar magnetic potential φ and the magnetic field intensity H have the following relationship [2]:

$$\vec{H} = -\nabla\varphi = -\left(\frac{\partial\varphi}{\partial x}, \frac{\partial\varphi}{\partial y}, \frac{\partial\varphi}{\partial z}\right) \tag{2.15}$$

Assuming a conductor with N_c turns and a constant current I_c, the magnetic potential difference between any two points (points 1 and 2) in the plane, as illustrated in Fig. 2.6, can be calculated using the line integral of the magnetic field:

$$\varphi_1 - \varphi_2 = \int_{12} \vec{H}\,\mathrm{d}s \tag{2.16}$$

However, if the integration path is a complete loop, we can derive:

$$0 = \varphi_1 - \varphi_1 = \oint \vec{H}\,\mathrm{d}s = N_c I_c \tag{2.17}$$

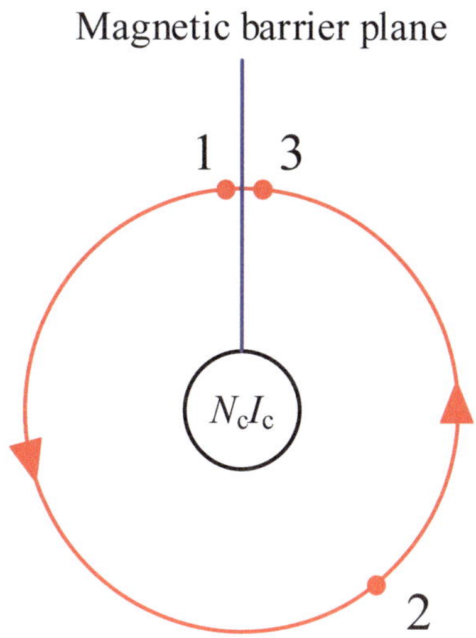

Fig. 2.6 Scalar magnetic potential

However, this approach is invalid due to the fundamental differences between magnetic and electric fields. The magnetic field is inherently a swirling field and does not theoretically have a potential function. However, for computational convenience, the magnetic potential is artificially defined in regions without current. Consequently, when using magnetic potential, a magnetic barrier plane must be specified, and the integral path cannot cross this plane. In the case it occurs, a magnetic potential difference will arise between the two sides, causing a discontinuity in the magnetic potential across the barrier plane. In Fig. 2.6, the magnetic potential difference between points 1 and 3 is:

$$\varphi_1 - \varphi_3 = \int_{13} \vec{H}\,\mathrm{d}s = N_c I_c \tag{2.18}$$

Subsequently, the distribution of scalar magnetic potentials in the armature field is examined. Initially, a starting region on the stator side is chosen, and its magnetic potential is set to zero, as shown in Fig. 2.7. In the absence of a magnetic barrier plane, the iron core is equipotential if its magnetic resistance is neglected. However, due to the presence of the magnetic barrier plane, the stator core is divided into separate Z_s regions. Each time the magnetic barrier plane is crossed, the stator magnetic potential changes by the ampere-turns of the corresponding conductors. Therefore, the magnetic potential difference within the stator core and on its inner surface is exactly equivalent to the previously defined function F_{AB}. As for the rotor core, which lacks a magnetic barrier plane, it remains equipotential. From the preceding discussion, it is evident that the magnetic potential is equivalent to the MMF. Henceforth, the magnetic potential will be denoted as F.

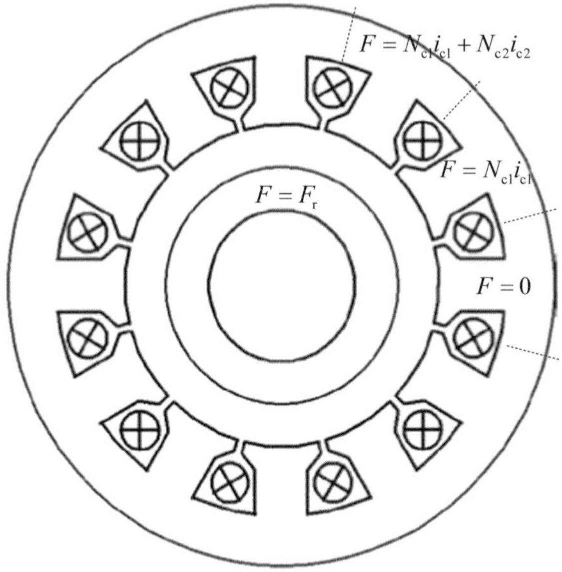

Fig. 2.7 Magnetic potential generated by the armature

Assuming the rotor's magnetic potential in Fig. 2.7 is F_r, then according to the principle of magnetic flux continuity:

$$\int_0^{2\pi} (F_{AB} - F_r)\mathrm{d}\theta = 0$$

$$F_r = \frac{1}{2\pi}\int_0^{2\pi} F_{AB}(\theta)\mathrm{d}\theta \tag{2.19}$$

By comparing Eqs. (2.13) and (2.19), it is clear that the armature MMF is the magnetic potential difference between the stator and rotor, $F_{AB} - F_r$. This implies that if the rotor's magnetic potential is set as the reference zero, the armature winding MMF F_a is identical to the stator magnetic potential F_s. Assuming that the MMF $F_a(\theta)$ is positive at a given spatial position θ, the magnetic potential increases uniformly along the radial direction inside the air gap, as depicted in Fig. 2.8. Here, r_{ig} represents the air gap's inner radius, which is also the rotor's outer radius.

2.2.2 *Armature Magnetic Field with Open Stator Slots*

For the air gap structure with open stator slots, the armature flux direction and the magnetic field intensity differ from those with semi-closed stator slots. Taking the

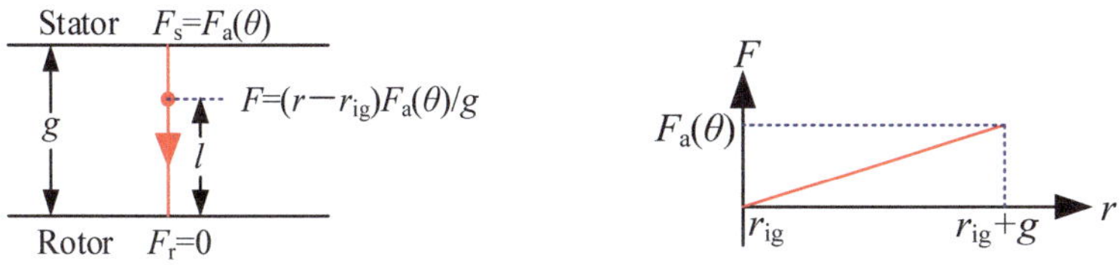

Fig. 2.8 Variation of the MMF (magnetic potential) in the air gap along the radial direction

structure shown in Fig. 2.9 as an example, at the position corresponding to the stator tooth, the flux lines pass through the air gap in the radial direction *AB*, similar to the semi-closed stator slots scenario. However, at the position corresponding to the stator slot, as the flux lines always take the shortest path, they will deflect in the slot and enter the tooth wall perpendicularly. In this case, the flux lines follow a longer path *CD* than the actual air gap. Furthermore, the magnetic field in the slots will be slightly weaker due to the influence of cogging. Additionally, if the energization condition of the armature winding changes, the path of the flux lines may vary accordingly.

To calculate the air gap magnetic field in this complex case, two assumptions are necessary:

(1) Unchanged flux line path: The path of the flux lines remains unchanged, regardless of the distribution of the magnetic potential difference between the stator and rotor.
(2) Constant magnetic field intensity: The magnetic field intensity along the same flux line does not change.

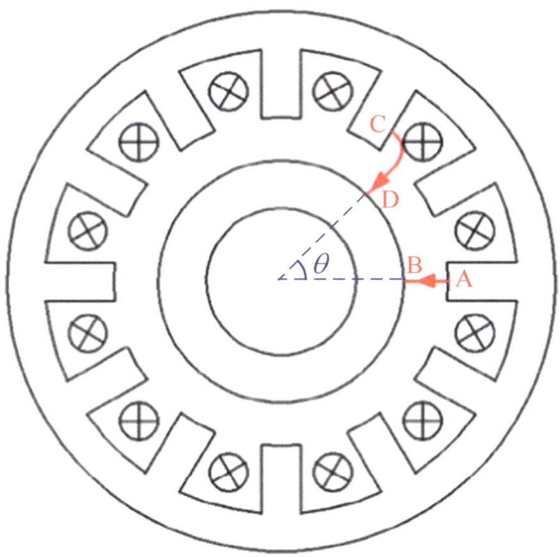

Fig. 2.9 Air gap flux lines with open stator slots

According to the first assumption, the length of the flux lines for the armature field is solely a function of the air gap dimensional parameters and can be expressed as $g(\theta)$. According to the second assumption, once the MMF is determined, the flux density and magnetic field intensity on each flux line are only related to θ, independent of the radial coordinate r (as depicted in Fig. 2.8), and can be represented as $B(\theta)$ and $H(\theta)$, respectively.

In the previous analysis of the uniform air gap magnetic field, if the zero point of the magnetic potential is considered at the rotor core, i.e., $F_r = 0$, then the stator magnetic potential F_s equals the winding MMF F_a. Conversely, if $F_s = F_a$ holds true, $F_r = 0$ can be deduced. Since the current distribution does not change in the open stator slot structure, $F_s = F_a$ remains valid. However, due to the altered magnetic field distribution, the rotor magnetic potential F_r may not be zero. Later, several MMFs with non-uniform air gap structures will also be encountered. To avoid ambiguity, F' is used to denote the MMF after slotting (i.e., non-uniform air gap), and F denotes the MMF before slotting (i.e., uniform air gap). The difference between the two, $F' - F$, represents the additional MMF introduced by the slotting.

$$F_a'(\theta) = \int_r H_a(\theta)\mathrm{d}r = H_a(\theta)g(\theta) \tag{2.20}$$

According to Eq. (2.20), the air gap flux density generated by the armature magnetic field can be computed as:

$$B_a(\theta) = \mu_0 H_a(\theta) = \frac{\mu_0}{g}\frac{g}{g(\theta)}F_a'(\theta) = \frac{\mu_0}{g}\frac{g}{g + \Delta g(\theta)}F_a'(\theta) \tag{2.21}$$

where $\Delta g(\theta)$ represents the increase in the length of the flux lines in the slots due to the slotting effect. The expression for flux density after slotting includes an additional term $g/[\Delta g(\theta) + g]$ compared to the flux density expression with the uniform air gap. According to the previous assumptions, it relates only to the machine's structural parameters. This function physically characterizes the change in air gap permeance before and after slotting and is termed "specific permeance," denoted by the symbol "λ" with the subscript "f" for modulation, i.e.,

$$\lambda_f(\theta) = \frac{g}{g(\theta)} = \frac{g}{g + \Delta g(\theta)} \tag{2.22}$$

The value of λ_f lies between 0 and 1. Since the non-uniform specific permeance is introduced through the stator slotting in Fig. 2.9, the subscript "f" can be replaced by "s". Then, Eq. (2.21) can be further written as:

$$B_a(\theta) = \frac{\mu_0}{g}\lambda_s(\theta)F_a'(\theta) \tag{2.23}$$

The analysis reveals that the specific permeance varies in a slotted air gap, exhibiting higher values (approaching 1) at the stator teeth positions and lower values (nearing 0) at the stator slot locations. Consequently, the waveform of the air gap magnetic field deviates from that of the MMF due to the slotting effect, introducing distortion. The air gap magnetic field decreases where the specific permeance is low and increases where it is high. This phenomenon resembles PWM in power electronics. PWM involves the intermittent alteration of a constant voltage source, converting a stable waveform into an irregular one with multiple square waves of varying widths. Similarly, Eq. (2.23) can be seen as a chopping mechanism, where the specific permeance function, influenced by a non-uniform air gap, modulates the MMF. This process of generating a magnetic field in a non-uniform air gap, described in Chap. 1 as flux modulation, will be further examined in the subsequent discussion. This exploration will delve into the characteristics of flux modulation, the specific permeance function, and the MMF in various structural configurations experiencing non-uniform air gaps.

2.2.2.1 The Structure of the Modulation Unit

Based on the preceding analysis, the crucial factor for the modulation unit to achieve its function is the creation of varied flux line lengths at different circumferential positions, implying the necessity of a non-uniform air gap structure. Such structures fall into three categories:

(1) Stator Slotting (Fig. 2.10a): The inner surface of the stator near the air gap incorporates open slots. This design exploits the shorter flux path facing the stator teeth, resulting in larger specific permeance, and the longer flux path facing the stator slots, resulting in smaller specific permeance. Therefore, a circumferentially varying specific permeance is generated using the stator's tooth-slot structure.

(2) Rotor Slotting (Fig. 2.10b): The outer surface of the rotor near the air gap features open slots, analogous to the stator's tooth-slot structure. Flux paths facing the rotor teeth are shorter, creating larger specific permeance, while those facing the rotor slots are longer, producing smaller specific permeance. The rotor's tooth-slot structure achieves varying specific permeance in the circumferential direction.

(3) Modulation Ring (Fig. 2.10c): Ferromagnetic blocks within the air gap alter the flux lines' paths as they move from the stator side to the rotor side. Flux lines passing through the blocks experience increased specific permeance due to a reduced effective path length, while those bypassing the blocks maintain smaller specific permeance. This structure employs ferromagnetic blocks to introduce variable specific permeance, often referred to as the modulation ring.

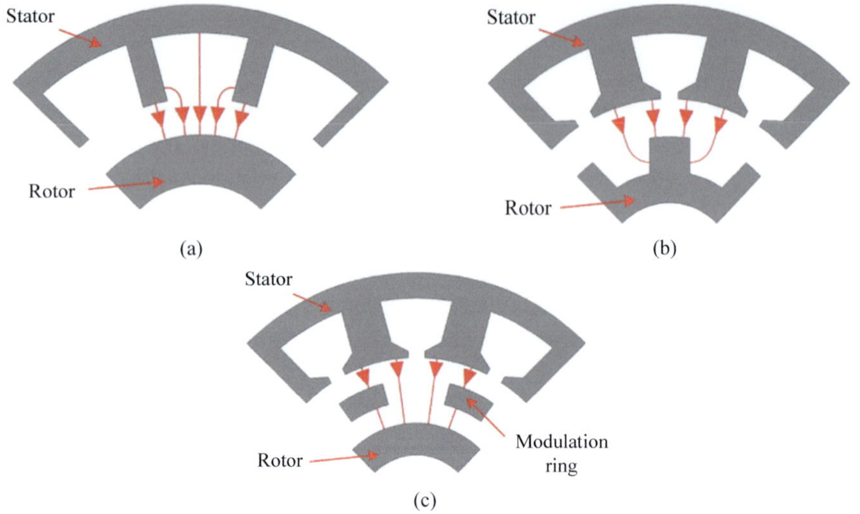

Fig. 2.10 Methods for generating a non-uniform air gap: **a** non-uniform air gap by stator slotting, **b** non-uniform air gap by rotor slotting, **c** non-uniform air gap by modulation ring

2.2.2.2 Specific Permeance Function

In these structures, whether formed by stator slotting, rotor slotting, or ferromagnetic blocks, the variable specific permeance functions are periodically distributed along the circumference. Consequently, specific permeance can be expressed as a Fourier series [3]:

$$\lambda_{\mathrm{f}}(\theta, t) = \lambda_{\mathrm{f0}} + \sum_{l=1}^{\infty} \lambda_{\mathrm{f}l} \cos lP_{\mathrm{f}}(\theta - \Omega_{\mathrm{f}}t) \tag{2.24}$$

where $\lambda_{\mathrm{f}l}$ represents the amplitude of the lth specific permeance harmonic, P_{f} denotes the pole pair number of specific permeance, and Ω_{f} is the rotation speed of the modulation unit.

Figure 2.10 demonstrates that P_{f} corresponds to the number of stator teeth, rotor teeth, and ferromagnetic blocks in configurations (a), (b), and (c) respectively. In the latter two configurations, both the rotor and the modulation ring can rotate, causing the specific permeance function to rotate as well, with a mechanical angular speed of Ω_{f}. In contrast to the MMF, the specific permeance of the air gap is always positive, necessitating a positive average term λ_{f0}, which results in even harmonics in its spatial spectrum. For subsequent analyses, only the average term and the fundamental term of the specific permeance function are typically considered [4], neglecting higher-order harmonics. This simplification reduces $\lambda_{\mathrm{f}}(\theta)$ to:

$$\lambda_{\mathrm{f}}(\theta, t) = \lambda_{\mathrm{f0}} + \lambda_{\mathrm{f1}} \cos P_{\mathrm{f}}(\theta - \Omega_{\mathrm{f}}t) \tag{2.25}$$

However, the amplitudes of the specific permeance harmonics remain unknown and need to be determined. Given that specific permeance is independent of the MMF, a constant MMF $F = 1$ A is assumed in the iron core for simplification (Fig. 2.11). Using this assumption, the specific permeance function can be derived as:

$$\lambda_f(\theta) = \frac{g}{\mu_0}\frac{B(\theta)}{F} = \frac{g}{\mu_0}B(\theta) \tag{2.26}$$

A qualitative observation of the specific permeance function, as depicted in Fig. 2.12, reveals that its waveform resembles that of the air gap flux density under constant MMF excitation. At the tooth centerline, the specific permeance is near 1, corresponding to the maximum air gap flux density. As the spatial position θ approaches the tooth edge, the specific permeance decreases due to the slotting effect, reaching a minimum at the slot centerline before gradually increasing again. This variation in specific permeance is periodic, with each period corresponding to one tooth-slot. Thus, the average term λ_{f0} depends on the average reduction caused by slotting, and the fundamental term λ_{f1} depends on the amplitude of the waveform fluctuation.

The quantitative methods for calculating the magnetic field are divided into two categories: analytical methods and finite element analysis. Analytical methods aim to determine the magnetic field in the air gap region by solving scalar magnetic potential boundary conditions. Given the periodicity and symmetry of the tooth-slot structure, it is sufficient to calculate the magnetic field for only half of a slot pitch, as illustrated in Fig. 2.13a. However, the numerous right angles present in this region complicate full analytical computation, necessitating further model simplification.

Initially, it is assumed that the slots are deep, which implies that even at the center, the flux lines reach the tooth wall rather than the slot's bottom. Consequently, slots are generally treated as infinitely deep. Moreover, when the distance between adjacent slots is sufficiently large, the magnetic field at the tooth center approximates the value as if slotting were absent (Note: this condition may not apply if the number of slots is large). Therefore, the influence of adjacent slots on the magnetic field distribution is minimal, permitting the left side of the solution area to be considered infinitely

Fig. 2.11 Calculation model of the specific permeance function

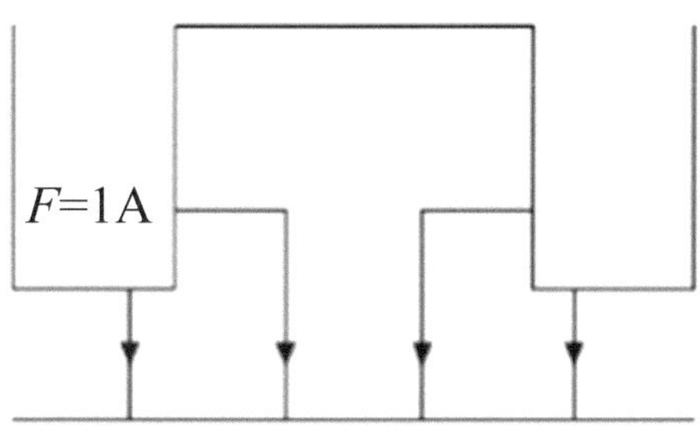

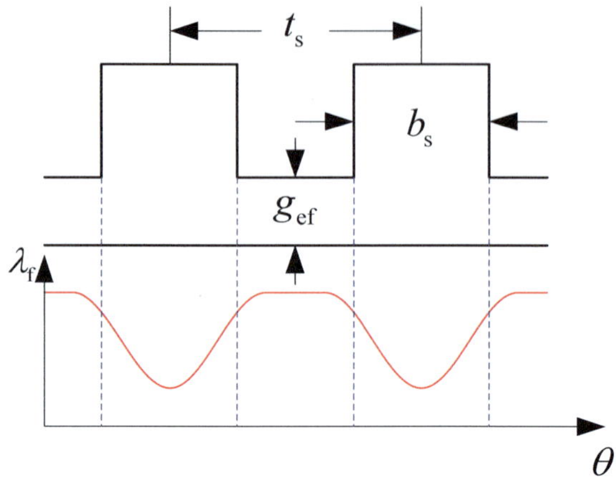

Fig. 2.12 Specific permeance waveform of a non-uniform air gap

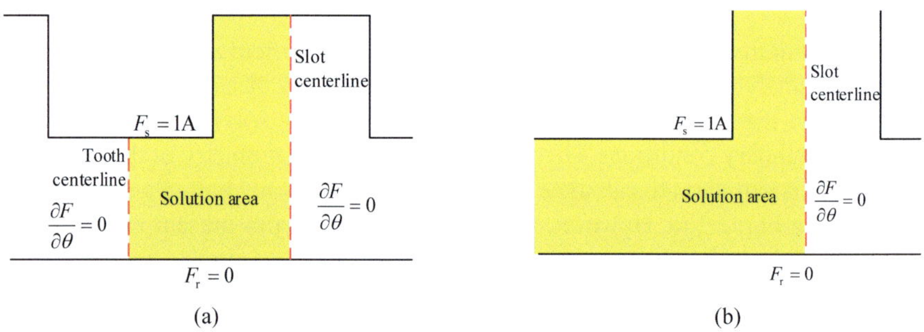

Fig. 2.13 Analytical method-based model for solving specific permeance: **a** exact model, **b** simplified model

long [5]. This assumption leads to the simplified model shown in Fig. 2.13b [6]. By employing the Schwarz-Christoffel mapping, the analytical expression for the specific permeance of the air gap can be derived, providing both the average term and the fundamental term [7]:

$$\begin{aligned} \lambda_{f0} &= 1 - 1.6\beta \frac{b_s}{t_s} \\ \lambda_{f1} &= \frac{4}{\pi}\beta \left[0.5 + \frac{\left(\frac{b_s}{t_s}\right)^2}{0.78125 - 2\left(\frac{b_s}{t_s}\right)^2} \right] \sin\left(1.6\pi \frac{b_s}{t_s}\right) \end{aligned} \quad (2.27)$$

where b_s represents the slot width, t_s denotes the slot pitch, and h_m indicates the thickness of the permanent magnet. The parameter β is determined by these variables, and its expression is given by:

$$\beta = 0.5 - \frac{1}{2\sqrt{1+\left(\frac{b_s}{2g_{ef}}\right)^2}} = 0.5 - \frac{1}{2\sqrt{1+\left(\frac{t_s}{2g_{ef}}\frac{b_s}{t_s}\right)^2}} \tag{2.28}$$

where g_{ef} is the equivalent air gap length, equal to g if no permanent magnets are present. It is calculated as follows for air gap structures with surface-mounted permanent magnets:

$$g_{ef} = g + \frac{h_m}{\mu_r} \tag{2.29}$$

From Fig. 2.12 and Eqs. (2.27)–(2.29), it is clear that three independent parameters—b_s, t_s, and g_{ef}—determine the specific permeance. However, analyzing their influences independently is not suitable owing to the proportional relationships within the equation (i.e., b_s/t_s and t_s/g_{ef}). Therefore, the ratios b_s/t_s and t_s/g_{ef}, which characterize the air gap's structure, will be used as variables for the analysis, as referenced in [8]. For illustrative purposes, a surface-mounted permanent magnet FMM is analyzed, with parameters detailed in Table 2.1.

1. Influence of b_s/t_s

The parameter b_s/t_s, known as the slot opening ratio, denotes the ratio of the slot opening width to the slot pitch. Figure 2.14 illustrates the variation in specific permeance with changes in the slot opening ratio. It is observed that the average term λ_{f0} diminishes as the slot opening ratio increases, reflecting a decrease in the average magnetic field value within the air gap due to a wider slot opening. Conversely, the fundamental term λ_{f1} initially increases and then decreases, indicating that the variations in the specific permeance waveform are less pronounced at extreme slot opening ratios. Thus, there exists an optimal b_s/t_s for maximizing λ_{f1}, which is approximately 0.6. Moreover, it is important to note that λ_{f1} is significantly smaller than λ_{f0}. For instance, at $b_s/t_s = 0.65$, λ_{f0} is measured at 0.75 while λ_{f1} is only 0.25, constituting merely 1/3 of the former value.

2. Influence of t_s/g_{ef}

The parameter t_s/g_{ef} can be manipulated through two distinct approaches. One method involves varying the slot number Z_s while keeping the slot opening ratio

Table 2.1 Parameters of the analytical model for the specific permeance function

Parameters	Value	Parameters	Value
Air gap radius	40 mm	Number of slots	12
Air gap length	0.8 mm	Thickness of permanent magnets	3.2 mm
Relative permeability of permanent magnets	1.05		

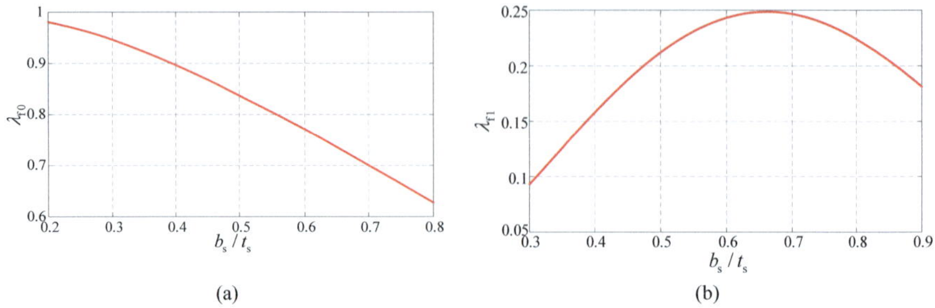

Fig. 2.14 Variation of specific permeance with b_0/t_s: **a** average term, **b** fundamental term

bs/ts constant. Under this scenario, t_s decreases as the number of slots increases. The changes in specific permeance relative to the slot number Z_s are illustrated in Fig. 2.15, with b_s/t_s maintained at 0.6. As the slot number rises, the average term λ_{f0} gradually increases, while the fundamental term λ_{f1} diminishes significantly when Z_s reaches 36, suggesting an almost negligible flux modulation effect at this configuration. This trend can be explained by noting that a reduction in tooth pitch t_s, caused by an increased number of slots while maintaining a constant slot opening ratio, leads to a narrower slot opening width. Consequently, this smoothens the fluctuation in specific permeance, resulting in an increase in the average term λ_{f0} and a decrease in the fundamental term λ_{f1}. This observation indicates that an excessive slot number might severely diminish the flux modulation effect, therefore the selection of slot number must be approached with caution [9].

The second approach to adjusting t_s/g_{ef} involves altering the equivalent air gap length. Given the challenges associated with the retaining sleeve and the magnetic suction during assembly, the physical air gap length in surface-mounted permanent magnet FMMs is typically larger. In contrast, electrically excited FMMs generally feature a smaller equivalent air gap length [10]. The influence of the equivalent air gap length on specific permeance is demonstrable by contrasting surface-mounted permanent magnet FMMs with electrically excited FMMs. Figure 2.14 displays the variation of specific permeance with b_s/t_s in a permanent magnet FMM, whereas

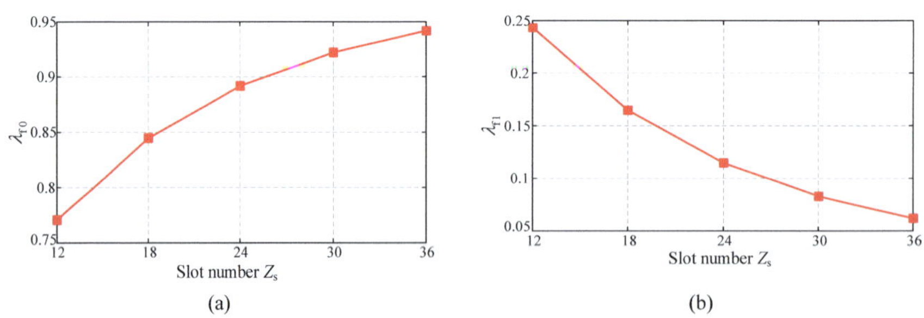

Fig. 2.15 Variation of specific permeance with the slot number Z_s: **a** average term, **b** fundamental term

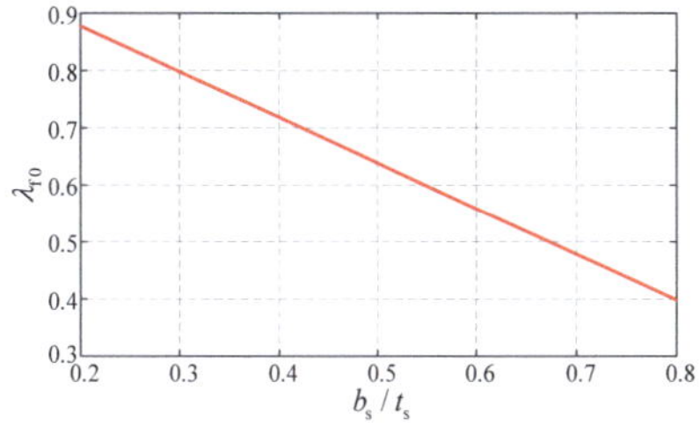

(a)

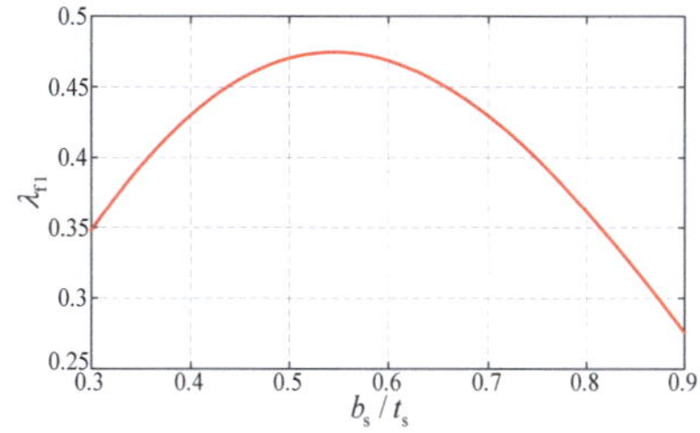

(b)

Fig. 2.16 Variation of specific permeance with b_s/t_s: **a** average term, **b** fundamental term in electrically excited FMM

Fig. 2.16 illustrates the corresponding changes in an electrically excited FMM, achieved by removing the permanent magnets and reducing the air gap length to 0.5 mm as per data in Table 2.1. Notably, in Fig. 2.16, λ_{f0} approaches 1 when the slot opening ratio b_s/t_s is small, mirroring observations in surface-mounted permanent magnet FMMs.

As the slot opening ratio increases, λ_{f0} declines more sharply in electrically excited FMMs than in surface-mounted permanent magnet FMMs. This more pronounced decrease is attributed to the smaller equivalent air gap length, where a wider slot width more significantly differentiates the magnetic fields between the slot and tooth regions. This results in a sharper decline in the average magnetic field intensity and an increase in magnetic field fluctuation. Consequently, λ_{f0} decreases, and λ_{f1} increases. The peak λ_{f1} is observed at a b_s/t_s ratio of 0.55, which is shifted leftward compared to the curve in Fig. 2.14. At $b_s/t_s = 0.55$, both λ_{f0} and λ_{f1} approximate 0.5, indicating that even in electrically excited FMMs, the utilization of specific permeance remains suboptimal [11].

2.2.2.3 The Additional Magneto-motive Force Introduced by Non-uniform Air Gap

The following is the derivation process of the additional MMF introduced via stator slotting, $F_a{}' - F_a$, which equals F_r. According to the principle of magnetic flux continuity:

$$0 = \int_0^{2\pi} B_a(\theta, t)\mathrm{d}\theta = \frac{\mu_0}{g}\int_0^{2\pi} F_a'(\theta, t)\lambda_s(\theta)\mathrm{d}\theta = \frac{\mu_0}{g}\int_0^{2\pi} [F_a(\theta, t) - F_r(t)]\lambda_s(\theta)\mathrm{d}\theta \tag{2.30}$$

Consequently, $F_r(t)$ can be solved as:

$$F_{\mathrm{r}}(t) = \frac{\int_0^{2\pi} F_{\mathrm{a}}(\theta, t)\lambda_{\mathrm{s}}(\theta)\mathrm{d}\theta}{\int_0^{2\pi} \lambda_{\mathrm{s}}(\theta)\mathrm{d}\theta} = \frac{\int_0^{2\pi} F_{\mathrm{a}}(\theta, t)\lambda_{\mathrm{s}}(\theta)\mathrm{d}\theta}{2\pi\lambda_{\mathrm{s}0}} \tag{2.31}$$

For simplicity, this analysis considers only the three-phase integer-slot winding condition, and its MMF can be written as:

$$F_{\mathrm{a}}(\theta, t) = \sum_{h=1}^{\infty} F_{\mathrm{a}(6h\pm1)} \cos\left[(6h \pm 1)P_{\mathrm{a}}\theta \mp \omega_{\mathrm{a}}t + \theta_{\mathrm{a}(6h\pm1)}\right] \tag{2.32}$$

where $F_{\mathrm{a}(6h\pm1)}$ represents the $(6h \pm 1)$th harmonic. Upon substituting Eqs. (2.24) and (2.32) into Eq. (2.31):

$$F_{\mathrm{r}}(t) = \frac{\sum_{(6h\pm1)P_{\mathrm{a}}=lP_{\mathrm{f}}} F_{\mathrm{a}(6h\pm1)}\lambda_{\mathrm{s}l} \cos\left(\theta_{\mathrm{a}(6h\pm1)} \mp \omega_{\mathrm{a}}t\right)}{2\lambda_{\mathrm{s}0}} \tag{2.33}$$

Any positive integers h and l that satisfy:

$$(6h \pm 1)P_{\mathrm{a}} = lP_{\mathrm{f}} \tag{2.34}$$

This condition can produce additional MMFs. When the stator is slotted, P_{f} equals Z_{s}, and therefore h and l do not satisfy Eq. (2.34). Consequently, $F_{\mathrm{r}}(t) = 0$, and no additional MMF is generated. Conversely, when the rotor is slotted, h and l might satisfy Eq. (2.34), resulting in $F_{\mathrm{r}}(t) \neq 0$. Therefore, the calculation of additional MMF should be performed under conditions of rotor slotting.

2.2.3 *Flux Modulation Theory of Surface-Mounted Permanent Magnet Structure*

Initially, the scenario with a uniform air gap is considered. Regardless of whether the permanent magnets are positioned on the stator or the rotor, they consistently rotate when observed from the reference system on the opposite side [12, 13]. Therefore, it is assumed by default that the permanent magnets are placed on the rotor. In the case of a uniform air gap, the stator and rotor are at equipotential under no-load conditions. Given that the permanent magnets operate in the second quadrant with a negative magnetic field intensity, the magnetic potential increases along the magnetization direction of the permanent magnets and subsequently decreases across the air gap, as depicted in Fig. 2.17.

The following equations can be obtained using Ampere's circuital law, the flux continuity principle, and the relationship between flux density and magnetic field intensity:

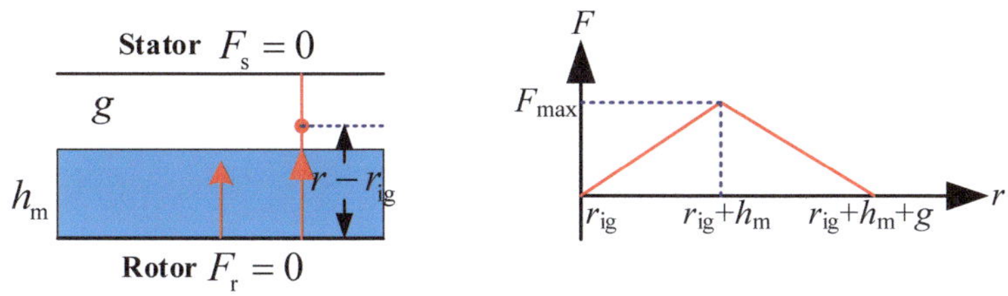

Fig. 2.17 Magnetic potential variation in the air gap and surface-mounted permanent magnets

$$\begin{aligned} B_{em} &= B_r + \mu_0\mu_r H_{em}, \quad B_{eg} = \mu_0 H_{eg} \\ B_{em} &= B_{eg}, \quad H_{em}h_m + H_{eg}g = 0 \end{aligned} \tag{2.35}$$

where B_{em} and H_{em} represent the flux density and magnetic field intensity in the permanent magnet, respectively; B_{eg} and H_{eg} indicate the flux density and magnetic field intensity in the air gap, respectively; and B_r denotes the remanence of the permanent magnets. By solving the above equations:

$$B_{eg} = \frac{B_r}{1 + \mu_r \frac{g}{h_m}} = \frac{\mu_0}{g + \frac{h_m}{\mu_r}} \frac{B_r h_m}{\mu_0 \mu_r} \tag{2.36}$$

A comparison of Eqs. (2.36) and (2.21) suggests that the air gap structure with permanent magnets in Fig. 2.17 is analogous to the air gap structure without permanent magnets in Fig. 2.18. In this latter configuration, the permanent magnets are substituted by excitation coils with a specific number of turns, N_c, and current, I_c, to ensure the air gap flux density remains constant:

$$\begin{aligned} g_{ef} &= g + \frac{h_m}{\mu_r} \\ N_c I_c &= \frac{B_r h_m}{\mu_0 \mu_r} \end{aligned} \tag{2.37}$$

Subsequently, the MMF generated by the permanent magnet is equated to the MMF produced by the excitation winding, which manifests as a symmetric square wave. Through Fourier decomposition, the MMF expression is derived as follows:

$$F_e(\theta, t) = \frac{4}{\pi} \frac{B_r h_m}{\mu_0 \mu_r} \sum_{h=1,3,5\ldots}^{\infty} \frac{1}{h} \sin\left(\frac{\pi h \alpha_m}{2}\right) \cos[h P_e(\theta - \Omega_e t)] \tag{2.38}$$

where α_m is the pole arc of the permanent magnets.

In Fig. 2.18, the permanent magnets are substituted with a closed-slot excitation winding, allowing the air gap to be considered uniform with an equivalent length denoted by g_{ef}. However, if the stator is slotted, the specific permeance function can

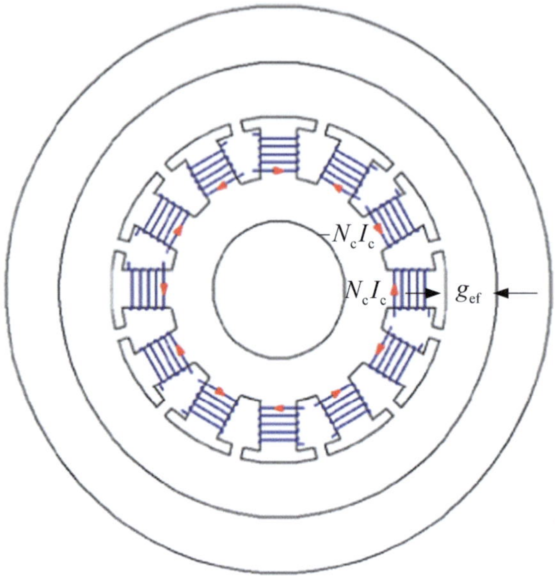

Fig. 2.18 Equivalent winding structure of the permanent magnets

be expressed as:

$$\lambda_s(\theta) = \frac{g + \frac{h_m}{\mu_r}}{g(\theta) + \frac{h_m}{\mu_r}} = \frac{g_{ef}}{g_{ef} + \Delta g(\theta)} \tag{2.39}$$

where $\Delta g(\theta)$ represents the additional air gap length introduced by the stator slotting. Due to the stator slotting, the air gap becomes non-uniform, and thus the stator and rotor are no longer at equipotential. Assuming the rotor's magnetic potential is zero, the stator's magnetic potential is derived using the flux continuity principle:

$$F_s(t) = -\frac{\sum_{hP_e=lP_f} F_{eh}\lambda_{sj}\cos(hP_e\Omega_e t)}{2\lambda_{s0}} \tag{2.40}$$

where F_{eh} and λ_{sl} represent the hth and lth harmonic amplitude of the MMF and specific permeance in Eqs. (2.38) and (2.39), respectively. Then, the air gap MMF under no-load condition and stator slotting becomes $F_e{}' = F_e - F_s$.

In most FMMs, the pole numbers P_f and P_e are closely aligned, resulting in large values of h and l that satisfy $hP_e = lP_f$ in Eq. (2.40), and the magnitude of the corresponding harmonic is minimal. Therefore, the stator magnetic potential F_s is generally approximated to zero. However, in specific "speed-increasing" FMMs where P_e significantly exceeds P_f, the values of h and l satisfying $hP_e = lP_f$ are smaller, and F_s cannot be disregarded.

2.2.4 Specific Permeance of Double-Sided Slotting

In certain FMM topologies such as flux switching machines, vernier reluctance machines, and flux reversal machines, both stators and rotors are slotted, as illustrated in Fig. 2.19a. The specific permeance functions for these air gaps differ from those with only single-sided slotting. As depicted in Fig. 2.19b, $g_s(\theta)$ indicates the air gap length with a slotted stator and a smooth rotor, whereas $g_r(\theta)$ represents the air gap length with a slotted rotor and a smooth stator. The actual air gap length with double-sided slotting is then determined by:

$$g(\theta) = g_s(\theta) + g_r(\theta) - g \tag{2.41}$$

Based on the definition of specific permeance:

$$\begin{aligned}\lambda_f(\theta) &= \frac{g}{g_s(\theta) + g_r(\theta) - g} = \frac{1}{\frac{1}{\lambda_s(\theta)} + \frac{1}{\lambda_r(\theta)} - 1} \\ &= \lambda_s(\theta)\lambda_r(\theta)\frac{1}{\lambda_s(\theta) + \lambda_r(\theta) - \lambda_s(\theta)\lambda_r(\theta)}\end{aligned} \tag{2.42}$$

where $\lambda_s(\theta)$ and $\lambda_r(\theta)$ denote the air gap's specific permeance functions when only the stator and rotor is slotted, respectively.

In Eq. (2.42), the specific permeance resulting from double-sided slotting is modeled as the product of the two specific permeances from single-sided slotting, adjusted by a coefficient. This coefficient is defined as:

$$k_\lambda = \frac{1}{\lambda_s(\theta) + \lambda_r(\theta) - \lambda_s(\theta)\lambda_r(\theta)} \tag{2.43}$$

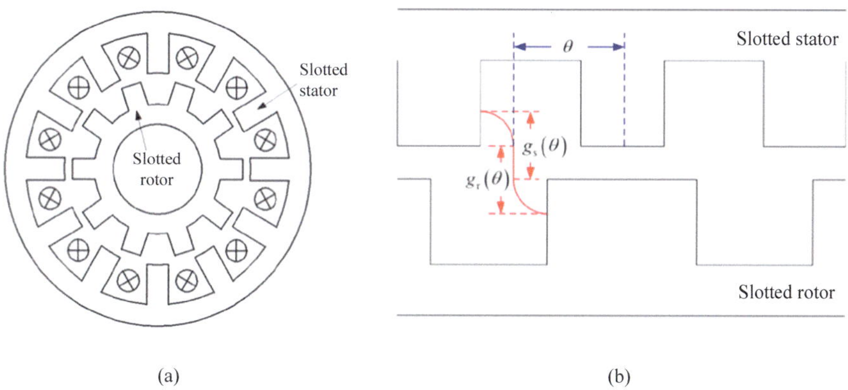

Fig. 2.19 FMM with double-sided slotting: **a** FMM structure, **b** equivalent air gap structure

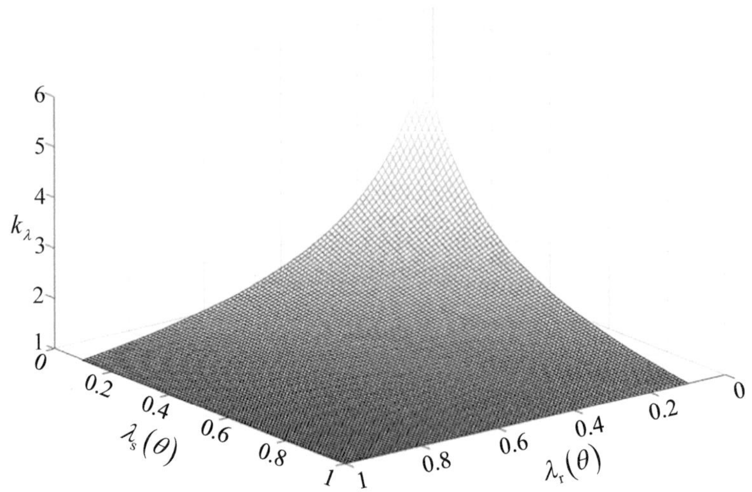

Fig. 2.20 Variation of the coefficient k_λ with specific permeance $\lambda_s(\theta)$ and $\lambda_r(\theta)$

The coefficient k_λ is a binary function of $\lambda_s(\theta)$ and $\lambda_r(\theta)$. When θ lies at the tooth centerline, $\lambda_s(\theta)$ is 1; and when θ lies at the slot centerline, $\lambda_s(\theta)$ reaches the minimum, close to 0. $\lambda_r(\theta)$ follows a similar trend. Figure 2.20 illustrates the variation of the coefficient k_λ with $\lambda_s(\theta)$ and $\lambda_r(\theta)$. When either $\lambda_s(\theta)$ and $\lambda_r(\theta)$ is close to 1, k_λ is on the verge of becoming 1. Only when both $\lambda_s(\theta)$ and $\lambda_r(\theta)$ are approximately 0 simultaneously, k_λ is much higher than 1. In this case, θ is positioned at both the stator and rotor slot centerlines, where $\lambda_s(\theta)$ and $\lambda_r(\theta)$ are very small. Even if k_λ is assumed to be 1, the absolute error for the specific permeance $\lambda_f(\theta)$ is negligible. Therefore, k_λ can be approximated as 1, and Eq. (2.42) can be rewritten as [3]:

$$\lambda_f(\theta) \approx \lambda_s(\theta)\lambda_r(\theta) \tag{2.44}$$

Considering the time-varying nature of the specific permeance introduced by the rotor rotation, the air gap's specific permeance functions $\lambda_s(\theta)$ and $\lambda_r(\theta)$ can be formulated as:

$$\begin{aligned} \lambda_s(\theta) &= \lambda_{s0} + \sum_{l_s=1}^{\infty} \lambda_{sl_s} \cos l_s Z_s \theta \\ \lambda_r(\theta, t) &= \lambda_{r0} + \sum_{l_r=1}^{\infty} \lambda_{rl_r} \cos l_r Z_r (\theta - \Omega_r t) \end{aligned} \tag{2.45}$$

Here, Z_s and Z_r represent the stator slot number and rotor slot number, respectively; λ_{s0} and λ_{r0} denote the average terms of the specific permeance with stator slotting and rotor slotting, respectively; λ_{sls} and λ_{rlr} indicate the amplitudes of the l_sth and l_rth harmonic of the specific permeance with stator slotting and rotor slotting, respectively.

By combining Eqs. (2.44) and (2.45):

$$\lambda_f(\theta, t) = \lambda_{s0}\lambda_{r0} + \lambda_{s0}\sum_{l_r=1}^{\infty}\lambda_{rl_r}\cos l_r Z_r(\theta - \Omega_r t)$$
$$+ \lambda_{r0}\sum_{l_s=1}^{\infty}\lambda_{sl_s}\cos l_s Z_s\theta + \frac{1}{2}\sum_{l_s,l_r}\lambda_{sl_s}\lambda_{rl_r}\cos[(l_r Z_r \pm l_s Z_s)\theta - l_r\Omega_r t] \tag{2.46}$$

For the double-sided slotting structure, new specific permeance harmonic components are generated in addition to the original harmonics caused by stator and rotor slotting. These new harmonics have pole pair numbers equal to the sum or difference of the stator slot number Z_s and the rotor slot number Z_r.

2.2.5 *Additional Magneto-Motive Force of Non-Surface-Mounted Flux Modulation Machine*

In Sects. 2.2.2 and 2.2.3, we investigate the flux modulation processes of the armature winding and the surface-mounted permanent magnet structure under slotting effects. Previous analyses indicate that the MMF in the air gap is not solely generated by the winding or the PMs; it also includes an additional MMF due to slotting. Although these additional MMFs are generally minor and can be disregarded for the armature winding and surface-mounted permanent magnet structures, they become significant in non-surface-mounted permanent magnet structures, when the air gap is non-uniform due to slotting effects. This subsection examines the flux modulation process and additional MMF in non-surface-mounted permanent magnet FMMs, focusing on two typical topologies: the consequent-pole vernier permanent magnet machine and the spoke-magnetization vernier permanent magnet machine.

2.2.5.1 Consequent-Pole Vernier Permanent Magnet Machine

Figure 2.21a illustrates the structure of the consequent-pole vernier permanent magnet machine. Unlike the surface-mounted permanent magnet FMM, the permanent magnets here share the same polarity and are evenly distributed along the rotor surface, with the number of permanent magnet blocks matching the rotor pole pairs. The iron core, located between adjacent permanent magnets, is depicted in Fig. 2.21b. Notably, the iron core acts as the opposite polarity, hence it is referred to as the ferromagnetic pole [14].

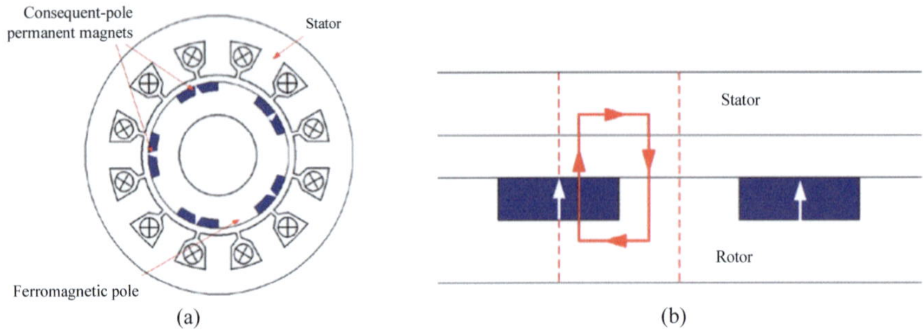

Fig. 2.21 Consequent-pole vernier permanent magnet machine: **a** structure, **b** magnetic circuit

To elucidate the working principle of the consequent-pole vernier permanent magnet machine, we employ the MMF/specific permeance model. Ideally, flux lines are perpendicular to the tangential direction, whereas actual flux lines are perpendicular to the iron core surface, as depicted in Fig. 2.22. Assuming ideal flux lines, the variation in magnetic potential in the permanent magnet and air gap is illustrated in Fig. 2.17. Consequently, the MMF generated by the permanent magnets can be equated to the MMF of the excitation winding. For the ferromagnetic pole, the MMF is the magnetic potential difference between the stator and rotor. Assuming the rotor magnetic potential is F_r, the MMF produced by the consequent-pole permanent magnet rotor is illustrated in Fig. 2.23, where α_m represents the pole arc of the permanent magnets.

The air gap specific permeance is a segmented function, which can be expressed as:

$$\lambda_r = \begin{cases} 1, & \frac{\pi}{2P_e}\alpha_m < \theta_r < \frac{\pi}{2P_e}(4-\alpha_m) \\ \frac{1}{1+\frac{h_m}{g\mu_r}}, & -\frac{\pi}{2P_e}\alpha_m < \theta_r < \frac{\pi}{2P_e}\alpha_m \end{cases} \tag{2.47}$$

According to the flux continuity principle:

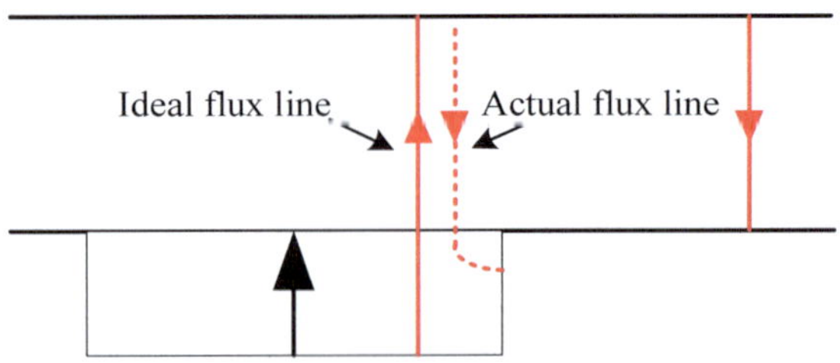

Fig. 2.22 No-load flux line distribution in a consequent-pole vernier permanent magnet machine

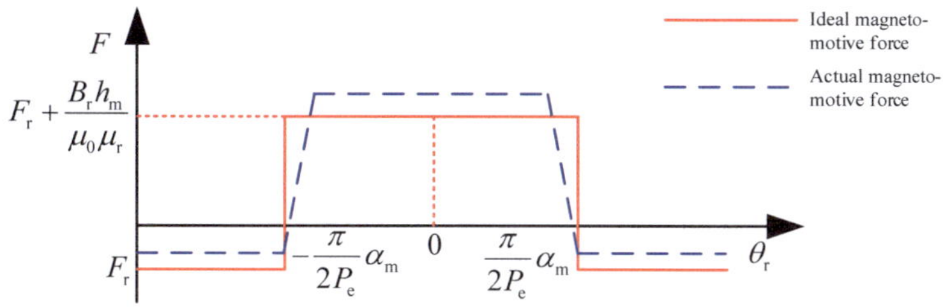

Fig. 2.23 MMF excited by the consequent-pole permanent magnet rotor

$$\int_0^{2\pi} B\mathrm{d}\theta_r = \int_0^{2\pi} \frac{\mu_0}{g} F_e \lambda_r \mathrm{d}\theta_r = 0 \tag{2.48}$$

Consequently, F_r can be obtained as:

$$F_r = -\frac{B_r h_m}{\mu_0 \mu_r} \frac{\alpha_m}{2 + \frac{h_m}{\mu_r g}(2 - \alpha_m)} \tag{2.49}$$

Therefore, the MMF excited by the consequent-pole permanent magnet rotor in Fig. 2.23 can be given by:

$$F_{e1} = \begin{cases} -\frac{B_r h_m}{\mu_0 \mu_r} \frac{\alpha_m}{2+\frac{h_m}{\mu_r g}(2-\alpha_m)}, & \frac{\pi}{2P_e}\alpha_m < \theta_r < \frac{\pi}{2P_e}(4 - \alpha_m) \\ \frac{B_r h_m}{\mu_0 \mu_r} \frac{\left(\frac{h_m}{\mu_r g}+1\right)(2-\alpha_m)}{2+\frac{h_m}{\mu_r g}(2-\alpha_m)}, & -\frac{\pi}{2P_e}\alpha_m < \theta_r < \frac{\pi}{2P_e}\alpha_m \end{cases} \tag{2.50}$$

Based on the MMF/specific permeance model, the air gap flux density can be calculated as:

$$B(\theta_r) = \frac{\mu_0}{g} F_{e1} \lambda_r = \begin{cases} -\frac{B_r h_m}{g \mu_r} \frac{\alpha_m}{2+\frac{h_m}{\mu_r g}(2-\alpha_m)}, & \frac{\pi}{2P_e}\alpha_m < \theta_r < \frac{\pi}{2P_e}(4 - \alpha_m) \\ \frac{B_r h_m}{g \mu_r} \frac{(2-\alpha_m)}{2+\frac{h_m}{\mu_r g}(2-\alpha_m)}, & -\frac{\pi}{2P_e}\alpha_m < \theta_r < \frac{\pi}{2P_e}\alpha_m \end{cases} \tag{2.51}$$

According to Eq. (2.50), the excitation MMF depends on parameters such as the permanent magnet pole arc (α_m) and $h_m/\mu_r g$. As α_m increases, the MMF of the permanent magnets decreases, whereas the MMF of the ferromagnetic poles increases. Additionally, since $h_m/\mu_r g$ is significantly greater than 1, the MMF amplitude of the permanent magnets is considerably larger than that of the ferromagnetic poles. These analyses are based on ideal flux line distributions. However, considering the actual flux line distribution, as shown in Fig. 2.22, the flux lines at the junction of the permanent magnet and the ferromagnetic pole bend and become perpendicular to the iron core surface. Consequently, the MMF at the junction does not immediately reach its maximum value but gradually rises, as illustrated in Fig. 2.23. This behavior is similar to a decrease in the pole arc coefficient of the permanent magnets, leading

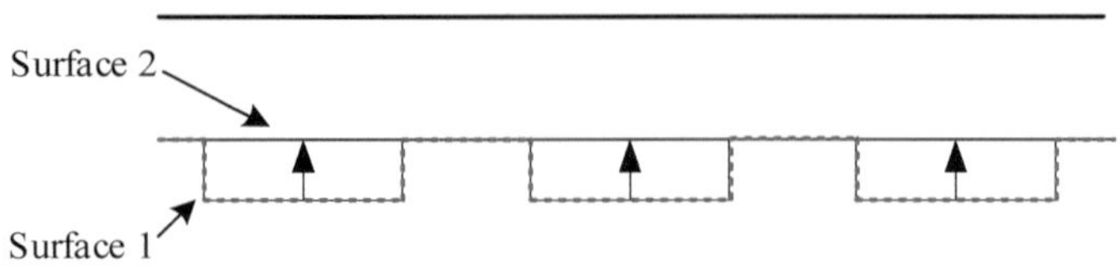

Fig. 2.24 Rotor surfaces from different viewpoints

to an overall increase in the average MMF value. Therefore, Eqs. (2.47), (2.50), and (2.51), based on ideal flux lines, entail some errors when compared to the actual situation. However, due to the complexity of calculating actual flux line distributions and the negligible impact on subsequent flux modulation process analyses, no further derivation is provided here. Interested readers can explore this aspect further.

The above analysis treats the rotor as a slotted structure, considering the permanent magnets and the ferromagnetic poles as two separate parts. In this context, F_{e1} corresponds to the magnetic potential on rotor surface 1, as depicted in Fig. 2.24. Alternatively, when the rotor is regarded as a non-slotted structure and the permanent magnets and ferromagnetic poles are considered as a whole, the equivalent rotor excitation MMF F_{e2} will be equal to the magnetic potential on rotor surface 2.

From Eq. (2.51), the equivalent rotor excitation MMF, F_{e2}, with a uniform air gap can be determined as:

$$F_{e2} = \frac{g}{\mu_0} B(\theta_r) = \begin{cases} -\frac{B_r h_m}{\mu_0 \mu_r} \frac{\alpha_m}{2 + \frac{h_m}{\mu_r g}(2 - \alpha_m)}, & \frac{\pi}{2P_e}\alpha_m < \theta_r < \frac{\pi}{2P_e}(4 - \alpha_m) \\ \frac{B_r h_m}{\mu_0 \mu_r} \frac{(2 - \alpha_m)}{2 + \frac{h_m}{\mu_r g}(2 - \alpha_m)}, & -\frac{\pi}{2P_e}\alpha_m < \theta_r < \frac{\pi}{2P_e}\alpha_m \end{cases} \tag{2.52}$$

Both F_{e1} and F_{e2} manifest as square waves, maintaining identical values within the ferromagnetic pole area, yet differing in the permanent magnet area. These calculations of F_{e1} and F_{e2} assume an unslotted stator. Subsequent analysis will assess the effects of stator slotting on these values.

With the introduction of stator slotting, alterations occur in the rotor magnetic potential. Assuming the post-slotting rotor magnetic potential as F_r', the relationship between F_{e1}' and F_{e1} is delineated as follows:

$$F_{e1}' = F_{e1} + F_r' - F_r \tag{2.53}$$

Combining Eqs. (2.42) and (2.53), the air gap flux density is derived as:

$$B = \frac{\mu_0}{g} F_{e1}' \frac{\lambda_s \lambda_r}{\lambda_s + \lambda_r - \lambda_s \lambda_r} = \frac{\mu_0}{g} F_{e1}' \lambda_s \lambda_r k_\lambda \tag{2.54}$$

Notably, in Sect. 2.2.4, the coefficient k_λ, defined as $1/(\lambda_s + \lambda_r - \lambda_s \lambda_r)$, is approximated as 1. However, in Eq. (2.54), treating k_λ as 1 is inappropriate due to the high value of F_{e1}'. This discrepancy indicates that the approximation of k_λ as 1 is solely applicable to specific permeance calculations and not when determining the air gap flux density in Eq. (2.54).

The procedural steps for calculating the air gap flux density in a consequent-pole vernier permanent magnet machine are as follows:

1. Assuming a smooth stator, calculate the rotor MMF, F_{e1}, and rotor specific permeance, λ_{r}. Subsequently, calculate the specific permeance, λ_{s}, for the slotted stator.
2. Utilize Eq. (2.53) and the flux continuity principle to compute $F_{\text{r}}{}'$ and $F_{\text{e1}}{}'$.
3. Employ Eq. (2.54) to determine the air gap flux density.

This analysis of the consequent-pole vernier permanent magnet machine underscores the physical concepts involved. Typically, researchers calculate the no-load air-gap flux density of an MMF under a uniform air gap without slotting. They then apply the specific permeance function introduced by slotting to compute the flux density distribution under a non-uniform air gap. This method is particularly applicable to the armature magnetic field and surface-mounted permanent magnet structures discussed in Sects. 2.2.2 and 2.2.3, where the additional MMF induced by slotting is considered negligible. However, for consequent-pole permanent magnet structures, following this conventional approach yields a flux density expression ($F_{\text{e1}}\mu_0/g\lambda_{\text{s}}\lambda_{\text{r}}$) that substantially deviates from Eq. (2.54). This discrepancy arises from differences between the MMF $F_{\text{e1}}{}'$ and F_{e1}, as well as the neglect of k_λ. Consequently, this method proves unsuitable for consequent-pole FMMs.

Subsequent investigations will focus on MMF F_{e2} as depicted in Fig. 2.24. According to the principle of flux modulation, the air gap flux density may also be represented as:

$$B = \frac{\mu_0}{g} F'_{\text{e2}} \lambda_{\text{s}} \tag{2.55}$$

Combining Eqs. (2.55) and (2.54), the $F_{\text{e2}}{}'$ can be derived as:

$$F'_{\text{e2}} = F'_{\text{e1}} \frac{\lambda_{\text{r}}}{\lambda_{\text{s}} + \lambda_{\text{r}} - \lambda_{\text{s}}\lambda_{\text{r}}} \tag{2.56}$$

In scenarios featuring a uniform air gap, the excitation MMF is F_{e2}. However, following stator slotting, the magnetic potential on the rotor surface is altered, resulting in an air gap flux density post-slotting that is not equivalent to $F_{\text{e2}}\lambda_{\text{s}}\mu_0/g$. This conclusion is applicable to other FMMs; the feasibility of obtaining the post-slotting air gap flux density by multiplying the excitation MMF by a uniform air gap and the specific permeance introduced by slotting hinges on the equality of excitation MMFs pre and post-slotting. For the consequent-pole vernier permanent magnet machine, $F_{\text{e2}}{}' \neq F_{\text{e2}}$. The difference between $F_{\text{e2}}{}'$ and F_{e2} originates from the additional MMF induced by changes in the rotor magnetic potential before and after slotting, coupled with the factor k_λ, as specified in Eq. (2.43). Notably, k_λ is a time-variant function that introduces additional harmonic content into the MMF post-slotting.

To analyze F_{e2}' thoroughly, examining the harmonics in k_λ is essential. However, as both stator and rotor specific permeance variables in k_λ are situated in the denominator, decoding the harmonic contents within k_λ presents challenges. Simplification may be achieved by transposing these variables to the numerator. According to Eq. (2.47), only two possible values for λ_r exist. When θ_r is within the ferromagnetic pole area, λ_r equals 1, and k_λ remains 1 without further adjustment. Conversely, when θ_r is in the permanent magnet area, λ_r attains its minimum value, denoted $\lambda_{r\text{-min}}$, rendering k_λ a univariate function of λ_s, which can be modeled using a quadratic function. The initial step for quadratic function fitting involves selecting three points of (λ_s, k_λ), namely:

$$\begin{aligned}
&\lambda_s = 1 && k_\lambda = 1, \\
&\lambda_s = \tfrac{\lambda_{s\text{-min}}+1}{2} && k_\lambda = \tfrac{2}{\lambda_{s\text{-min}}+\lambda_{r\text{-min}}-\lambda_{s\text{-min}}\lambda_{r\text{-min}}+1}, \\
&\lambda_s = \lambda_{s\text{-min}} && k_\lambda = \tfrac{1}{\lambda_{s\text{-min}}+\lambda_{r\text{-min}}-\lambda_{s\text{-min}}\lambda_{r\text{-min}}}.
\end{aligned} \tag{2.57}$$

where $\lambda_{s\text{-min}}$ represents the minimum value of stator specific permeance. The corresponding quadratic function is then derived as follows:

$$k_\lambda \approx k_1\lambda_s^2 + k_2\lambda_s + k_3 \tag{2.58}$$

where the coefficients k_1, k_2 and k_3 are defined as:

$$\begin{aligned}
k_1 = {} & \frac{2}{(1-\lambda_{s\text{-min}})^2}\left(1 - \frac{4}{\lambda_{s\text{-min}}+\lambda_{r\text{-min}}-\lambda_{s\text{-min}}\lambda_{r\text{-min}}+1}\right. \\
& \left. + \frac{1}{\lambda_{s\text{-min}}+\lambda_{r\text{-min}}-\lambda_{s\text{-min}}\lambda_{r\text{-min}}}\right) \\
k_2 = {} & -\frac{1}{(1-\lambda_{s\text{-min}})^2}\left[3\lambda_{s\text{-min}}+1 - \frac{8(\lambda_{s\text{-min}}+1)}{\lambda_{s\text{-min}}+\lambda_{r\text{-min}}-\lambda_{s\text{-min}}\lambda_{r\text{-min}}+1}\right. \\
& \left. + \frac{\lambda_{s\text{-min}}+3}{\lambda_{s\text{-min}}+\lambda_{r\text{-min}}-\lambda_{s\text{-min}}\lambda_{r\text{-min}}}\right] \\
k_3 = {} & \frac{1}{(1-\lambda_{s\text{-min}})^2}\left[(\lambda_{s\text{-min}}+1)\lambda_{s\text{-min}} - \frac{8\lambda_{s\text{-min}}}{\lambda_{s\text{-min}}+\lambda_{r\text{-min}}-\lambda_{s\text{-min}}\lambda_{r\text{-min}}+1}\right. \\
& \left. + \frac{\lambda_{s\text{-min}}+1}{\lambda_{s\text{-min}}+\lambda_{r\text{-min}}-\lambda_{s\text{-min}}\lambda_{r\text{-min}}}\right]
\end{aligned} \tag{2.59}$$

As Eq. (2.58) is valid only when $\lambda_r = \lambda_{r\text{-min}}$, to make it equally valid when $\lambda_r = 1$, the function can be rewritten as:

$$k_\lambda \approx \frac{(1-\lambda_r)\left(k_1\lambda_s^2 + k_2\lambda_s + k_3\right) + (\lambda_r - \lambda_{r\text{-min}})}{1-\lambda_{r\text{-min}}} \tag{2.60}$$

In Eq. (2.60), the terms $(1 - \lambda_r)$ and $(\lambda_r - \lambda_{r\text{-min}})$ encompass harmonics with lP_e pole pairs, while the term $(k_1\lambda_s^2 + k_2\lambda_s + k_3)$ incorporates harmonics with lZ_s pole pairs. Consequently, k_λ contains harmonics with pole pair numbers $|l_sZ_s \pm l_eP_e|$. When k_λ is multiplied by F_{e1}' in Eq. (2.56), a series of harmonics with the same pole pairs will be generated in F_{e2}'.

The harmonic contents in F_{e2}' can be categorized as follows:

(1) Harmonics with l_eP_e $(l_e > 0)$ pole pairs: These harmonics are present in the rotor excitation MMF F_{e2} even without stator slotting. However, their magnitudes differ before and after stator slotting. Among these harmonics, the one with P_e pole pairs is considered the primary source for generating the fundamental wave of back electromotive force.

(2) Harmonics with lZ_s $(l_r > 0)$ pole pairs: These are new components, primarily resulting from the product of the DC bias term in F_{e1}' and the l_sZ_s pole pair harmonics in k_λ. While these harmonics are stationary with respect to the stator and do not directly affect the winding flux linkage and back electromotive force, their interaction with the stator-specific permeance generates unipolar air gap magnetic fields, indirectly influencing the main magnetic field.

(3) Harmonics with $|l_sZ_s \pm l_eP_e|$ pole pairs: These are also new components, with particular focus on the harmonic with $|Z_s - P_e|$ pole pairs. This is crucial because both its pole pairs and electrical frequency match the main working magnetic field. Additionally, the $|Z_s - P_e|$ pole pair harmonic, when interacting with the average term of the specific permeance, can produce additional working flux density harmonics. Further explanation of its origin and influence on the working magnetic field is provided below.

Equation (2.52) provides the expression for F_{e2}, and Fig. 2.25 depicts the waveforms of F_{e2}. The waveform is a periodic square wave with positive and negative symmetry, where the positive region corresponds to the permanent magnet region, and the negative region corresponds to the ferromagnetic pole region. After stator slotting, F_{e2} undergoes changes. The magnetic potential of the rotor core transitions from F_r to F_r', causing the waveform in the ferromagnetic pole region to shift downward. The magnetic potential on the permanent magnet surface is then examined. Due to the unequal number of stator teeth and rotor permanent magnets, the relative positions of each permanent magnet to the stator teeth differ. In Fig. 2.25, the central permanent magnet aligns with the stator tooth centerline. At this position, the magnetic resistance between the stator and the rotor is minimized, resulting in the strongest magnetic field generated by the permanent magnet and the greatest magnetic pressure drop within it. Consequently, the magnetic potential on the surface of the permanent magnet is nearly identical to that without stator slotting. As the permanent magnet centerline gradually offsets from the stator tooth centerline, the magnetic resistance increases, the magnetic field generated by the permanent magnet weakens, and its internal magnetic pressure drop decreases. This leads to an increase in the magnetic potential on the surface of the permanent magnet, as indicated by the dashed line in Fig. 2.25. Therefore, stator slotting alters the rotor magnetic potential, introducing additional MMF.

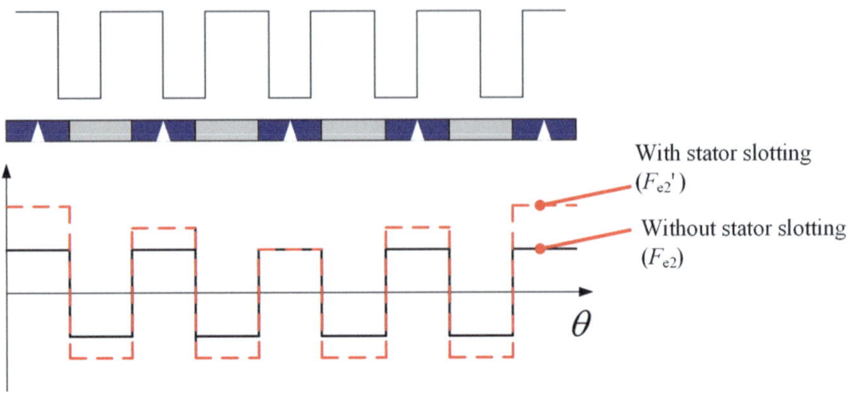

Fig. 2.25 Change in the magnetic potential on the rotor surface before and after stator slotting

Referencing the methodologies described in Sect. 2.1, further exploration into the nature of the additional MMF and its impact on the air gap magnetic field is warranted. As depicted in Fig. 2.26, the centerline of the middle permanent magnet aligns with the stator tooth centerline, resulting in zero additional MMF. The displacement of the left permanent magnet from the centerline of the adjacent stator tooth is given by:

$$\theta_{ef} = 2\pi\left(\frac{1}{P_e} - \frac{1}{Z_s}\right) \tag{2.61}$$

Given that $\theta_{ef} > 0$, the left permanent magnet produces additional MMF. As this displacement from the stator tooth centerline increases, the additional MMF from subsequent permanent magnets escalates sequentially until the centerline of the permanent magnet nears the stator tooth centerline, at which point the additional MMF begins to diminish. If the centerline of the n_{ef}th stator tooth realigns with that of the permanent magnet, the additional magnetic potential reverts to zero, as illustrated by the leftmost permanent magnet in Fig. 2.26.

From the relationship of structural parameters, n_{ef} can be calculated as:

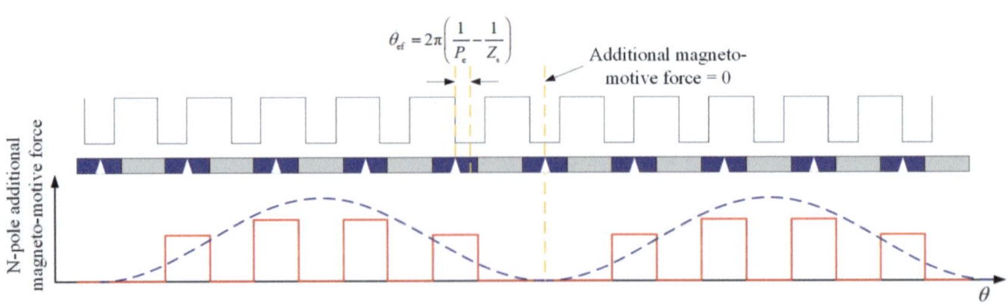

Fig. 2.26 Additional MMF on the surface of the permanent magnets

$$n_{\text{ef}} = \frac{2\pi}{Z_{\text{s}}\theta_{\text{ef}}} = \frac{P_{\text{e}}}{Z_{\text{s}} - P_{\text{e}}} \tag{2.62}$$

Consequently, the number of pole pairs of the fundamental component of the additional MMF is determined as:

$$P = \frac{P_{\text{e}}}{n_{\text{ef}}} = Z_{\text{s}} - P_{\text{e}} \tag{2.63}$$

Subsequently, the rotational speed of the additional MMF is examined. As shown in Fig. 2.27, when the rotor rotates counterclockwise by θ_{ef} degrees, the centerline of the right permanent magnet aligns with the stator tooth centerline, inducing the additional MMF to rotate clockwise by $2\pi/Z_{\text{s}}$ degrees. Therefore, the rotational speed of the additional MMF is defined as:

$$\Omega = -\frac{2\pi}{Z_{\text{s}}\theta_{\text{ef}}}\Omega_{\text{e}} = -\frac{P_{\text{e}}}{Z_{\text{s}} - P_{\text{e}}}\Omega_{\text{e}} \tag{2.64}$$

Assuming the rotor's position in Fig. 2.26 as the initial condition, the rotor excitation MMF is expressed as:

$$F'_{\text{e2}} \approx F_{P_{\text{e}}} \cos(P_{\text{e}}\theta - \omega_{\text{e}}t) - F_{P_{\text{a}}} \cos[(Z_{\text{s}} - P_{\text{e}})\theta + \omega_{\text{e}}t] \tag{2.65}$$

where $F_{P\text{e}}$ and $F_{P\text{a}}$ represent the harmonic amplitudes of the magnetomotive forces with P_{e} and $P_{\text{a}} = Z_{\text{s}} - P_{\text{e}}$ pole pairs, respectively.

If considering only the average and fundamental terms of the specific permeance function λ_{s}, the air gap flux density can be calculated, i.e.

$$B_{\text{e}}(\theta, t) = \frac{\mu_0}{g}F'_{\text{e2}}\lambda_{\text{s}} = \frac{\mu_0}{g}\left(F_{P_{\text{e}}}\lambda_{\text{s0}} - \frac{F_{P_{\text{a}}}\lambda_{\text{s1}}}{2}\right)\cos(P_{\text{e}}\theta - \omega_{\text{e}}t)$$

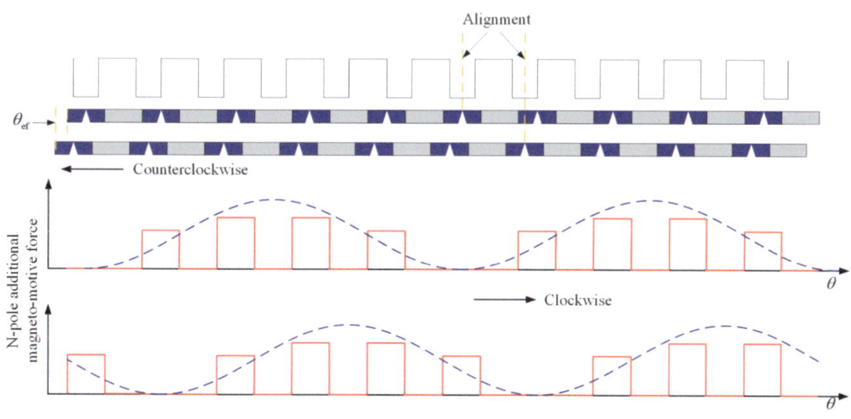

Fig. 2.27 Rotation of additional MMF at different rotor positions

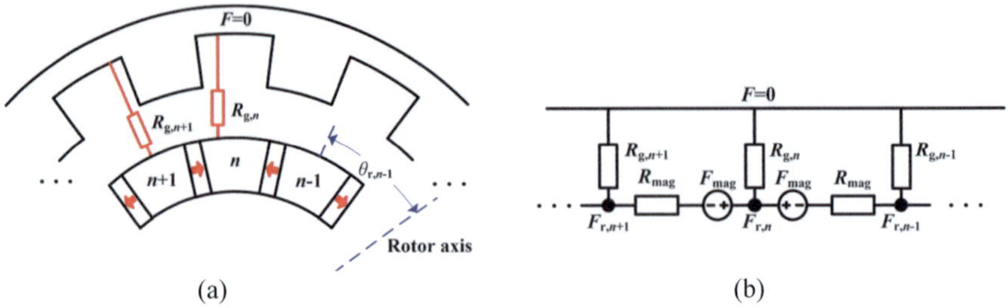

Fig. 2.28 Spoke-magnetization vernier permanent magnet machine: **a** topology, **b** equivalent magnetic circuit

$$+\frac{\mu_0}{g}\left(\frac{F_{P_e}\lambda_{s1}}{2}-F_{P_a}\lambda_{s0}\right)\cos(P_a\theta+\omega_e t)$$
$$+\frac{\mu_0}{g}\frac{F_{P_e}\lambda_{s1}}{2}\{\cos[(Z_s+P_e)\theta+\omega_e t]-\cos[(2Z_s-P_e)\theta+\omega_e t]\} \quad (2.66)$$

It is observed that the additional MMF introduces new flux density harmonics in the air gap. Notably, it attenuates the magnetic field harmonics associated with P_e and P_a pole pairs, where the latter denotes the working magnetic field harmonic of FMMs. Therefore, the additional MMF detrimentally impacts the performance of FMMs.

2.2.5.2 Spoke-Magnetization Vernier Permanent Magnet Machine

The previous subsection introduced the calculation of MMF for consequent-pole permanent magnets, which belongs to the inset type. In this subsection, we will derive and analyze the MMF of spoke-magnetization permanent magnets, which is more complex than that of the inset permanent magnet type. The basic structure of the spoke-magnetization vernier permanent magnet machine is shown in Fig. 2.28a.

In calculating the air gap flux density using the MMF/specific permeance model, the process begins by selecting the two boundaries of the air gap. Next, the magnetic potential difference and specific permeance between these boundaries are computed. Finally, multiplying the magnetic potential difference by the specific permeance yields the air gap flux density. For the spoke-magnetization permanent magnet structure, the air gap boundaries are the rotor's outer surface and the stator's inner surface. While the specific permeance in this region is straightforward to calculate, determining the magnetic potential on the rotor's outer surface is more complex. This complexity arises because the rotor surface comprises two distinct areas: permanent magnets and ferromagnetic poles. Each ferromagnetic pole is equipotential, thus having a constant magnetic potential. Conversely, the magnetic potential distribution of the permanent magnets is intricate. Additionally, after stator slotting, the air gap structure of each ferromagnetic pole changes, resulting in a lack of periodicity and symmetry in the magnetic potential of the ferromagnetic poles.

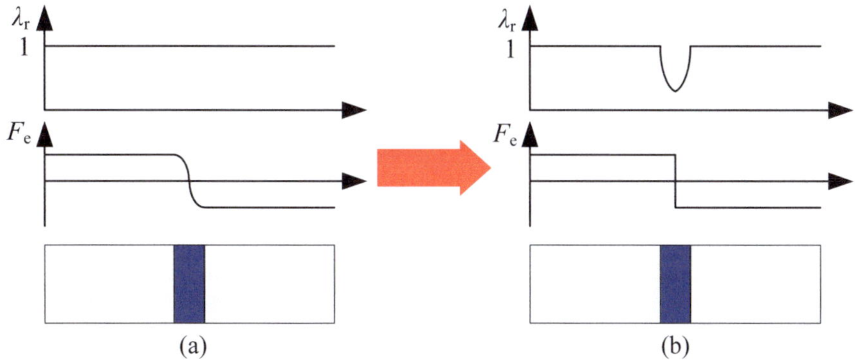

Fig. 2.29 Equivalent MMF and rotor specific permeance: **a** actual MMF versus rotor specific permeance, **b** equivalent MMF versus rotor specific permeance

To address this problem, an equivalent approach is employed, as explained below. Theoretically, the rotor's magnetic potential F_e and the specific permeance λ_r are depicted in Fig. 2.29a. The magnetic potentials of adjacent ferromagnetic poles are opposite, and the magnetic potential of the permanent magnet between two ferromagnetic poles gradually transitions from one ferromagnetic pole to the other. Closer to the centerline of the permanent magnet, the absolute value of its magnetic potential decreases. This drop in magnetic potential within the permanent magnet reduces the air gap flux density at the same position. This phenomenon is similar to the reduction of air gap flux density due to slotting, hence these permanent magnets can be equated to rotor slots. Furthermore, the change in the magnetic potential of the permanent magnets can be viewed as analogous to the change in specific permeance due to slotting, as shown in Fig. 2.29b. This equivalence simplifies the subsequent calculation of the MMF and specific permeance.

To address the second problem, we adopt a lumped parameter magnetic circuit model, as shown in Fig. 2.28b. In this model, each ferromagnetic pole is treated as a magnetic circuit node. Here, $F_{r,n}$ represents the magnetic potential of the nth ferromagnetic pole, while F_{mag} and R_{mag} represent the MMF and magnetic resistance of the permanent magnet, respectively. $R_{g,n}$ denotes the magnetic resistance between the nth ferromagnetic pole and the stator core. Without stator slotting, $R_{g,n}$ remains constant for different values of n. According to the symmetry of the magnetic circuit, all ferromagnetic poles with odd numbers have the same magnetic potential $F_{r,2n+1}$ (where $n = 1, 2, \ldots, P_e$), and all ferromagnetic poles with even numbers have the same magnetic potential $F_{r,2n}$ ($n = 1, 2, \ldots, P_e$). These potentials are opposite to each other. Therefore, the MMF on the rotor surface forms a square wave with positive and negative symmetry when there is no stator slotting. However, in practical applications, the stator is slotted, resulting in variable $R_{g,n}$ values. Considering the average and fundamental terms of the specific permeance of both the stator and rotor, $R_{g,n}$ can be calculated as:

$$\begin{aligned} R_{g,n} &= \frac{F_{r,n}}{\phi_{g,n}} = \frac{F_{r,n}}{r_g l_{stk} \int_{\theta_{r,n}-\frac{\pi}{2P_e}}^{\theta_{r,n}+\frac{\pi}{2P_e}} \frac{\mu_0}{g} F_{r,n} \lambda_s(\theta) \lambda_r(\theta) d\theta} \\ &= \frac{1}{\frac{\mu_0}{g} r_g l_{stk}} \frac{1}{\lambda_{lump\_0} + \lambda_{lump\_1} \cos Z_s(\theta_{r,n} + \theta_m)} \end{aligned} \tag{2.67}$$

where $\Phi_{g,n}$ represents the air gap flux produced by the nth ferromagnetic pole, θ_m denotes the angle between the stator and rotor axes, and λ_{lump_0} and λ_{lump_1} represent the average and fundamental terms of the specific permeance in the lumped parameter magnetic circuit model, respectively. These can be expressed as:

$$\begin{cases} \lambda_{lump\_0} = \frac{\pi}{P_e} \lambda_{s0} \lambda_{r0} \\ \lambda_{lump\_1} = \frac{\pi}{P_e} \lambda_{s1} \sin\left(\frac{Z_s}{P_e} \times \frac{\pi}{2}\right) \left\{ \frac{\lambda_{r0}}{\frac{Z_s}{P_e} \times \frac{\pi}{2}} + \frac{\lambda_{r1}}{\left[4\left(\frac{P_e}{Z_s}\right)^2 - \left(\frac{Z_s}{P_e}\right)^2\right] \frac{\pi}{2}} \right\} \end{cases} \tag{2.68}$$

Based on this model and as shown in Fig. 2.28b, the nodal equation can be expressed as:

$$\frac{F_{r,n} - F_{r,n-1} - (-1)^n F_{mag}}{R_{mag}} + \frac{F_{r,n} - F_{r,n+1} - (-1)^n F_{mag}}{R_{mag}} + \frac{F_{r,n}}{R_{g,n}} = 0 \tag{2.69}$$

Since n ranges from 1 to $2P_e$, Eq. (2.69) forms a non-homogeneous linear equation system composed of $2P_e$ equations, which can be solved through matrix inversion. This method, however, has drawbacks. It does not intuitively reveal the corresponding physical laws, and $R_{g,n}$ includes rotor position variables, complicating the solution at different positions. To derive an intuitive analytical solution, we first determine the rotor magnetic potential $F_{r,n}$. For any ferromagnetic pole, due to the periodic nature of the stator core's structure, $F_{r,n}$ should be a periodic function of the relative position between the stator and rotor axes θ_m. As θ_m rotates through one full cycle, $F_{r,n}$ will alternate Z_s times. Therefore, the fundamental period of $F_{r,n}$ is Z_s. Taking $F_{r,1}$ as an example, its Fourier expansion can be presented as:

$$F_{r,1} = (-1)\left[F_{r0} + \sum_{h=1}^{\infty} F_{rh} \cos hZ_s(\theta_{r,1} + \theta_m)\right] \tag{2.70}$$

where F_{r0} and F_{rh} represent the average term and the hth harmonic amplitude, respectively. The factor of 1 in the leading position of Eq. (2.70) corresponds to the 1st ferromagnetic pole, which is the S-pole, and exhibits a negative magnetic potential. Furthermore, the initial phase in Eq. (2.70) is designated as $\theta_{r,1}$, which alignswith the stator tooth when $\theta_m = -\theta_{r,1}$. The magnetic potentials of subsequent ferromagnetic poles, denoted as $F_{r,n}$, mirror the waveform of $F_{r,1}$. For odd values of n, $F_{r,n}$ and $F_{r,1}$ are phase-shifted relative to each other; for even values of n, they not only differ in phase but also require a reversal in magnitude because the ferromagnetic pole corresponds to the N polarity.

$$F_{\mathrm{r},n}(\theta_{\mathrm{m}}) = (-1)^{n-1} F_{\mathrm{r},1}\left(\theta_{\mathrm{m}} + \theta_{\mathrm{r},n} - \theta_{\mathrm{r},1}\right) \tag{2.71}$$

By combining Eqs. (2.70) and (2.71):

$$F_{\mathrm{r},n} = (-1)^n \left[F_{\mathrm{r}0} + \sum_{h=1}^{\infty} F_{\mathrm{r}h} \cos hZ_{\mathrm{s}}\left(\theta_{\mathrm{r},n} + \theta_{\mathrm{m}}\right) \right] \tag{2.72}$$

Consequently, all ferromagnetic poles share identical Fourier coefficients and differ only in phase. This is further corroborated by combining Eqs. (2.69) and (2.72):

$$\begin{aligned}
&\left[4F_{\mathrm{r}0} - 2F_{\mathrm{mag}} + \frac{\mu_0}{g} r_{\mathrm{g}} l_{\mathrm{stk}} R_{\mathrm{mag}} \left(F_{\mathrm{r}0}\lambda_{\mathrm{lump_0}} + \frac{F_{\mathrm{r}1}\lambda_{\mathrm{lump_1}}}{2} \right)\right] \\
&+ \left\{ 2F_{\mathrm{r}1}\left[1 + \cos\left(\frac{Z_{\mathrm{s}}}{P_{\mathrm{e}}}\pi\right)\right] + \frac{\mu_0}{g} r_{\mathrm{g}} l_{\mathrm{stk}} R_{\mathrm{mag}} \right. \\
&\left. \left(F_{\mathrm{r}1}\lambda_{\mathrm{lump_0}} + F_{\mathrm{r}0}\lambda_{\mathrm{lump_1}} + \frac{F_{\mathrm{r}2}\lambda_{\mathrm{lump_1}}}{2} \right) \right\} \cos Z_{\mathrm{s}}\left(\theta_{\mathrm{r},n} + \theta_{\mathrm{m}}\right) \\
&+ \sum_{h=2}^{\infty} \left\{ 2F_{\mathrm{r}h}\left[1 + \cos\left(h\frac{Z_{\mathrm{s}}}{P_{\mathrm{e}}}\pi\right)\right] \right. \\
&\left. + \frac{\mu_0}{g} r_{\mathrm{g}} l_{\mathrm{stk}} R_{\mathrm{mag}} \lambda_{\mathrm{lump_1}} \frac{F_{\mathrm{r}(h+1)} + F_{\mathrm{r}(h-1)}}{2} \right\} \cos hZ_{\mathrm{s}}\left(\theta_{\mathrm{r},n} + \theta_{\mathrm{m}}\right) = 0
\end{aligned} \tag{2.73}$$

According to Eq. (2.73), the following relationships can be deduced:

$$\begin{aligned}
&4F_{\mathrm{r}0} - 2F_{\mathrm{mag}} + \frac{\mu_0}{g} r_{\mathrm{g}} l_{\mathrm{stk}} R_{\mathrm{mag}} \left(F_{\mathrm{r}0}\lambda_{\mathrm{lump_0}} + \frac{F_{\mathrm{r}1}\lambda_{\mathrm{lump_1}}}{2} \right) = 0, \\
&2F_{\mathrm{r}1}\left[1 + \cos\left(\frac{Z_{\mathrm{s}}}{P_{\mathrm{e}}}\pi\right)\right] + \frac{\mu_0}{g} r_{\mathrm{g}} l_{\mathrm{stk}} R_{\mathrm{mag}} \left(F_{\mathrm{r}1}\lambda_{\mathrm{lump_0}} + F_{\mathrm{r}0}\lambda_{\mathrm{lump_1}} + \frac{F_{\mathrm{r}2}\lambda_{\mathrm{lump_1}}}{2} \right) = 0, \\
&2F_{\mathrm{r}h}\left[1 + \cos\left(h\frac{Z_{\mathrm{s}}}{P_{\mathrm{e}}}\pi\right)\right] + \frac{\mu_0}{g} r_{\mathrm{g}} l_{\mathrm{stk}} R_{\mathrm{mag}} \lambda_{\mathrm{lump_1}} \frac{F_{\mathrm{r}(h+1)} + F_{\mathrm{r}(h-1)}}{2} = 0 \quad \forall h \geq 2
\end{aligned} \tag{2.74}$$

As indicated in Eq. (2.74), each harmonic amplitude $F_{\mathrm{r}h}$ is linked to the magnitude of higher-order harmonics, which are challenging to determine accurately. To simplify, it is assumed that $F_{\mathrm{r}0}$ is substantially greater than $F_{\mathrm{r}1}$, and $F_{\mathrm{r}1}$ significantly exceeds the amplitudes of other harmonics. This assumption relies on the observation that the amplitude of $F_{\mathrm{r}h}$ decreases as the harmonic order h increases. Thus, the analysis primarily considers only $F_{\mathrm{r}0}$ and $F_{\mathrm{r}1}$:

$$\begin{aligned}
F_{\mathrm{r}0} &\approx \frac{2F_{\mathrm{mag}}}{4 + \frac{\mu_0}{g} r_{\mathrm{g}} l_{\mathrm{stk}} R_{\mathrm{mag}} \lambda_{\mathrm{lump_0}}}, \\
F_{\mathrm{r}1} &\approx -\frac{2F_{\mathrm{mag}}}{4 + \frac{\mu_0}{g} r_{\mathrm{g}} l_{\mathrm{stk}} R_{\mathrm{mag}} \lambda_{\mathrm{lump_0}}} \times \frac{\frac{\mu_0}{g} r_{\mathrm{g}} l_{\mathrm{stk}} R_{\mathrm{mag}} \lambda_{\mathrm{lump_1}}}{2\left[1 + \cos\left(\frac{Z_{\mathrm{s}}}{P_{\mathrm{e}}}\pi\right)\right] + \frac{\mu_0}{g} r_{\mathrm{g}} l_{\mathrm{stk}} R_{\mathrm{mag}} \lambda_{\mathrm{lump_0}}}
\end{aligned} \tag{2.75}$$

In Eq. (2.75), F_{r0} represents the magnetic potential component arising from the average term of the ferromagnetic pole's specific permeance, which is independent of the stator slotting and the pole ratio. Conversely, F_{r1} is a component introduced by the stator slotting and is influenced by the ratio Z_s/P_e, determined by the pole ratio. In systems with smaller pole ratios, such as fractional-slot concentrated-winding vernier machines, Z_s/P_e approximates 2, leading to $\cos(Z_s\pi/P_e)$ reaching its maximum value of 1. Consequently, F_{r1} is minimal. Conversely, as the pole ratio increases, Z_s/P_e approaches 1, and $\cos(Z_s\pi/P_e)$ reaches its minimum value of − 1, which results in a larger F_{r1}.

Based on the above calculations of the magnetic potentials of the ferromagnetic poles, the rotor's MMF can be determined using Fourier expansion:

$$\begin{aligned} F_e = {} & \frac{4F_{r0}}{\pi}\sum_{h=1,3\ldots}^{\infty}\cos h(\theta-\theta_m) \\ & + F_{r1}\sum_{P=1}^{\infty}\frac{\sin\left(\frac{P\pi}{2P_e}\right)}{P\pi}\sum_{n=1}^{2P_e}(-1)^n\Big\{\cos\big[P\theta+(Z_s-P)\theta_{r,n}+(Z_s-P)\theta_m\big] \\ & +\cos\big[P\theta-(Z_s+P)\theta_{r,n}-(Z_s+P)\theta_m\big]\Big\} \end{aligned} \tag{2.76}$$

The rotor MMF consists of two components: the first produced by the average term of the ferromagnetic pole's magnetic potential F_{r0}, and the second generated by the fundamental term F_{r1}, which is the additional MMF. This additional MMF is non-zero when $P = hP_e \pm Z_s$ (where h is odd). Substituting $P = hP_e \pm Z_s$ into Eq. (2.76) yields:

$$\begin{aligned} F_e = {} & \frac{4F_{r0}}{\pi}\sum_{h=1,3\ldots}^{\infty}\cos h(\theta-\theta_m) + F_{r1}\sum_{h=1,3\ldots}^{\infty}\frac{\sin\frac{(hP_e+Z_s)\pi}{2P_e}}{\frac{(hP_e+Z_s)\pi}{2P_e}}\cos[(hP_e+Z_s)\theta-hP_e\theta_m] \\ & + F_{r1}\sum_{h=1,3\ldots}^{\infty}\frac{\sin\frac{(hP_e-Z_s)\pi}{2P_e}}{\frac{(hP_e-Z_s)\pi}{2P_e}}\cos[(hP_e-Z_s)\theta-hP_e\theta_m] \end{aligned} \tag{2.77}$$

In Eq. (2.77), the spoke-magnetization vernier permanent magnet machine shares similarities with the consequent-pole vernier permanent magnet machine. Both structures include a component with $(Z_s - P_e)$ pole pairs in the additional MMF. Since the magnitude of the additional MMF F_{r1}, is negative, it negatively impacts the main working magnetic field of $(Z_s - P_e)$ pole pairs [15]. Similarly, for other interior permanent magnet structures, such as flat-shaped or V-shaped configurations, the MMF will also contain additional components after slotting on the opposite side of the air gap. These components suppress the working magnetic field and cannot be ignored in the analysis. The solution process for the additional MMF in these structures is similar to that of the spoke-magnetization permanent magnet structure, and thus the derivation process is not included in this book. Interested readers are encouraged to complete it on their own.

2.3 Air Gap Flux Density Under Flux Modulation

Section 2.2 introduces the theory and analysis method of flux modulation. This section explores the characteristics of the air gap magnetic field in FMMs, providing a foundation for subsequent chapters. Using the basic model of an FMM illustrated in Fig. 1.5 as an example, where the excitation unit is a surface-mounted permanent magnet structure, and its excitation MMF is given in Eq. (2.38). Considering only the average and fundamental terms, the specific permeance function of the FMM is expressed as:

$$\lambda_{\mathrm{f}}(\theta) \approx \lambda_{\mathrm{f}0} + \lambda_{\mathrm{f}1} \cos P_{\mathrm{f}}(\theta - \Omega_{\mathrm{f}} t) \tag{2.78}$$

Based on the MMF/specific permeance model, the air gap flux density under no-load conditions can be derived as follows:

$$\begin{aligned} B_{\mathrm{e}}(\theta, t) = & \sum_{h=1,3,5,\ldots} \frac{\mu_0}{g} \lambda_{\mathrm{f}0} F_{\mathrm{e}h} \cos hP_{\mathrm{e}}(\theta - \Omega_{\mathrm{e}} t) \\ & + \sum_{h=1,3,5,\ldots} \frac{\mu_0}{g} \frac{\lambda_{\mathrm{f}1} F_{\mathrm{e}h}}{2} \cos[(hP_{\mathrm{e}} - P_{\mathrm{f}})\theta - (hP_{\mathrm{e}}\Omega_{\mathrm{e}} - P_{\mathrm{f}}\Omega_{\mathrm{f}})t] \\ & + \sum_{h=1,3,5,\ldots} \frac{\mu_0}{g} \frac{\lambda_{\mathrm{f}1} F_{\mathrm{e}h}}{2} \cos[(hP_{\mathrm{e}} + P_{\mathrm{f}})\theta - (hP_{\mathrm{e}}\Omega_{\mathrm{e}} + P_{\mathrm{f}}\Omega_{\mathrm{f}})t] \end{aligned} \tag{2.79}$$

Observations indicate that by neglecting the high-order harmonics of specific permeance, the pole pair numbers of the no-load air gap flux density are hP_{e}, $|P_{\mathrm{f}} - hP_{\mathrm{e}}|$, and $P_{\mathrm{f}} + hP_{\mathrm{e}}$, where h is a positive odd number. Table 2.2 delineates the characteristics of each harmonic component in the air gap flux density. In the first row, the flux density associated with hP_{e} pole pairs results from the interaction of the excitation MMF with the corresponding pole pairs and the average term $\lambda_{\mathrm{f}0}$ of the specific permeance. This component, prevalent in conventional permanent magnet machines, is not affected by the flux modulation effect. Conversely, the harmonic components listed in the second and third rows are generated through the interaction between the excitation MMF and the fundamental term $\lambda_{\mathrm{f}1}$ of the specific permeance, hence influenced by the flux modulation effect. Consequently, the pole pair numbers of these flux density harmonics differ from those of the excitation MMF, being determined by both the excitation MMF and the specific permeance. With respect to the amplitudes of the air gap flux density harmonics, they diminish as h increases, rendering the high-order harmonics often negligible. The harmonics associated with $|P_{\mathrm{f}} - P_{\mathrm{e}}|$ pole pairs ($h = 1$) are considered the primary operational harmonics. Moreover, as detailed in Sect. 2.2.2, for surface-mounted permanent magnet structures, the fundamental term of the specific permeance function is significantly lower than the average term. Therefore, the amplitude of the flux density generated by the flux modulation effect (i.e., the modulated flux density) is substantially lower than that of the unmodulated flux density. Additionally, the dynamics of the working magnetic

Table 2.2 Characteristics of no-load air gap flux density harmonics

Pole pair	Magnitude	Speed	Source	
			Specific permeance	Permanent magnet MMF
hP_e	$\frac{4\lambda_0}{\pi h}\frac{\mu_{f0}}{g}\frac{B_r h_m}{\mu_0\mu_r}\sin\frac{\alpha_m}{2}$	Ω_e	λ_{f0}	hth MMF harmonic
$\lvert P_f - hP_e\rvert$	$\frac{2\lambda_{f1}}{\pi h}\frac{\mu_0}{g}\frac{B_r h_m}{\mu_0\mu_r}\sin\frac{\alpha_m}{2}$	$\frac{hP_e\Omega_e - P_f\Omega_f}{hP_e - P_f}$	λ_{f1}	
$P_f + hP_e$	$\frac{2\lambda_{f1}}{\pi h}\frac{\mu_0}{g}\frac{B_r h_m}{\mu_0\mu_r}\sin\frac{\alpha_m}{2}$	$\frac{hP_e\Omega_e + P_f\Omega_f}{hP_e + P_f}$	λ_{f1}	

field's harmonics are notable, warranting further discussion as outlined in reference [16].

2.3.1 Magnetic Field Motion with Stationary Modulation Unit

When the modulation unit is stationary, i.e., $\Omega_f = 0$, the speed of the modulated flux density harmonic described in the second row of Table 2.2 is as follows:

$$\Omega = \frac{P_e\Omega_e}{P_e - P_f} \tag{2.80}$$

The magnetic field harmonic with $\lvert P_f - P_e\rvert$ pole pairs is the additional magnetic field introduced in Sect. 2.1, and Eq. (2.80) aligns with Eq. (2.6). Assuming P_e is close to P_f, the speed of the modulated flux density is significantly higher than that of the excitation unit Ω_e. Consequently, the FMM enhances the speed of the working magnetic field [16].

2.3.2 Magnetic Field Motion with Stationary Excitation Unit

When the excitation unit is stationary, i.e., $\Omega_e = 0$, the speed of the modulated flux density harmonic in the second row of Table 2.2 is:

$$\Omega = \frac{P_f\Omega_f}{P_e - P_f} \tag{2.81}$$

In conventional permanent magnet machines, stationary permanent magnets result in no rotating magnetic field in the air gap. To generate armature reaction, the armature winding cannot be an AC winding; it must be a DC winding with brushes and slip rings. However, in FMMs, even with a stationary excitation unit (i.e., permanent magnets), the rotating modulation unit can still produce a rotating magnetic field in the air gap. Therefore, the armature winding can be an AC winding. This principle underlies the operation of stator-excitation FMMs, which will be detailed in Chap. 3 [17–19].

2.3.3 Magnetic Field Motion with Rotating Excitation and Modulation Units

When both the excitation and modulation units are rotating, the speed of the modulated flux density harmonic in the second row of Table 2.2 is:

$$\Omega = \frac{P_e\Omega_e - P_f\Omega_f}{P_e - P_f} \tag{2.82}$$

The rotation of the modulated magnetic field depends on the relative speeds of the two units. The magnetic field is stationary only if $P_e\Omega_e = P_f\Omega_f$; otherwise, it rotates [20].

2.4 Summary of This Chapter

In this chapter, the physical process of flux modulation is qualitatively examined using a basic magnetic circuit model and the MMF/specific permeance model is developed to analyze the air gap magnetic field. Based on this study, several conclusions can be drawn:

1. *Fundamental Cause of Flux Modulation*: The flux modulation effect arises from the non-uniformity of the air gap structure, which enhances or weakens the air gap magnetic field at different positions. This causes distortion in the magnetic field's waveform, introducing new magnetic field harmonics with different pole pairs in the air gap.
2. *Air Gap Flux Density in FMMs*: The air gap flux density is the product of the magnetic potential difference between the two boundaries of the air gap (i.e., air gap MMF) and the specific permeance. The specific permeance represents the ratio of the uniform air gap length to the non-uniform air gap length. In surface-mounted permanent magnet FMMs with a large equivalent air gap length, the fundamental term of the specific permeance is only 1/5–1/3 of the average term, peaking when the slot opening ratio is 0.6. In electrically excited FMMs, which have a smaller equivalent air gap length, both the average and fundamental terms of the specific permeance can reach 0.5.
3. *MMF/Specific Permeance Models for Different FMM Topologies*: Various MMF/specific permeance models are established for different FMM topologies. In armature winding and surface-mounted permanent magnet structures, the MMF

and specific permeance are generally independent, with the MMF remaining unchanged before and after slotting. For double-sided slotting structures, the air gap specific permeance is approximately the product of the single-sided specific permeance when either the stator or rotor is slotted. This introduces new specific permeance harmonic components whose pole pair numbers are the sum or difference of the stator and rotor teeth numbers. In consequent-pole and spoke magnetization permanent magnet structures, the MMF is influenced by the distribution of the specific permeance function, which must be considered in calculations.

4. *Flux Modulation Effect*: The flux modulation effect generates a new flux density harmonic with $|P_f - P_e|$ pole pairs in the air gap, which serves as the main working harmonic in all types of FMMs. However, its magnitude is significantly smaller than the non-modulated flux density component. Additionally, regardless of whether the excitation unit or the modulation unit is stationary, the flux modulation effect results in a speed-increasing phenomenon for the working magnetic field, making its speed much higher than the rotor speed.

By exploring these concepts, this chapter lays a solid foundation for the subsequent chapters, which will delve deeper into the practical applications and further theoretical developments of flux modulation in various FMM configurations.

References

1. Gu CL, Chen QF, Xiong YC (2018) Electrical machine (in Chinese). Huazhong University of Science and Technology Press, Wuhan
2. Ye QZ, Chen DZ (2012) Electromagnetic field tutorial (in Chinese). Higher Education Press, Beijing
3. Li HM, Li QF (1982) Electromagnetic reduction machine (in Chinese). Machinery Industry Press, Beijing
4. Fang L, Li DW, Ren X et al (2022) IEEE Trans Ind Electron 69(6):6058
5. Shi CJ (2020) Research on the electromagnetic characteristics and topologies of linear permanent magnet vernier machines (in Chinese). Huazhong University of Science and Technology Press, Wuhan
6. Tang YQ (1998) Electromagnetic fields in electrical machines (in Chinese). Science Press, Beijing
7. Zhu ZQ, Howe D (1993) IEEE Trans Magn 29(1):143
8. Li DW, Qu RH, Li J et al (2016) IEEE Trans Ind Appl 52(1):125
9. Wu LL, Qu RH, Li DW et al (2015) IEEE Trans Ind Appl 51(5):3707
10. Jia SF (2017) Research on stator DC-current excited vernier reluctance machines (in Chinese). Huazhong University of Science and Technology Press, Wuhan
11. Jia SF, Qu RH, Li J et al (2016) IEEE Trans Energy Convers 31(3):935
12. Zhao Y, Li DW, Ren X et al (2022) IEEE J Emerg Sel Top Power Electron 10(3):2934
13. Xie KF, Li DW, Qu RH et al (2019) IEEE Trans Ind Electron 55(2):1420
14. Li DW, Qu RH, Li J et al (2015) IEEE Trans Ind Appl 51(6):4470
15. Ren X, Li DW, Qu RH et al (2018) IEEE Trans Energy Convers 33(4):2153
16. Li DW (2015) Research on flux modulation permanent magnet machine (in Chinese). Huazhong University of Science and Technology Press, Wuhan
17. Gao YT, Qu RH, Li DW et al (2017) IEEE Trans Ind Appl 53(3):2110
18. Gao YT, Qu RH, Li DW et al (2017) IEEE Trans Ind Appl 53(5):4232

19. Gao YT (2017) Theoretical analysis and topology investigation of flux reversal permanent magnet machines (in Chinese). Huazhong University of Science and Technology Press, Wuhan
20. Xie KF (2020) Permanent magnet machines theory and topology study based on electromagnetic superimposition (in Chinese). Huazhong University of Science and Technology Press, Wuhan

Chapter 3
Classification of Flux Modulation Machines

In contrast to conventional electrical machines, which typically consist of two components—excitation and armature units—FMMs incorporate an additional element, the modulation unit. This tripartite structure introduces multiple degrees of freedom in design, leading to a diverse range of FMM configurations. Despite the extensive exploration of various FMM types, including flux switching permanent magnet machines, flux reversal permanent magnet machines, and electrically excited doubly-salient machines, the underlying flux modulation mechanisms remain elusive. Consequently, the electromagnetic characteristics of these machines, characterized by high torque densities and low power factors, lack theoretical elucidation. Research findings applicable to one machine type do not necessarily translate to others, and a consistent framework for topological laws and design methodologies has yet to be established. This chapter elucidates five viable FMM types based on fundamental magnetic field constraints and the dynamics of the functional units. It further categorizes existing FMM topologies through the selection and integration of these units, thereby standardizing the framework and proposing novel FMM configurations, such as the flux modulation induction machine, for additional research.

3.1 Structural Constraint and Classification

Section 1.2 delineates the basic structure of FMMs. As illustrated in Fig. 1.5, the fundamental FMM model comprises three segments: the excitation, modulation, and armature units. The MMF produced by the excitation unit is modulated to generate a no-load air gap magnetic field, the specifics of which, including pole pair number and rotational speed, are specified in Table 2.2. FMMs employ a specific harmonic of the magnetic field, with $|P_f - P_e|$ pole pairs, as the operational magnetic field. This harmonic engages with the armature MMF to facilitate electromechanical energy conversion. According to principles of electrical machine theory, the interacting

R. Qu et al., *Flux Modulation Machine*, https://doi.org/10.1007/978-981-95-4746-3_3

harmonics must share identical pole pair numbers and rotational speeds. Based on these considerations, FMMs are categorized according to the motions of their functional units [1].

(1) In configurations where both the MMF of the excitation unit and the modulation unit are stationary, the resultant no-load air gap magnetic field is also stationary. Consequently, the armature MMF must remain stationary as well. This configuration is termed the stationary three-unit FMM. It is crucial to clarify that "stationary" here pertains to the MMF, not the physical units themselves. If all three units were stationary, no mechanical energy would be converted. For instance, in a DC machine, although the armature MMF is stationary, the armature itself rotates, allowing the stationary MMF of the armature to interact with the stationary modulated magnetic field for electromechanical energy conversion.
(2) When the excitation MMF rotates while the modulation unit remains stationary, their interaction produces a rotating no-load magnetic field. The armature MMF must also rotate synchronously to align with this field. This configuration is identified as the stationary modulation unit FMM.
(3) Conversely, if the modulation unit rotates and the excitation MMF is stationary, a rotating no-load magnetic field is generated. The armature MMF must then rotate in synchrony with this magnetic field. This type is designated as the stationary excitation unit FMM.
(4) In scenarios where both the excitation MMF and the modulation unit rotate, the behavior of the armature MMF depends on the relative speeds of the excitation and modulation units. If the speed of the excitation MMF Ω_e, and the speed of the modulation unit Ω_f, fulfill the conditions outlined in the following equation:

$$P_f\Omega_f = P_e\Omega_e \tag{3.1}$$

In Eq. (2.82), the resultant no-load magnetic field remains stationary. The armature MMF must also be stationary to facilitate electromechanical energy conversion. Machines meeting these criteria are called stationary armature MMF FMMs.
(5) If the speeds of the excitation MMF and modulation unit both vary and their relationship does not satisfy Eq. (3.1), the no-load magnetic field rotates. Consequently, the armature MMF must also rotate. This configuration is referred to as the non-stationary unit FMM.

In addition to the motion of each unit, the diversity among FMMs is also evident in the structure of the excitation unit. The variations include: (1) Permanent magnet FMMs, which utilize permanent magnets for excitation; (2) Electrically excited FMMs, where excitation is provided by direct current (DC) in the excitation windings; (3) Brushless doubly-fed machines, typically characterized by dual armature windings—the power winding and the control winding. In this discourse, the control winding is treated as the AC excitation winding to maintain coherence with flux modulation theory; (4) Induction machines, where the rotor current and magnetic

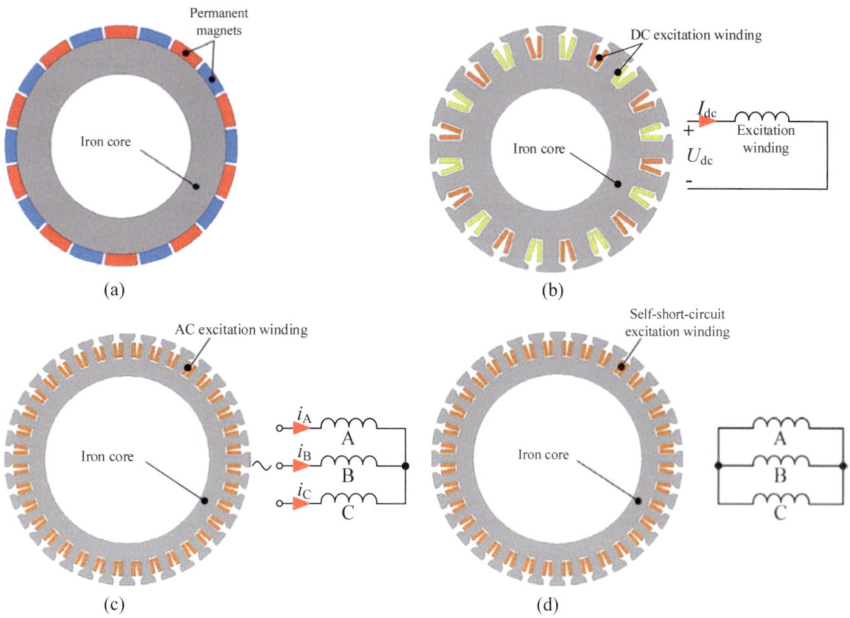

Fig. 3.1 Feasible excitation structures in FMMs: **a** permanent magnet excitation unit, **b** DC excitation winding, **c** AC excitation winding, **d** self-short-circuit excitation winding

field are induced by the rotating armature MMF, typically lacking a distinct excitation unit. In these machines, the self-short-circuit rotor winding is considered a specialized form of the excitation unit. Consequently, four distinct excitation unit structures are viable for FMMs, as depicted in Fig. 3.1. FMMs are classified based on the motion of each unit and the type of excitation, as outlined in Table 3.1.

As indicated in Table 3.1, all predominant FMM types discussed in current literature are categorized based on the aforementioned criteria. Additionally, two novel FMM configurations—the flux modulation DC machine and the flux modulation induction machine—which have not been previously proposed, are identified. However, some FMM types, such as the stationary modulation unit FMM with AC excitation winding, lack practical functionality or present implementation challenges, and are therefore excluded from Table 3.1. Furthermore, although theoretical derivations suggest the possibility of a stationary armature MMF FMM, a viable machine topology for this configuration has not yet been identified, leading to its omission in this chapter.

Table 3.1 Classification of FMMs

Machine type	Stationary three-unit FMM	Stationary modulation unit FMM	Stationary excitation unit FMM	Non-stationary unit FMM
Permanent magnet	Flux modulation permanent magnet DC machine	Magnetic geared machine, vernier permanent magnet machine, transverse-flux permanent magnet machine	Flux reversal permanent magnet machine, flux switching permanent magnet magnetic, doubly-salient permanent magnet machine	Flux modulation brushless dual-mechanical-port dual-electrical-port machine
DC excitation winding	Flux modulation electrically excited DC machine	Electrically excited vernier machine	Flux modulation electrically excited machine	–
AC excitation winding	–	–	–	Vernier reluctance machine, brushless doubly-fed reluctance machine
Self-short-circuit winding	–	Flux modulation induction machine	–	Asynchronous induction machine

3.2 Flux Modulation Machine with Stationary Three Units

In the stationary three-unit FMM, both the excitation MMF and the modulation unit remain stationary. For electromechanical energy conversion to occur, the armature unit must rotate; otherwise, the transfer of mechanical energy is unachievable. Since the armature MMF is stationary, brushes and commutators are employed to align the magnetic field. Consequently, the stationary three-unit FMM bears structural similarities to the conventional DC machine, hence it is also termed the flux modulation DC machine. The primary distinction between the two lies in the generation of the no-load magnetic field; in a standard DC machine, it is produced solely by the excitation MMF without any modulation, whereas in the flux modulation DC machine, it results from the combined effects of the excitation MMF and the flux modulation from the modulation unit.

3.2.1 *Flux Modulation Permanent Magnet DC Machine*

Regarding the flux modulation DC machine, when the excitation unit consists of permanent magnets, this variant is identified as the flux modulation permanent

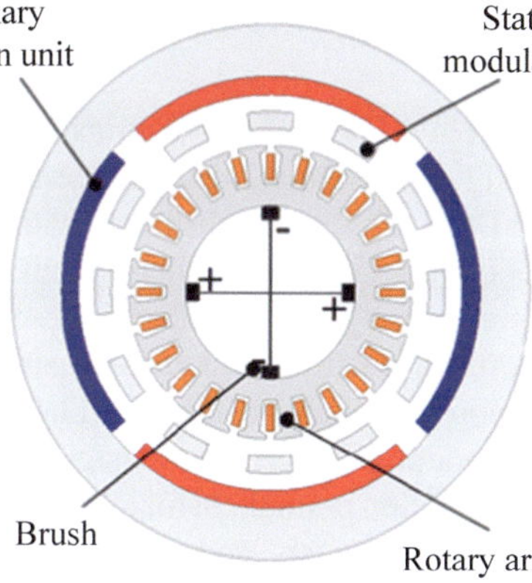

Fig. 3.2 Flux modulation permanent magnet DC machine

magnet DC machine, as illustrated in Fig. 3.2. The configuration encompasses an outer permanent magnet excitation unit, a middle modulation unit, and an inner armature unit. The AC in the armature winding is converted to DC output via the brushes and slip rings.

According to Eq. (2.79), the relationship between the pole pair number of the armature unit (P_a), the number of ferromagnetic blocks in the modulation unit (P_f), and the pole pair number of the excitation unit P_e need to satisfy the following relationship:

$$P_a = |P_e \pm P_f| \tag{3.2}$$

As outlined in Sect. 2.3, the amplitude of the modulated component of the no-load air gap magnetic field in FMMs is significantly lower than that of the non-modulated component. In conventional permanent magnet DC machines, which lack a modulation unit, the no-load air gap magnetic field consists solely of the non-modulated component. Conversely, the flux modulation permanent magnet DC machine operates primarily through the modulated component of the no-load air gap magnetic field, which is inherently weaker. Additionally, the dual air gap structure of this machine type further diminishes the magnetic field, resulting in substantially lower magnetic loading compared to the regular permanent magnet DC machine. Furthermore, the pole ratio of the flux modulation permanent magnet DC machine is set at 1, eliminating any torque amplification effect. (The pole ratio refers to the ratio of the pole pair number of the mechanical rotating part to the pole pair number of the armature winding; a higher pole ratio enhances torque amplification within a specific range.) Therefore, while the flux modulation principle in permanent magnet DC machines is technically sound, its performance does not offer advantages over the conventional permanent magnet DC machine.

3.2.2 Electrically Excited Flux Modulation DC Machine

The development process of the electrically excited flux modulation DC machine is depicted in Fig. 3.3. Initially, the permanent magnets in the flux modulation permanent magnet DC machine are substituted with an excitation winding. If the number of ferromagnetic blocks in the modulation unit, P_f, satisfies the condition $P_f = 2n_f P_e$, where n_f represents any positive integer, these blocks can be affixed to the stator teeth, leading to a transformation into a single air gap structure. The electrically excited flux modulation DC machine shares similar operating principles and characteristics with the flux modulation permanent magnet DC machine. The key distinction is that in the flux modulation permanent magnet DC machine, the excitation MMF is generated by permanent magnets, and thus, its amplitude remains unaffected by the pole pair number of the permanent magnet.

In the electrically excited flux modulation DC machine, the excitation MMF is generated by the excitation winding. The number of pole pairs in this winding influences the amplitude of the excitation MMF. Specifically, under equivalent electrical loading of the excitation winding, a smaller number of winding pole pairs results in a higher excitation MMF. Thus, the excitation MMF in the electrically excited flux modulation DC machine can be optimized by adjusting the pole pair number of the excitation winding.

The subsequent discussion will evaluate the practicality of utilizing the AC excitation winding or the self-short-circuit winding in the flux modulation DC machine. In the stationary three-unit FMM, the excitation MMF is stationary. However, if either of the aforementioned excitation methods is implemented, the excitation unit must be mechanically rotary. This requirement for rotational excitation adds structural complexity without conferring performance benefits over the permanent magnet or DC winding excitations. Consequently, employing the AC excitation winding or the self-short-circuit winding in the stationary three-unit FMM is deemed impractical.

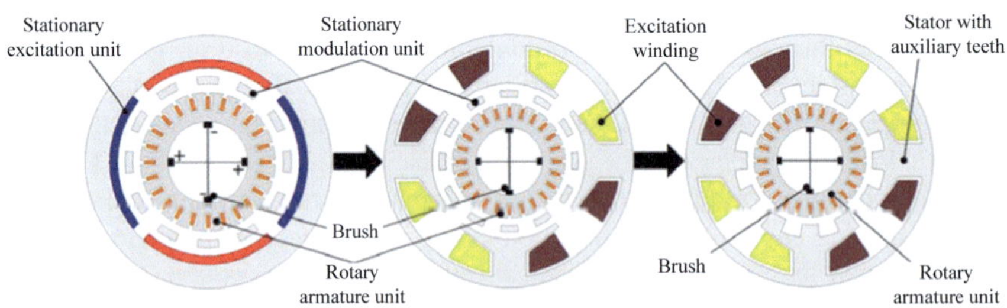

Fig. 3.3 Evolution process of electrically excited flux modulation DC machine

3.3 Flux Modulation Machine with Stationary Modulation Unit

In Fig. 1.5b, when the modulation unit remains stationary and the excitation unit rotates, this configuration aligns with the traditional topology of the stationary modulation unit FMM, also referred to as the magnetic geared permanent magnet machine. However, the complexity of its dual air gap structure necessitates a simplification of the topology through the integration of functional units.

3.3.1 Vernier Permanent Magnet Machine

The developmental process of the vernier permanent magnet machine is illustrated in Fig. 3.4. In this magnetic geared machine, both the modulation and armature units are stationary, allowing for the elimination of the gap between them. The ferromagnetic blocks of the modulation unit can then be attached to the inner surface of the stator. If the number of stator teeth matches the number of ferromagnetic blocks in the modulation unit, the original stator teeth can be converted to an open-slot structure. This modification enables the stator teeth to simultaneously fulfill the roles of traditional iron teeth and flux modulation elements [2–4].

In the vernier permanent magnet machine, the pole pair number of the modulation unit (P_f) is equivalent to the number of stator teeth (Z_s). For effective electromechanical energy conversion, the pole pair number of the armature winding must meet the following criteria:

$$Z_s = P_e \pm P_a \tag{3.3}$$

As previously mentioned, the pole ratio is defined as the ratio of the pole pair number of the mechanical rotating part to the pole pair number of the armature winding. In vernier permanent magnet machines, where the excitation unit is rotary, this ratio can be defined as:

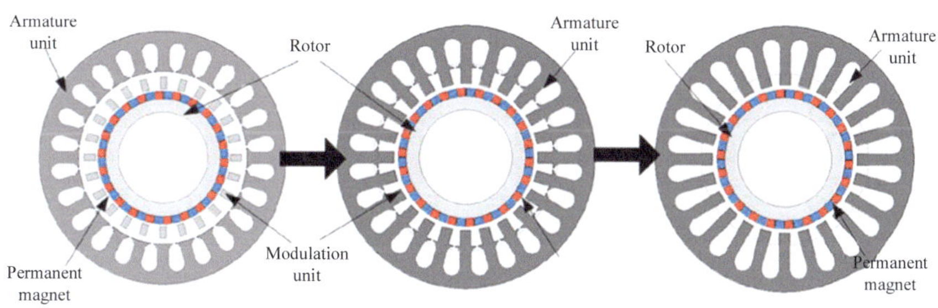

Fig. 3.4 Evolution process from magnetic geared machine to vernier permanent magnet machine

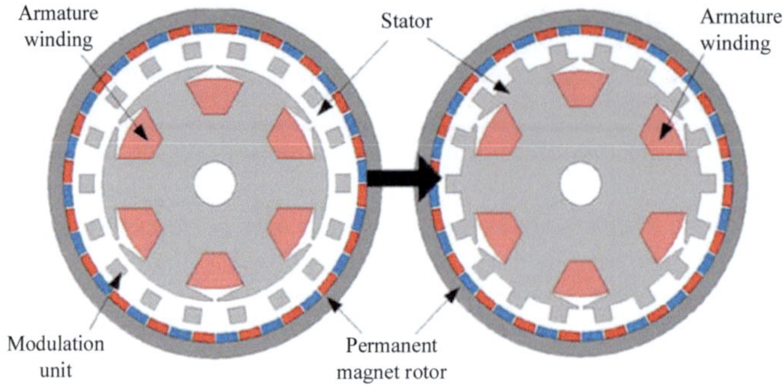

Fig. 3.5 Evolution process from outer-rotor magnetic geared machine to auxiliary-tooth vernier permanent magnet machine

$$\mathrm{PR} = \frac{P_\mathrm{e}}{P_\mathrm{a}} = \frac{Z_\mathrm{s}}{P_\mathrm{a}} \mp 1 \tag{3.4}$$

To enhance the pole ratio and torque density, the pole pair number of the armature winding (P) should be considerably lower than the number of stator teeth (Z_s). This configuration results in a larger winding pitch and an extended end winding length, which increases the overall volume of the machine and the consumption of copper, thereby diminishing the advantages in torque density. To address this issue, the introduction of a vernier permanent magnet machine with an auxiliary-tooth structure has been proposed [5]. The evolution process of this machine topology is depicted in Fig. 3.5. Unlike the configuration shown in Fig. 3.4, in Fig. 3.5, the number of ferromagnetic blocks in the modulation unit is a multiple of the number of stator teeth. Consequently, several auxiliary teeth are affixed to the surface of each main stator tooth, and these auxiliary teeth play a pivotal role in modulating the excitation MMF.

According to flux modulation theory, the slot-pole combination of the auxiliary-tooth vernier permanent magnet machine satisfies the following condition:

$$\begin{aligned} &P_\mathrm{f} = P_\mathrm{e} \pm P_\mathrm{a}, \\ &P_\mathrm{f} = n_\mathrm{f} Z_\mathrm{s} \quad n_\mathrm{f} = 2, 3, \ldots \end{aligned} \tag{3.5}$$

where n_f represents the number of auxiliary teeth on each main tooth. Its pole ratio is:

$$\mathrm{PR} = \frac{P_\mathrm{e}}{P_\mathrm{a}} = \frac{n_\mathrm{f} Z_\mathrm{s}}{P_\mathrm{a}} \mp 1 \tag{3.6}$$

Tables 3.2 and 3.3 detail the feasible slot-pole combinations for straight-tooth and auxiliary-tooth vernier permanent magnet machines, respectively, where k_wa represents the winding factor of the armature winding, and q denotes the number of slots per pole per phase. For the straight-tooth vernier machine, a shorter end winding length is achieved only with a low pole ratio (P_e/P_a). Conversely, a high pole ratio results in a longer end winding length. In the case of the auxiliary-tooth vernier machine, even with a high pole ratio, the presence of a large number of auxiliary teeth (n_f) on each main tooth allows the use of non-overlapping concentrated windings with shorter ends, effectively addressing the issue of long end windings. However, the auxiliary-tooth vernier machine also presents drawbacks. The auxiliary teeth and their junctions with the main tooth consume additional space, thus reducing the slot area, decreasing the air gap radius, and impacting the split ratio. Therefore, in practical engineering applications, selecting an appropriate topology based on specific design requirements is essential. For instance, when a low aspect ratio (i.e., the ratio of length to radius) is critical, the auxiliary-tooth topology is preferable due to its shorter end winding. Alternatively, the straight-tooth configuration is advantageous when minimizing the machine's outer diameter is desired.

Overall, apart from the open slot stator and the unequal number of stator and rotor poles, the architecture of vernier permanent magnet machines closely resembles that of conventional permanent magnet machines. This similarity ensures that the processing, manufacturing, and assembly techniques for vernier machines are well-established and straightforward. Owing to these advantages, the vernier permanent magnet machine has emerged as a prominent area of research within the field of electrical machines.

Table 3.2 Slot-pole combinations of straight-tooth vernier permanent magnet machine

	$Z_s = 6$				$Z_s = 12$				$Z_s = 18$				$Z_s = 24$			
P_a	P_e	k_{wa}	PR	q	P_e	k_{wa}	PR	q	P_e	k_{wa}	PR	q	P_e	k_{wa}	PR	q
1	5	1	5	1	11	0.97	11	2	17	0.96	17	3	23	0.96	23	4
2	4	0.87	2	0.5	10	1	5	1	16	0.95	8	1.5	22	0.97	11	2
3	–				–				15	1	5	1	–			
4	–				8	0.87	2	1/2	14	0.95	3.5	3/4	20	1	5	1
5	–				7	0.93	1.4	2/5	13	0.95	2.6	3/5	19	0.93	3.8	4/5
6	–				–				12	0.86	2	1/2	–			
7	–				–				11	0.90	1.57	3/7	17	0.93	2.43	4/7
8	–				–				10	0.95	1.25	3/8	16	0.87	2	1/2
10	–				–				–				14	0.93	1.4	2/5
11	–				–				–				13	0.95	1.18	4/11

Integer-slot distributed winding
Fractional-slot distributed winding
Non-overlapping concentrated winding

Table 3.3 Slot-pole combination of auxiliary-tooth vernier permanent magnet machine

	$Z_s = 6, P_f = 12$				$Z_s = 6, P_f = 18$				$Z_s = 6, P_f = 24$			
P_a	P_e	k_w	PR	q	P_e	k_w	PR	q	P_e	k_w	PR	q
2	10	0.87	5	0.5	16	0.87	8	0.5	22	0.87	11	0.5
	$Z_s = 12, P_s = 24$				$Z_s = 12, P_s = 36$				$Z_s = 12, P_s = 48$			
P_a	P_e	k_w	PR	q	P_e	k_w	PR	q	P_e	k_w	PR	q
5	19	0.93	3.8	2/5	31	31	6.2	2/5	43	0.93	8.6	2/5

Non-overlapping concentrated winding

3.3.2 Transverse-Flux Permanent Magnet Machine

The topology of the transverse-flux permanent magnet machine is illustrated in Fig. 3.6 [6]. The stator core comprises several U-shaped structures evenly distributed along the circumference and is wound with a toroidal armature winding. The rotor adopts a surface-mounted permanent magnet configuration, segmented into two parts in the axial direction. The magnetization directions of the permanent magnets on the two rotor segments are opposed. The pole pair number of the rotor permanent magnets (P_e) is equal to the number of U-shaped stator cores.

The flux modulation principle of the transverse-flux permanent magnet machine is delineated as follows. Initially, the excitation field of the permanent magnet is computed. Given that the U-shaped stator core's iron tooth functions as the modulation unit, and its number of pole pairs matches that of the rotor, the resultant no-load air gap flux density can be expressed as:

$$\begin{aligned} B_e(\theta, t) &= \frac{\mu_0}{g} F_e(\theta, t)\lambda_s(\theta) \\ &= \frac{\mu_0}{g} F_{e1}\left\{\lambda_{s0} cos P_f(\theta - \Omega_e t) + \frac{1}{2}\lambda_{s1}[\cos(2P_f\theta - P_f\Omega_e t) + \cos(P_f\Omega_e t)]\right\} \end{aligned} \tag{3.7}$$

where P_f denotes the pole pair number of the modulation unit, equivalent to P_e; Ω_e represents the mechanical rotation speed of the excitation unit. As indicated in Eq. (3.7), the flux density comprises three principal components: the first is the non-modulated magnetic field, arising from the interaction of the fundamental MMF and the average specific permeance; the second is the modulated magnetic field with a relatively higher pole number; the third is the modulated magnetic field with a relatively lower pole number. In the transverse-flux machine, since the pole pair

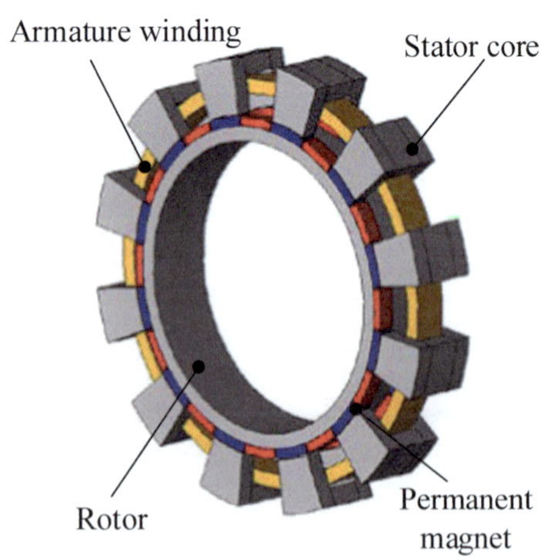

Fig. 3.6 Transverse-flux permanent magnet machine

numbers of the excitation and modulation units are identical, the pole pair number of the third term is zero, resulting in a unipolar magnetic field. This unipolar magnetic field constitutes an average term in the air gap flux density, yet it still links with the armature winding due to its toroidal characteristic. As the rotor rotates, the direction of the unipolar magnetic field continuously alternates, thereby inducing a back-EMF in the toroidal winding.

From this analysis, two critical observations emerge regarding the transverse-flux permanent magnet machine. First, to facilitate the flux path for the unipolar magnetic field, the machine must adopt a three-dimensional (3D) structure with the U-shaped stator core. Second, considering flux modulation, the pole pair number of the armature winding is zero, rendering the pole ratio of the transverse-flux machine infinite. This accounts for its superior torque density compared to other FMMs. Due to this high torque density, the transverse-flux machine emerged as a focal point of research in the 1990s. However, its relatively low power factor, typically between 0.3 and 0.4, has impeded its widespread adoption.

3.3.3 Electrically Excited Vernier Machine

By substituting the permanent magnets in the vernier machine with an excitation winding, the electrically excited vernier machine is created, as depicted in Fig. 3.7. This modification eliminates the need for permanent magnets, thereby reducing costs. Additionally, the excitation winding allows for flexible adjustment of the magnetic field. However, the drawbacks of electrically excited vernier machines are also pronounced. Firstly, to achieve a higher pole ratio, the pole pair number of the excitation winding is intentionally increased, which in turn reduces the excitation capacity. Second, the air gap length in electrically excited vernier machines is considerably shorter than that in vernier permanent magnet machines. This reduction leads to increased synchronous inductance and a lowered power factor. Moreover, the shorter air gap length predisposes the iron core to saturation, limiting the machine's application in environments requiring robust cooling and high power. To address the challenges of low excitation capability and power factor, some researchers have proposed the development of superconducting electrically excited vernier machines, which significantly enhance excitation capability and power factor through the use of superconducting excitation windings with high current density [7].

3.3.4 Flux Modulation Induction Machine

In the magnetic geared machine, replacing permanent magnets with a self-short-circuit winding gives rise to the flux modulation induction machine, as illustrated in Fig. 3.8a. The operational principle is as follows: the armature winding receives alternating current (AC) to produce an armature MMF. This MMF is then modulated

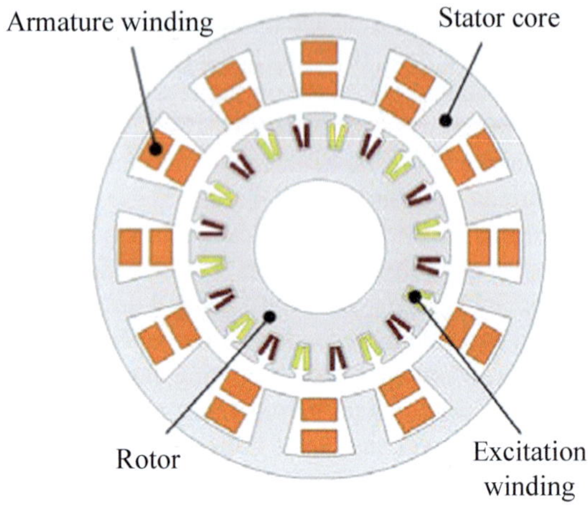

Fig. 3.7 Electrically excited vernier machine

by the modulation unit, creating a rotating no-load magnetic field in the inner air gap. The self-short-circuit winding within the rotor induces back-EMF and current from this rotating magnetic field. Subsequently, the rotor current also produces a rotor MMF, which, after modulation by the modulation unit, interacts with the stator armature MMF. The operational principle of the flux modulation induction machine closely resembles that of the conventional induction machine, with the key difference being the additional requirement for a modulation unit to modulate both the stator and rotor MMFs.

Figure 3.8b demonstrates the flux line distribution in the flux modulation induction machine. It shows how the 5-pole-pair stator armature MMF, modulated by the 6 ferromagnetic blocks of the modulation unit, generates a 1-pole-pair magnetic field in the rotor. Thus, the pole pair number relationship of the magnetic fields in the flux

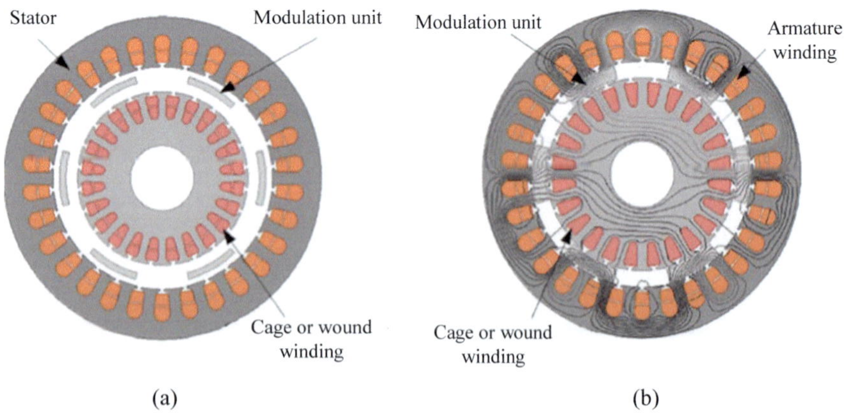

Fig. 3.8 Flux modulation induction machine: **a** structure. **b** Flux lines

modulation induction machine adheres to Eq. (2.79), and consequently, its slot-pole combination must also conform to Eq. (3.3).

The discussion now turns to the topology of the flux modulation induction machine. Regarding the stator structure, when the number of ferromagnetic blocks in the modulation unit equals or is an integer multiple of the number of stator teeth, these blocks can be affixed to the stator teeth to form either a straight or auxiliary teeth structure. This configuration not only simplifies the mechanical design but also reduces the equivalent air gap length, enhancing the flux modulation effect and the overall output torque. In terms of rotor structure, regular induction machines typically utilize either cage type or wound winding type rotors. Due to manufacturing ease, most induction machines feature cage rotors. However, within the flux modulation induction machine, a multitude of magnetic field harmonics with varied rotation speeds and directions exist. Utilizing a cage rotor could induce back-EMF and current in the cage bars from various harmonics, effectively making the machine a composite of multiple induction machines. This multitude of harmonics, some rotating in directions opposite to or at speeds slower than the rotor's motion, generates considerable braking torque, which not only diminishes torque density but also contributes to additional rotor losses. Thus, it is advisable for the flux modulation induction machine to use a wound winding rotor. Moreover, if an AC excitation winding is employed in the excitation unit rather than a self-short-circuit winding, the rotor current in the flux modulation induction machine becomes controllable. However, as both the modulation and armature units are stationary, only the excitation unit is rotational. Therefore, the use of brushes and slip rings is necessary to facilitate the rotor's AC excitation winding. This configuration of the flux modulation induction machine is also referred to as the brush-type doubly-fed FMM.

3.4 Flux Modulation Machine with Stationary Excitation Unit

In Fig. 1.5b, a configuration where the excitation unit is stationary and the modulation unit is rotary characterizes the stationary excitation unit FMM. Given that both the excitation and armature units are stationary, integration of these components simplifies the machine's topology. Depending on the pole pair number of the excitation unit and the arrangement of the permanent magnets, the stationary excitation unit FMM can assume various topologies, as detailed below.

3.4.1 Flux Reversal Permanent Magnet Machine

When the stationary excitation unit employs permanent magnets, this variant is referred to as the stator permanent magnet machine. There are three distinct types of

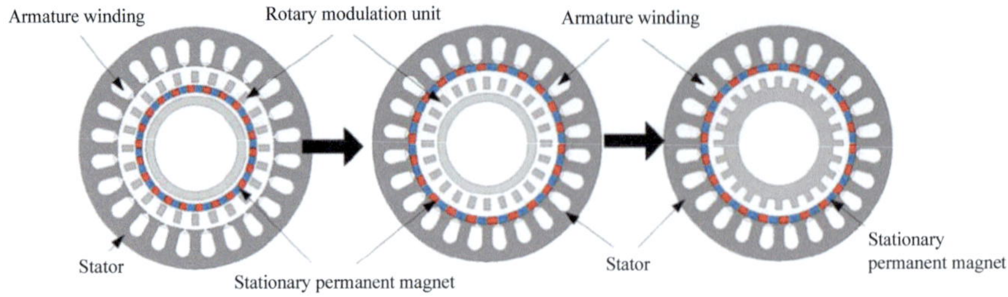

Fig. 3.9 Evolution process of evenly distributed flux reversal permanent magnet machine

stator permanent magnet machines: the flux reversal permanent magnet machine [5], the flux switching permanent magnet machine [6], and the doubly-salient permanent magnet machine [7]. This section will focus on the flux reversal permanent magnet machine.

Figure 3.9 illustrates the evolutionary process from the classical topology shown in Fig. 1.5b to the evenly distributed flux reversal permanent magnet machine. Initially, the permanent magnets of the stationary excitation unit are detached from the inner stator yoke and positioned between the armature and modulation units. As the permanent magnets are stationary, they are affixed to the stator surface. Subsequently, the remaining inner stator, now a smooth iron core, does not affect the machine's performance, regardless of whether it is stationary or rotary. Thus, the rotary modulation unit can be attached to the inner stator, creating a reluctance rotor. This configuration culminates in the realization of the flux reversal machine with evenly distributed permanent magnets [8]. Since the rotor slot number (Z_r) equals the number of ferromagnetic blocks in the modulation unit (P_f), the slot-pole combinations for the evenly distributed flux reversal machine must satisfy $Z_r = P_e \pm P_a$.

However, the machine topology depicted in Fig. 3.9 diverges from that of the conventional flux reversal permanent magnet machine and requires further evolution or modification. Initially, the pole pair number of the permanent magnets (P_e) and the stator teeth number (Z_s) need to conform to the relationship outlined in references [9, 10]:

$$P_e = \frac{3}{2} Z_s \tag{3.8}$$

Assuming that the sum of the widths of the stator teeth occupies two-thirds of the machine's circumference, the topology shown in Fig. 3.10a can be achieved. In this arrangement, several permanent magnets are positioned at the stator slot openings, contributing minimally to the back electromotive force (EMF) and torque. Consequently, these permanent magnets can be omitted, leading to the topology presented in Fig. 3.10b, which represents the traditional flux reversal permanent magnet machine as documented in the literature.

The explanation of the flux modulation effect in the flux reversal permanent magnet machine is as follows: The excitation MMF of the permanent magnets,

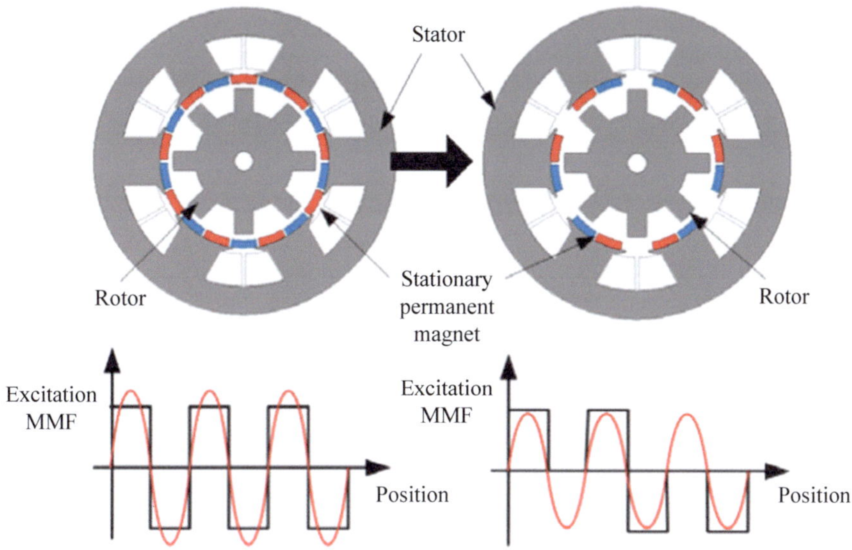

Fig. 3.10 Further evolution of flux reversal permanent magnet machine

depicted in Fig. 3.10, is noteworthy. Owing to the removal of the permanent magnets at the stator slot openings, the excitation MMF does not manifest as a square wave with positive and negative symmetry. It is observed that the pole pair number of the fundamental waveform of the excitation MMF is half the number of stator teeth ($Z_s/2$), and the Fourier series of the excitation MMF can be articulated as follows [11, 12]:

$$F_e = \sum_{h=1,3,5...}^{\infty} F_{eh} \sin\left(\frac{hZ_s}{2}\theta\right) \tag{3.9}$$

where F_{eh} represents the amplitude of the hth harmonic of the excitation MMF. Continuing, the calculation of air gap specific permeance is addressed. Given that the stator slot openings are already accounted for in the excitation MMF, the specific permeance introduced by stator slotting can be disregarded. Thus, a smooth stator surface is assumed, and only the rotor slotting is considered relevant. The air gap specific permeance due to rotor slotting is expressed as:

$$\lambda_r(\theta, t) = \lambda_{r0} + \sum_{l=1}^{\infty} \lambda_{rl} \cos lZ_r(\theta - \Omega_r t) \tag{3.10}$$

Finally, based on the principles of flux modulation theory, the no-load air gap magnetic field is determined using Eqs. (3.9) and (3.10). The harmonic contents of this field are summarized in Table 3.4.

Table 3.4 No-load air gap flux density harmonics of flux reversal machine

Order	Pole pair number	Magnitude	Angular frequency
1	$\frac{hZ_s}{2}$	$\frac{\mu_0}{g}\lambda_{r0}F_{mh}$	0
2	$\frac{hZ_s}{2}+lZ_r$	$\frac{\mu_0}{2g}\lambda_{rl}F_{eh}$	$l\Omega_r t$
3	$\left\lvert\frac{hZ_s}{2}-lZ_r\right\rvert$	$\frac{\mu_0}{2g}\lambda_{rl}F_{eh}$	$l\Omega_r t$

As detailed in Table 3.4, the no-load air gap flux density in flux reversal machines can be categorized into two groups. The first category comprises stationary harmonics, where the pole pair number of each harmonic can be expressed as:

$$P = \frac{h}{2}Z_s, \quad h = 1, 3, 5\ldots \tag{3.11}$$

These harmonics do not participate in electromechanical energy conversion. The second category consists of rotating harmonics, characterized by pole pair numbers that can be delineated as:

$$P = \left\lvert \frac{h}{2}Z_s \pm lZ_r \right\rvert, \quad h = 1, 3, 5\ldots, \quad l = 1, 2, 3\ldots \tag{3.12}$$

The electrical frequencies of these harmonics are exclusively associated with the order of the specific permeance l. Consequently, when $l = 1$, all the corresponding flux density harmonics share the same fundamental electrical frequency. Analysis of their pole pair numbers reveals that these are tooth harmonics, meaning their sum or difference is an integer multiple of the number of stator teeth. Therefore, they are capable of inducing fundamental back-EMF in the same armature winding. Typically, the pole pair number P_a of the armature winding is selected as the lowest positive integer that satisfies Eq. (3.12) and $l = 1$.

$$P_a = \min\left\{ \left\lvert \frac{h}{2}Z_s \pm Z_r \right\rvert \right\}, \quad h = 1, 3, 5\ldots \tag{3.13}$$

The pole ratio of the flux reversal permanent magnet machine is defined as:

$$\mathrm{PR} = \frac{Z_r}{P_a} \tag{3.14}$$

The feasible slot-pole combinations for flux reversal permanent magnet machines are summarized in Table 3.5. Initially, the flux reversal machine was perceived as a type of variable reluctance machine, meaning that as each rotor tooth rotates, the magnitude and direction of the magnetic flux passing through each tooth vary. Therefore, many of the early designs for flux reversal machines employed fractional-slot concentrated windings. However, according to Eq. (3.14), when the winding pole pair number is small, adopting distributed winding significantly increases the

Table 3.5 Slot-pole combination of flux reversal permanent magnet machine

	$Z_s = 6, P_e = 3$				$Z_s = 12, P_e = 6$				$Z_s = 18, P_e = 9$				$Z_s = 24, P_e = 12$			
P_a	Z_r	k_{wa}	PR	q	Z_r	k_{wa}	PR	q	Z_r	k_{wa}	PR	q	Z_r	k_{wa}	PR	q
1	4	1	4	1	7	0.97	7	2	10	0.96	10	3	13	0.96	13	4
2	5	0.87	2.5	0.5	8	1	4	1	11	0.95	5.5	1.5	14	0.97	7	2
3	–				–				12	1	4	1	–			
4	–				10	0.87	2	1/2	13	0.95	3.25	3/4	16	1	4	1
5	–				11	0.93	1.4	2/5	14	0.95	2.8	3/5	17	0.93	3.4	4/5
6	–				–				15	0.86	2.5	1/2	–			
7	10	0.87	5	0.5	13	0.93	2.6	2/5	16	0.90	2.29	3/7	19	0.93	2.71	4/7
8	–				14	0.87	3.5	1/2	17	0.95	2.13	3/8	20	0.87	2.5	1/2
9	–				–				–				21	0.93	2.33	2/5
10	–				–				19	0.95	2.38	3/8	22	0.95	2.2	4/11
11	–				–				20	0.90	2.86	3/7	–			
12	–				–				21	0.86	3.5	1/2	–			
13	–				–				–				–			
14	–				–				–				26	0.95	2.6	4/11
15	–				–				–				27	0.93	3	2/5
16	–				–				–				28	0.87	3.5	1/2

Integer-slot distributed winding
Fractional-slot distributed winding
Non-overlapping concentrated winding

winding factor. As indicated in Table 3.5, high pole ratios suggest the use of either integer-slot or fractional-slot distributed windings.

3.4.2 Flux Switching Permanent Magnet Machine

The evolutionary process from a flux reversal permanent magnet machine to a flux switching permanent magnet machine is illustrated in Fig. 3.11. In the flux reversal configuration, each stator tooth is associated with two permanent magnets, each set having opposite excitation directions. Additionally, each stator slot opening is flanked by two permanent magnets that share the same excitation direction. This arrangement allows each stator tooth of the flux reversal machine to be bisected by a gap. The permanent magnets, initially paired due to their identical excitation direction, can be merged into a single unit and positioned within the stator yoke, as shown in the middle topology of Fig. 3.11. Subsequently, the permanent magnets in the stator yoke can be replaced with spoke-magnetized permanent magnets positioned in the stator gaps, culminating in the flux switching permanent magnet machine, as depicted in the rightmost part of Fig. 3.11 [13].

In comparison with the flux reversal machine, the excitation MMF and specific permeance of the flux switching machine demonstrate notable differences. For the excitation MMF, the permanent magnets in the flux switching machine are uniformly distributed along the circumference. Assuming the absence of slotting effects, the excitation MMF resembles an ideal square wave with a pole pair number of $Z_s/2$, as

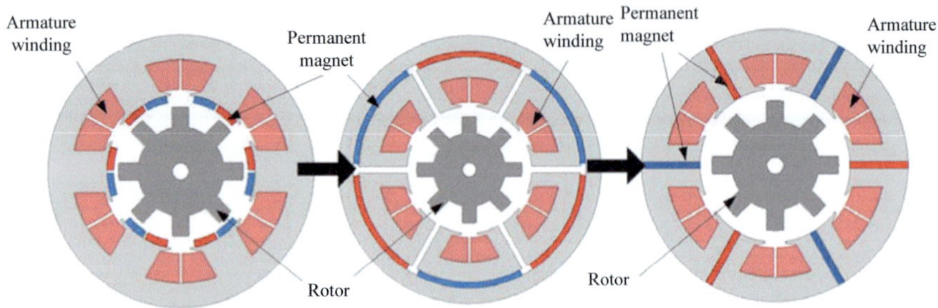

Fig. 3.11 Evolution process from flux reversal machine to flux switching machine

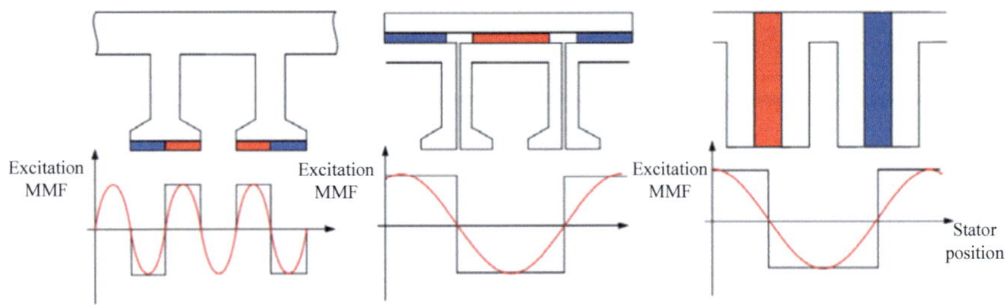

Fig. 3.12 Excitation MMF of flux switching permanent magnet machine

depicted in Fig. 3.12. The Fourier series representation of the excitation MMF can be expressed as:

$$F_e(\theta) = \sum_{h=1,3,5\ldots}^{\infty} F_{eh} \cos\left(\frac{h}{2} Z_s \theta\right) \tag{3.15}$$

It is crucial to acknowledge, based on the analysis in Sect. 2.2.5, that rotor slotting introduces additional MMF harmonics, which could introduce inaccuracies in quantitative analytical calculations. However, as this chapter primarily addresses qualitative analysis, the impact of rotor slotting on the excitation MMF is not considered here.

The next step involves calculating the air gap specific permeance. In the flux switching machine, each stator tooth is split into two smaller teeth by a permanent magnet. To achieve optimal configuration, the widths of the permanent magnet, the smaller tooth, and the slot opening should be equal. Consequently, the pole pair number of the specific permeance due to stator slotting is $2Z_s$, and the specific permeance function can be articulated as follows:

$$\lambda_s(\theta) = \lambda_{s0} + \sum_{l_s=1}^{\infty} \lambda_{sl_s} \cos 2l_s Z_s \theta \tag{3.16}$$

In Eq. (3.16), λ_{s1} is negative because the reference axis is positioned at the centerline of the stator tooth, where the specific permeance is minimal.

The rotor-specific permeance in the flux switching machine is maintained as outlined in Eq. (3.10), similar to that of the flux reversal machine. By combining Eqs. (3.10) and (3.16), the overall air gap specific permeance, considering both stator and rotor slotting, can be established as follows:

$$\lambda_f(\theta, t) \approx \lambda_s\lambda_r = \lambda_{s0}\lambda_{r0} + \lambda_{s0}\sum_{l_r=1}^{\infty}\lambda_{rl_r}\cos l_r Z_r(\theta - \Omega_r t) + \lambda_{r0}\sum_{l_s=1}^{\infty}\lambda_{sl_s}\cos 2l_s Z_s\theta + \frac{1}{2}\sum_{l_s,l_r}\lambda_{sl_s}\lambda_{rl_r}\cos[(l_r Z_r \pm 2l_s Z_s)\theta - l_r\Omega_r t] \quad (3.17)$$

In Eq. (3.17), to distinguish between the stator and rotor specific permeance, the harmonic order of the specific permeance on the rotor side is denoted as l_r. By employing the excitation MMF and specific permeance outlined in Eqs. (3.15) and (3.17), the no-load flux density of the flux-switching permanent magnet machine can be determined, with its harmonic contents detailed in Table 3.6. The no-load air gap flux density harmonics are divided into two categories. The first category encompasses the stationary magnetic field, which does not engage in electromechanical energy conversion. The second category comprises the rotating magnetic field, where the electrical frequency of these fields depends exclusively on the harmonic order of the rotor-specific permeance l_r. When $l_r = 1$, the corresponding air gap flux density harmonics are capable of inducing fundamental back-EMF in the armature winding. The pole pair numbers of these harmonics are expressed as follows:

$$P = \left|\left(\frac{h}{2} \pm 2l_s\right)Z_s \pm Z_r\right|, \quad h = 1, 3, 5\ldots, \quad l_s = 0, 1, 2\ldots \quad (3.18)$$

It is important to note that the sum or difference of the aforementioned harmonics is an integer multiple of the stator teeth number. Consequently, all harmonics that satisfy Eq. (3.18) are classified as tooth harmonics. To attain a higher pole ratio ($= Z_r/P_a$), the minimum positive integer that fulfills Eq. (3.18) is chosen as the pole pair number for the armature winding, i.e.,

$$P_a = \min\left|\left(\frac{h}{2} \pm 2l_s\right)Z_s \pm Z_r\right|, \quad h = 1, 3, 5\ldots, \quad l_s = 0, 1, 2\ldots \quad (3.19)$$

The feasible slot-pole combinations for flux-switching permanent magnet machines are cataloged in Table 3.7. Analogous to the flux reversal machine, the majority of flux-switching machines initially utilize fractional-slot concentrated winding. However, when the pole pair number of the winding is small, the adoption of distributed winding results in a notably higher winding factor. As demonstrated in Table 3.7, it is recommended to employ integer-slot or fractional-slot distributed windings when P_a is small.

Table 3.6 No-load air gap flux density harmonics of flux switching permanent magnet machine

Order	Pole pair number	Magnitude	Angular frequency
1	$\frac{hZ_s}{2}$	$\frac{\mu_0}{g}\lambda_{s0}\lambda_{r0}F_{eh}$	0
2	$\left(2l_s + \frac{h}{2}\right)Z_s$	$\frac{g}{2\mu_0}\lambda_{sl_s}\lambda_{r0}F_{eh}$	0
3	$\left\|2l_s - \frac{h}{2}\right\|Z_s$	$\frac{g}{2\mu_0}\lambda_{sl_s}\lambda_{r0}F_{eh}$	0
4	$\left(2l_s + \frac{h}{2}\right)Z_s + l_rZ_r$	$\frac{g}{4\mu_0}\lambda_{sl_s}\lambda_{rl_r}F_{eh}$	$l_r\Omega_rt$
5	$\left(2l_s + \frac{h}{2}\right)Z_s + l_rZ_r$	$\frac{g}{4\mu_0}\lambda_{sl_s}\lambda_{rl_r}F_{eh}$	$l_r\Omega_rt$
6	$\left(2l_s + \frac{h}{2}\right)Z_s + l_rZ_r$	$\frac{g}{4\mu_0}\lambda_{sl_s}\lambda_{rl_r}F_{eh}$	$l_r\Omega_rt$
7	$\left\|\left(2l_s - \frac{h}{2}\right)Z_s + l_rZ_r\right\|$	$\frac{g}{4\mu_0}\lambda_{sl_s}\lambda_{rl_r}F_{eh}$	$l_r\Omega_rt$
8	$\left\|\left(2l_s + \frac{h}{2}\right)Z_s - l_rZ_r\right\|$	$\frac{g}{4\mu_0}\lambda_{sl_s}\lambda_{rl_r}F_{eh}$	$l_r\Omega_rt$
9	$\left\|\left(2l_s - \frac{h}{2}\right)Z_s - l_rZ_r\right\|$	$\frac{g}{4\mu_0}\lambda_{sl_s}\lambda_{rl_r}F_{eh}$	$l_r\Omega_rt$

Table 3.7 Slot-pole combination of flux switching permanent magnet machine

	$Z_s = 6, P_e = 3$				$Z_s = 12, P_e = 6$				$Z_s = 18, P_e = 9$				$Z_s = 24, P_e = 12$			
P_a	Z_r	k_w	PR	q	Z_r	k_w	PR	q	Z_r	k_w	PR	q	Z_r	k_w	PR	q
1	16	1	16	1	17	0.97	17	2	26	0.96	26	3	35	0.96	35	4
2	17	0.87	8.5	0.5	16	1	8	1	25	0.95	12.5	1.5	34	0.97	17	2
3	–				–				24	1	8	1	–			
4	–				14	0.87	3.5	1/2	23	0.95	5.75	3/4	32	1	8	1
5	–				13	0.93	2.6	2/5	22	0.95	4.4	3/5	31	0.93	6.2	4/5
6	–				–				21	0.86	3.5	1/2	–			
7	–				11	0.93	2.2	2/5	20	0.90	2.86	3/7	29	0.93	4.14	4/7
8	–				10	0.87	2.5	1/2	19	0.95	2.38	3/8	28	0.87	3.5	1/2
9	–				–				–				27	0.93	3	2/5
10	–				–				17	0.95	2.13	3/8	26	0.95	2.6	4/11
11	–				–				16	0.90	2.29	3/7	–			
12	–				–				15	0.86	2.5	1/2	–			
13	–				–				–				–			
14	–				–				–				22	0.95	2.2	4/11
15	–				–				–				21	0.93	2.33	2/5
16	–				–				–				20	0.87	2.5	1/2

Integer-slot distributed winding
Fractional-slot distributed winding
Non-overlapping concentrated winding

3.4.3 Doubly-Salient Permanent Magnet Machine

The evolution from a flux reversal permanent magnet machine to a doubly-salient permanent magnet machine is elucidated as follows. A slot-pole combination that satisfies the expression:

$$P_{\mathrm{e}} = \frac{Z_{\mathrm{s}}}{6}, \quad Z_{\mathrm{r}} = \frac{2}{3}Z_{\mathrm{s}} \tag{3.20}$$

Under this specific slot-pole combination, each set of three adjacent stator teeth is covered by one permanent magnet, as depicted in the leftmost image of Fig. 3.13. Subsequently, these permanent magnets can be repositioned to the stator yoke while preserving the excitation field integrity, leading to the formation of a radial-magnetization doubly-salient permanent magnet machine, as illustrated in the middle image of Fig. 3.13. Furthermore, the radial-magnetization permanent magnets can be replaced with spoke-magnetization permanent magnets, which are evenly distributed every three teeth in the stator yoke. This adjustment maintains an unchanged excitation field for each stator tooth, culminating in the realization of the doubly-salient permanent magnet machine, as shown in the rightmost image of Fig. 3.13 [14].

In line with Eq. (3.20), the air gap flux density of the doubly-salient permanent magnet machine can be deduced, employing a process analogous to that used for the flux reversal and flux switching machines. Similarly, when the harmonic order of rotor-specific permeance $l_{\mathrm{r}} = 1$, the corresponding air gap flux density harmonics are capable of inducing the fundamental back-EMF in the armature winding. The pole pair numbers of these harmonics are specified as follows:

$$P = \left|\frac{h}{6} \pm l_{\mathrm{s}} \pm \frac{2}{3}\right| Z_{s}, \quad \begin{cases} h = 1, 3, 5 \ldots \\ l_{\mathrm{s}} = 0, 1, 2 \ldots \end{cases} \tag{3.21}$$

Subsequently, the minimum positive integer that satisfies Eq. (3.21) is selected as the pole pair number for the armature winding:

$$P_{\mathrm{a}} = \min\left\{\left|\frac{h}{6} \pm l_{\mathrm{s}} \pm \frac{2}{3}\right| Z_{s}\right\} = \frac{Z_{\mathrm{s}}}{6} \tag{3.22}$$

Consequently, the pole ratio of the doubly-salient permanent magnet machine is calculated as:

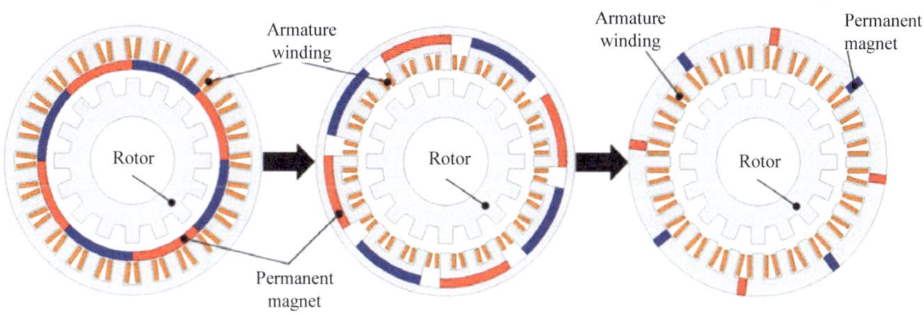

Fig. 3.13 Evolution process from flux reversal machine to doubly-salient permanent magnet machine

Table 3.8 Slot-pole combination of doubly-salient permanent magnet machine

	$Z_s = 6, P_e = 1$				$Z_s = 12, P_e = 2$				$Z_s = 18, P_e = 3$				$Z_s = 24, P_e = 4$			
P_a	Z_r	k_w	PR	q	Z_r	k_w	PR	q	Z_r	k_w	PR	q	Z_r	k_w	PR	q
1	4	1	16	1	–				–				–			
2	–				8	1	4	1	–				–			
3	–				–				12	1	4	1	–			
4	–				–				-				16	1	4	1

Integer-slot distributed winding

$$\mathrm{PR} = \frac{Z_r}{P_a} = 4 \tag{3.23}$$

This ratio reflects the machine's unique structural characteristics. Table 3.8 lists the feasible slot-pole combinations for the doubly-salient permanent magnet machine. Similar to its predecessors, the early models of doubly-salient permanent magnet machines typically utilized fractional-slot concentrated windings. However, with a pitch factor of only 0.5, such configurations significantly lower the winding factor and torque density. Therefore, to achieve an enhanced winding factor, the adoption of integer-slot distributed winding with the number of slots per pole per phase ($q = 1$) is recommended for doubly-salient permanent magnet machines. This approach optimizes both the magnetic field distribution and the machine's overall performance.

3.5 Electrically Excited Flux Modulation Machine

The subsequent analysis focuses on stationary excitation unit FMMs utilizing DC winding excitations. This configuration circumvents the complications associated with rotary excitation windings and allows the excitation MMF to be flexibly adjusted by tuning the current in the DC winding. Such stationary excitation unit FMMs are particularly promising for aviation starting and power generation systems. Three specific categories of stationary excitation unit FMMs with DC winding excitations are examined: electrically excited vernier reluctance machines [15–17], electrically excited flux switching machines [18], and electrically excited doubly-salient machines [19].

The electrically excited vernier reluctance machine can be traced back to the evolution from a flux reversal permanent magnet machine, as depicted in Fig. 3.14.

In the leftmost topology of Fig. 3.14, assume the pole pair number of permanent magnets in the flux reversal machine satisfies the following equation:

$$P_e = \frac{Z_s}{2} \tag{3.24}$$

Under these conditions, each stator tooth is excited by a single permanent magnet, and adjacent stator teeth are excited in opposite directions. Each permanent magnet

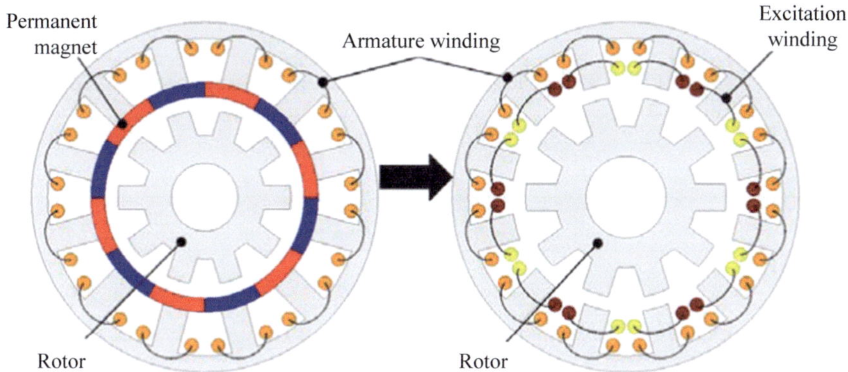

Fig. 3.14 Evolution from flux reversal machine to electrically excited vernier reluctance machine

can be replaced with an excitation coil, resulting in the electrically excited vernier reluctance machine, as shown in the rightmost part of Fig. 3.14. Similar to permanent magnet excitations, the pole pair number of the no-load air gap flux density for the electrically excited vernier reluctance machine can be derived as:

$$P = \left| \left(\frac{h}{2} \pm l_s \right) Z_s \pm Z_r \right| \tag{3.25}$$

Subsequently, the armature winding's pole pair number and pole ratio of the electrically excited vernier reluctance machine can be defined similarly to Eqs. (3.13) and (3.14). The feasible slot-pole combinations are summarized in Table 3.9. Given its stationary excitation unit, the utilization of superconducting excitation windings can enhance its excitation capacity. Stationary superconducting excitation windings offer advantages such as easy cooling and brushless operation compared to conventional superconducting synchronous machines [20].

The electrically excited flux switching machine, illustrated in Fig. 3.15, is derived directly from the flux switching permanent magnet machine. In the latter, the stator is divided into several U-shaped cores, each excited by two adjacent permanent magnets. These permanent magnets can be replaced by excitation coils wound around the U-shaped cores. Additionally, the previously separated stator yoke in the flux switching permanent magnet machine can be connected to reduce the magnetic resistance of the excitation field, resulting in the electrically excited flux switching machine. It is important to note that since the slot initially occupied by the permanent magnet now contains the excitation winding, both the excitation and armature windings have a winding pitch of two, causing them to overlap.

The electrically excited doubly-salient machine, depicted in Fig. 3.16, is a direct transformation of the doubly-salient permanent magnet machine. In the doubly-salient permanent magnet machine, each permanent magnet provides an excitation field for three adjacent stator teeth. These permanent magnets can be replaced by excitation coils spanning three stator teeth. This transformation requires approximately

Table 3.9 Slot-pole combination of electrically excited vernier reluctance machine

	$Z_s = 6, P_e = 3$				$Z_s = 12, P_e = 6$				$Z_s = 18, P_e = 9$				$Z_s = 24, P_e = 12$			
P_a	Z_r	k_w	PR	q	Z_r	k_w	PR	q	Z_r	k_w	PR	q	Z_r	k_w	PR	q
2	11	0.87	5.5	0.5	–				–				–			
4	–				22	0.87	5.5	1/2	–				–			
5	–				23	0.93	4.6	2/5	–				–			
6	–				–				21	0.86	3.5	1/2	–			
7	–				25	0.93	3.57	2/5	20	0.90	2.86	3/7	–			
8	–				26	0.87	3.25	1/2	19	0.95	2.38	3/8	28	0.87	3.5	1/2
9	–				–				–				27	0.93	3	2/5
10	–				–				17	0.95	2.13	3/8	26	0.95	2.6	4/11
11	–				–				16	0.90	2.29	3/7	–			
12	–				–				15	0.86	2.5	1/2	–			
14	–				–				–				22	0.95	2.2	4/11
15	–				–				–				21	0.93	2.33	2/5
16	–				–				–				20	0.87	2.5	1/2

(shaded) Non-overlapping concentrated winding

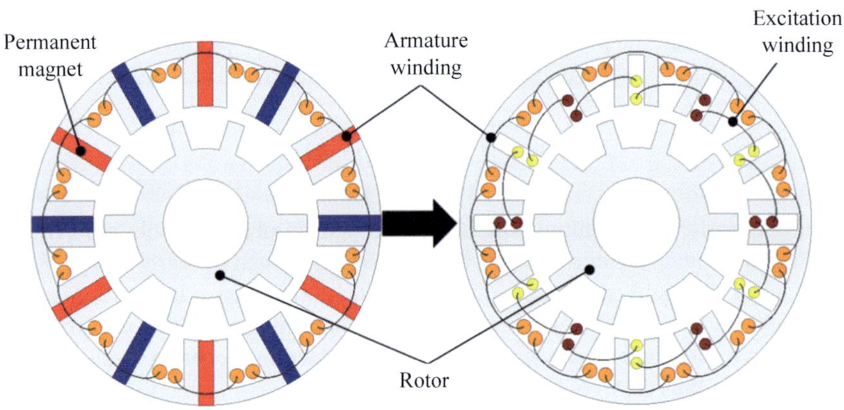

Fig. 3.15 Evolution process from flux switching permanent magnet machine to electrically excited flux switching machine

one-fourth of the stator slots to house additional excitation coils, necessitating an increase in the slot area. Therefore, optimizing the stator core shape is essential, ensuring that the slots accommodating the excitation winding are larger than the others. However, the electrically excited doubly-salient machine is not widely used due to its low armature winding factor and the extended length of the excitation winding ends. The low winding factor of the armature winding results in reduced torque density, and the increased volume and weight are due to the long end length of the excitation winding.

The preceding subsections delineate stationary excitation unit FMMs with permanent magnet and DC winding excitations. Finally, the feasibility of AC winding excitation and self-short-circuit winding excitation is discussed. Implementing AC

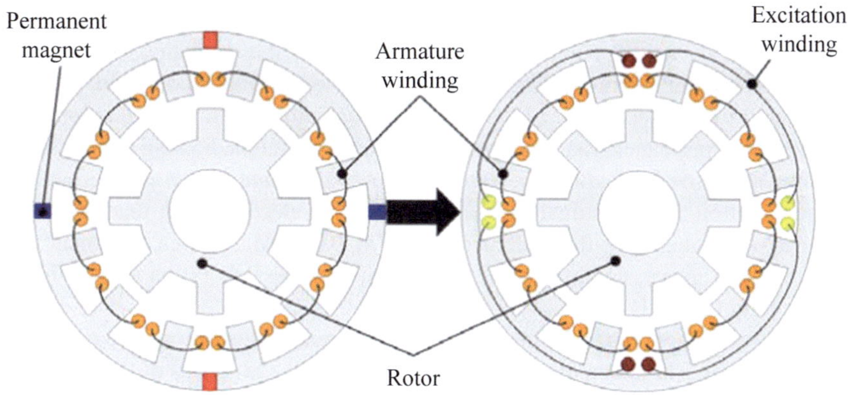

Fig. 3.16 Evolution process from doubly-salient permanent magnet machine to electrically excited doubly-salient machine

winding or self-short-circuit winding necessitates rotary excitation units to generate a stationary excitation MMF. However, this introduces mechanical complexity to the excitation unit and compromises the machine's overall performance. Consequently, AC winding excitation and self-short-circuit winding excitation are deemed impractical in this context.

3.6 Flux Modulation Machine with Non-Stationary Units

This section delves into non-stationary unit FMMs, where the excitation MMF, armature MMF, and modulation unit are all rotary. This category of FMM exhibits a relatively complex operating principle and encompasses a diverse range of topologies, including the flux modulation brushless dual-mechanical-port dual-electrical-port machine, brushless doubly-fed reluctance machine, vernier reluctance machine, and asynchronous induction machine. Detailed descriptions of these configurations are provided below.

3.6.1 Flux Modulation Brushless Dual-Mechanical-Port Dual-Electrical-Port Machine

When all three units of an FMM rotate, and the excitation unit comprises permanent magnets, the fundamental topology of the FMM with non-stationary units is illustrated in Fig. 1.5. Due to the independent rotation of the excitation and modulation units, it is classified as a dual-mechanical-port machine. However, the speed control of the two rotors is only partially constrained by Eq. (2.82), necessitating an additional set of armature windings in the stator. This newly added winding, characterized by a pole pair number equal to that of the excitation unit (P_e), is termed the regular winding, distinguished from the existing modulation winding represented by the subscript "a1." These windings, along with the excitation unit, form a regular permanent magnet machine. This configuration, hosting two armature windings and two rotors within the stator without brushes and slip rings, is termed the flux modulation brushless dual-mechanical-port dual-electrical-port machine (FM-BLDDM). By integrating these elements, the FM-BLDDM achieves a sophisticated level of performance, combining the benefits of brushless operation and dual-port functionality [21, 22].

Based on the topology in Fig. 3.17a, two evolved configurations are presented in Fig. 3.17b, c. These evolutions involve exchanging the positions of the inner and outer rotors to enhance the coupling between the permanent magnets and the regular winding. Subsequently, surface-mounted permanent magnets are replaced by spoke-magnetization permanent magnets, resulting in the topology shown in Fig. 3.18b. Here, the modulation winding and the inner and outer rotors constitute a dual-rotor FMM, while the regular winding and the outer rotor form a spoke-magnetization permanent magnet machine. Further optimization includes replacing the reluctance-type inner rotor with a surface-mounted permanent magnet rotor, which enhances torque density. The evolved topologies in Fig. 3.17b, c exhibit higher torque density than the basic topology in Fig. 3.17a due to changes in the magnetic circuit structure and excitation methods [23, 24].

Using the basic topology in Fig. 3.17a as an example, the derivations of torque and speed characteristics for FM-BLDDM are provided. Although other FM-BLDDM topologies offer performance improvements, their torque and speed characteristics align closely with those of the basic topology. Assuming $P_f > P_e$, when only the modulation winding is energized, the torque of the three units is proportional to their respective pole pair numbers. Thus, the torque expressions are as follows:

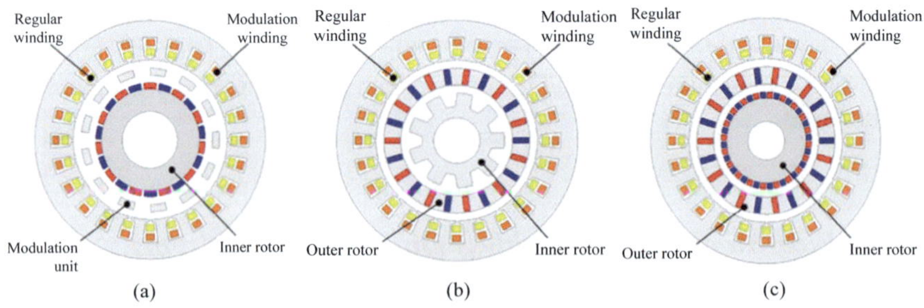

Fig. 3.17 FM-BLDDM topologies: **a** surface-mounted permanent magnet type, **b** spoke-magnetization permanent magnet type, **c** integrated excitation type

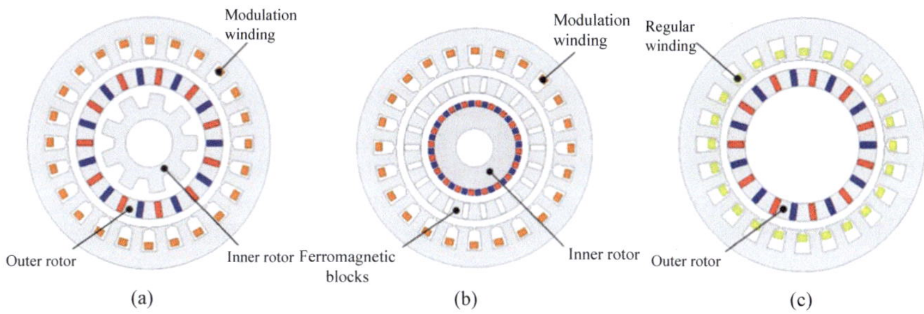

Fig. 3.18 Topology decomposition of FM-BLDDMs: **a** dual-rotor FMM in spoke-magnetization FM-BLDDM, **b** dual-rotor FMM in integrated excitation FM-BLDDM, **c** spoke-magnetization permanent magnet machine in two FM-BLDDMs

$$\begin{cases} T_{f1} = -k_{T1} I_{a1} P_f \\ T_{e1} = k_{T1} I_{a1} P_e \\ T_{a1} = k_{T1} I_{a1} P_{a1} \end{cases} \tag{3.26}$$

Except for the pole pair number, each torque is also proportional to the current amplitude I_{a1} in the modulation winding, with k_{T1} representing the torque coefficient. When only the regular winding is energized, the outer rotor (i.e., the modulation unit) remains inactive, and only the excitation unit and the stator are subjected to electromagnetic torque. Their torque expressions are:

$$\begin{cases} T_{f2} = 0 \\ T_{e2} = k_{T2} I_{a2} \\ T_{a2} = -k_{T2} I_{a2} \end{cases} \tag{3.27}$$

where I_{a2} denotes the current amplitude in the regular winding, and k_{T2} represents the corresponding torque coefficient. When both the modulation winding and regular winding are energized, the resultant torque is the superposition of both torques under single-winding energization, regardless of the saturation effect:

$$\begin{cases} T_f = -k_{T1} I_{a1} P_f \\ T_e = k_{T1} I_{a1} P_e + k_{T2} I_{a2} \end{cases} \tag{3.28}$$

Therefore, by adjusting the currents in the two windings, the torque of the two rotors can be independently controlled. For instance, if specific torques T_f and T_e need to be generated, one can first determine the required I_{a1} based on the first expression in Eq. (3.28). Using the values of T_e and I_{a1} the amplitude of I_{a2} can then be determined by substituting these into the second expression of Eq. (3.28). Furthermore, according to Eq. (2.82), combined with the fundamental relationship of electromechanical energy conversion:

$$\begin{cases} \omega_{a1} = P_f \Omega_f - P_e \Omega_e \\ \omega_{a2} = P_e \Omega_e \end{cases} \tag{3.29}$$

Here, ω_{a1} and ω_{a2} refer to the angular frequencies of the modulation and regular windings, respectively. Similar to the torque relationship described in Eq. (3.28), the rotation speeds of the two rotors can also be flexibly adjusted by tuning the angular frequencies of the two windings. In summary, the FM-BLDDM represents a highly integrated FMM where the rotation speed and torque of the two rotors can be controlled separately, making it highly suitable for applications requiring multiple mechanical ports, such as helicopter propeller drives and underwater propulsion systems.

Given that the FM-BLDDM features two electrical ports and two mechanical ports, the power flow within the system is somewhat complex and requires further analysis. When the modulation winding is energized, the mechanical powers of the two rotors are calculated as:

$$\begin{aligned} p_{f1} &= T_{f1} \Omega_f = -k_{T1} I_{a1} P_f \Omega_f \\ p_{e1} &= T_{e1} \Omega_e = k_{T1} I_{a1} P_e \Omega_e \end{aligned} \tag{3.30}$$

Notably, in this chapter, lowercase "p" denotes power, while uppercase "P" indicates pole pair number. As stated in Eq. (3.30), when both rotors rotate counterclockwise, the mechanical power of the modulation unit p_{f1} is negative, indicating that it absorbs mechanical power from an external source. Conversely, the mechanical power of the excitation unit p_{e1} is positive, signifying power output. Since, $p_{f1} \neq p_{e1}$, the residual power must originate from the stator side. Although the stationary stator does not generate mechanical power, it can supply electrical power through the modulation winding. According to the Conservation of Energy Law, the electrical power provided by the modulation winding can be calculated as:

$$p_{a1} = -(p_{f1} + p_{e1}) = k_{T1} I_{a1} (P_f \Omega_f - P_e \Omega_e) \tag{3.31}$$

When $P_f \Omega_f > P_e \Omega_e$, the electrical power is positive, indicating that the modulation winding outputs electrical power. Conversely, when $P_f \Omega_f < P_e \Omega_e$, the modulation winding absorbs electrical power. In this scenario, as shown in Fig. 3.19a, the mechanical power of the excitation unit is sourced from two components: the mechanical power of the modulation unit and the electrical power of the modulation winding. Subsequently, the excitation unit outputs the resultant mechanical power to the external system.

Similarly, when the regular winding is energized, the power dynamics of the regular winding and the excitation unit are as follows:

$$\begin{aligned} p_{e2} &= T_{e2} \Omega_e = k_{T2} I_{a2} \Omega_e \\ p_{a2} &= -p_{e2} = -k_{T2} I_{a2} \Omega_e \end{aligned} \tag{3.32}$$

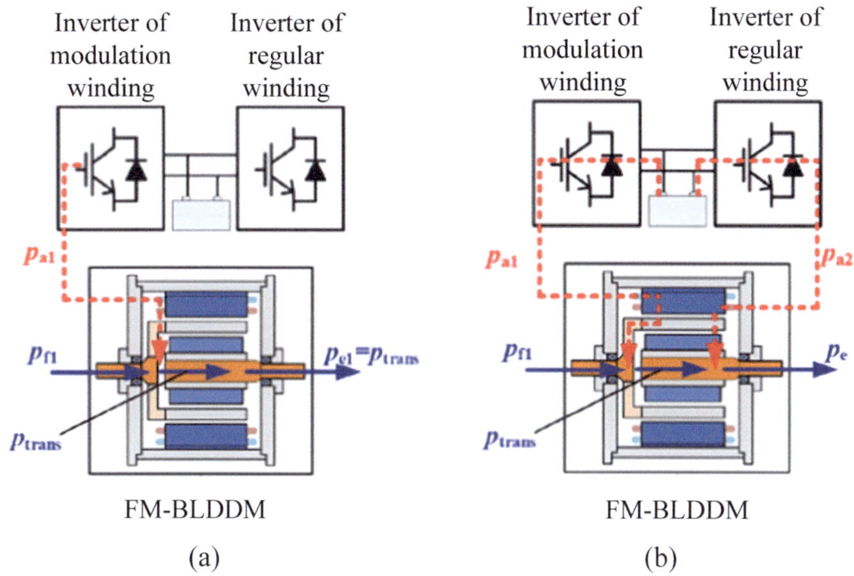

Fig. 3.19 Power flows of FM-BLDDM: **a** only the modulation winding is energized, **b** both the modulation winding and the excitation winding are energized

The direction of power flow for the regular winding depends on the polarity of its current I_{a2}. If I_{a2} is positive, the regular winding outputs electrical power, and the excitation unit absorbs mechanical power. Conversely, if I_{a2} is negative, the regular winding absorbs electrical power, and the excitation unit outputs mechanical power. In this scenario, depicted in Fig. 3.19b, the total output mechanical power of the excitation unit, p_e is derived from three sources: the mechanical power of the modulation unit p_{f1}, and the electrical powers from the modulation winding p_{a1} and the excitation winding p_{a2}.

The FM-BLDDM finds application in the drivetrain system of hybrid electric vehicles, as illustrated in Fig. 3.20. Traditionally, the energy generated by the internal combustion engine must be transmitted to the wheels through various mechanical devices, such as gears, to maintain a constant ratio of speed and torque between the engine and the wheels. However, varying road conditions cause fluctuations in the speed and torque requirements of the wheels, which in turn affect the operational conditions of the internal combustion engine. Traditional internal combustion engines operate efficiently only within a narrow speed and torque range, which contributes to overall low fuel efficiency in vehicles. To mitigate this issue, hybrid vehicles were developed, incorporating a special device known as the energy distribution device. This device enables the decoupling of the internal combustion engine from the wheels. Through strategic energy management, the internal combustion engine can operate in a high-efficiency zone. During conditions of low speed or light load, excess energy generated by the engine is temporarily stored in the battery, which is

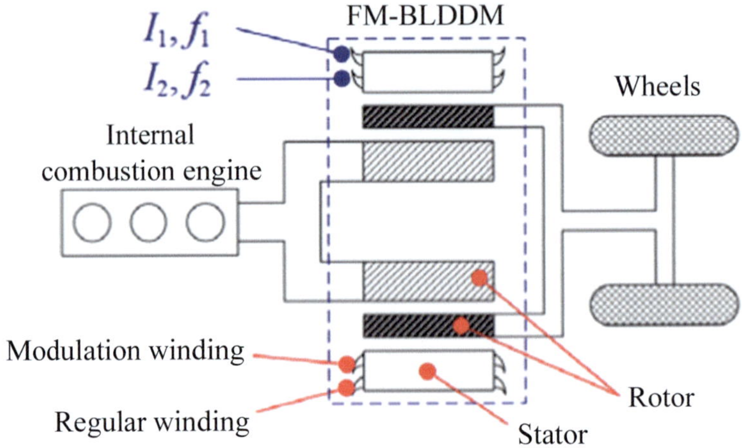

Fig. 3.20 FM-BLDDM-based hybrid drive system

then supplied to the wheels during conditions of high speed or heavy load. Thus, the hybrid vehicle efficiently manages energy delivery and dispatch through the energy distribution device, enhancing overall vehicle efficiency and performance.

The operational principle of the FM-BLDDM-based hybrid drive system involves connecting the two rotors of the FM-BLDDM—the modulation unit and the excitation unit—to the internal combustion engine and the wheels, respectively, as illustrated in Fig. 3.20. This configuration allows for the decoupling of the internal combustion engine from the wheels. Additionally, energy transfer and management between these units can be accomplished, as demonstrated in Fig. 3.19. Consequently, the FM-BLDDM serves effectively as an energy distribution device in hybrid electric vehicles.

If the excitation unit is transitioned from permanent magnets to a DC excitation winding, various multi-port FMM topologies can be realized. However, utilizing a DC excitation winding in FM-BLDDMs, whether located in the rotary excitation unit or modulation unit, requires brushes and slip rings, negating some of the advantages of multi-port FMMs over conventional dual-mechanical-port machines. Therefore, incorporating DC excitation windings in FM-BLDDMs is not advisable.

3.6.2 Brushless Doubly-Fed Reluctance Machine

The development process of the brushless doubly-fed reluctance machine involves several steps. Initially, based on the basic FMM model shown in Fig. 1.5, the positions of the excitation unit and the modulation unit are swapped, transforming the modulation unit into a reluctance inner rotor. Subsequently, to replace the rotating permanent magnets on the outer rotor, an AC excitation winding is added to the stator, resulting in the inclusion of two windings (the armature winding and the AC excitation winding) in the stator, as depicted in Fig. 3.21. This configuration of the brushless doubly-fed reluctance machine, featuring a reluctance rotor and two stationary windings, eliminates the need for brushes and slip rings, thereby earning it the designation of "brushless doubly-fed reluctance machine" [25].

Two options are available for the slot-pole combinations of the brushless doubly-fed reluctance machine. In the scenario of a semi-closed stator slot, the pole pair number P_f of the modulation unit is equal to the number of rotor teeth Z_r. Therefore, according to Table 2.2, the pole pair numbers of the two windings and the number of rotor teeth must satisfy the following relationship:

$$Z_r = |P_a \pm P_e| \tag{3.33}$$

where P_a and P_e are the pole pair numbers of the armature winding and the AC excitation winding, respectively. For an open-slot stator structure, as stipulated in Eq. (2.44), the air gap specific permeance includes a component with a pole pair number $|Z_s - Z_r|$, which represents the working permeance harmonic. Consequently, the slot-pole combination should meet the following condition:

$$|Z_s - Z_r| = |P_a \pm P_e| \tag{3.34}$$

Table 3.10 outlines the viable slot-pole combinations for the brushless doubly-fed reluctance machine. It is evident that when the slot-pole combination adheres to Eq. (3.33), the rotor teeth number (Z_r) closely aligns with either P_a or P_e, leading to a low pole ratio and, consequently, a relatively modest torque density. Conversely, if the slot-pole combination conforms to Eq. (3.34), the stator teeth number (Z_s) approaches the rotor teeth number (Z_r), with Z_r significantly larger than P_a and P_e. In this scenario, the pole ratio is high, yielding a relatively large torque density.

The power distribution of the brushless doubly-fed reluctance machine is a critical characteristic, and its derivation follows. The electromagnetic power of the two windings can be expressed as:

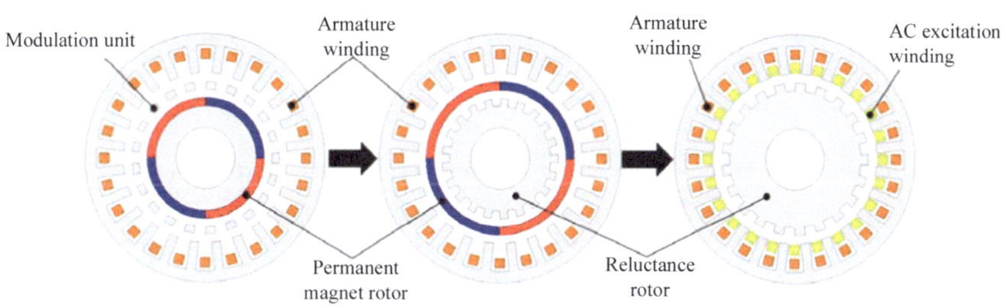

Fig. 3.21 Evolution process of brushless doubly-fed reluctance machine

Table 3.10 Slot-pole combination of brushless doubly-fed reluctance machine

	$Z_s = 12$				$Z_s = 18$				$Z_s = 24$			
P_a	P_e	Z_r	$k_{wa}k_{we}$	Z_r/P_aP_e	P_e	Z_r	$k_{wa}k_{we}$	Z_r/P_aP_e	P_e	Z_r	$k_{wa}k_{we}$	Z_r/P_aP_e
1	2	15	0.97	7.5	2	21	0.91	10.5	2	27	0.93	13.5
			–		3	22	1	7.33				
			–		4	23	0.95	5.75	4	29	0.96	7.25
			–		5	24	0.95	4.8	5	30	0.89	6
2			–		3	23	0.95	3.83			–	
			–		4	24	0.90	3	4	30	0.97	3.75
			–		5	25	0.90	2.5	5	31	0.90	3.1
			–				–		7	33	0.90	2.36
3			–		4	25	0.95	2.08			–	
			–		5	26	0.95	1.73			–	
4			–		5	27	0.90	1.35	5	33	0.93	1.65
			–				–		7	35	0.93	1.25
5			–				–		7	36	0.86	1.03

Integer-slot distributed winding
Fractional-slot distributed winding
Non-overlapping concentrated winding

$$\begin{cases} p_e = \Omega_e T_e \\ p_a = \Omega_a T_a \end{cases} \tag{3.35}$$

where Ω_e and Ω_a denote the rotation speeds of the excitation winding MMF and the armature winding MMF, respectively, while T_e and T_a represent the equivalent electromagnetic torques experienced by the excitation winding and the armature winding, respectively. Assuming that the slot-pole combination adheres to Eq. (3.36), the electromagnetic torque of the two windings is proportional to their respective pole pair numbers:

$$Z_s - Z_r = P_a + P_e \tag{3.36}$$

and by combining Eq. (2.82), the rotation speed relationship can be determined as:

$$P_e\Omega_e + P_a\Omega_a = -P_r\Omega_r \tag{3.37}$$

Similarly, the electromagnetic torque of the two windings is proportional to the corresponding pole pair number, i.e.,

$$\frac{T_e}{T_a} = \frac{P_e}{P_a} \tag{3.38}$$

In brushless doubly-fed reluctance machines, the synchronous speed is often defined as the rotor's speed when the excitation MMF is stationary (i.e., $\Omega_e = 0$). Thus, according to Eq. (3.37), the synchronous speed can be calculated as:

$$\Omega_{r0} = -\frac{P_a\Omega_a}{P_r} \tag{3.39}$$

The slip ratio is defined as:

$$s = \frac{\Omega_{r0} - \Omega_r}{\Omega_{r0}} \tag{3.40}$$

Then, by substituting Eqs. (3.39) and (3.40) into Eq. (3.37):

$$P_e \Omega_e = -sP_a \Omega_a \tag{3.41}$$

Afterward, substituting Eqs. (3.41) and (3.38) into Eq. (3.35):

$$p_e = -sp_a \tag{3.42}$$

Consequently, it can be deduced that the power of the AC excitation winding is equal to the product of the power of the armature winding and the slip ratio. While this relationship is derived under the assumption that the slot-pole combination satisfies Eq. (3.36), it has been demonstrated to apply to any brushless doubly-fed reluctance machine whose slot-pole combination conforms to Eqs. (3.33) or (3.34). This relationship highlights that brushless doubly-fed reluctance machines are particularly well-suited for fixed-frequency variable-speed applications, such as wind generators. In such applications, the armature winding is directly connected to the grid, and the AC excitation winding is linked to the grid through an inverter. As wind speed varies, the frequency of the excitation winding can be adjusted by the inverter according to Eq. (3.37), thus maintaining a constant frequency in the armature winding. Consequently, the machine speed can be flexibly adjusted to optimize the utilization of wind power. Moreover, due to the minor variation in blade speed, the slip ratio remains low, which results in a low power requirement for the excitation winding as per Eq. (3.42). Therefore, the inverter capacity required is substantially smaller than the power generation capacity, allowing the brushless doubly-fed reluctance wind generator to achieve rotor variable speed while minimizing the weight, volume, and cost of the inverter.

3.6.3 Vernier Reluctance Machine

The vernier reluctance machine evolves from the brushless doubly-fed reluctance machine, as depicted in Fig. 3.22. If the brushless doubly-fed reluctance machine satisfies the condition $P_a + P_e = |Z_s - Z_r|$, where Pe is the pole pair number of the excitation winding and P_a is the pole pair number of the armature winding, and Pe equals P_a, then the windings share the same configuration. Consequently, the excitation winding can be eliminated, and the excitation current can be injected into the armature winding to maintain a constant excitation MMF. Subsequently, as per Eq. (3.37), if the rotor speed satisfies:

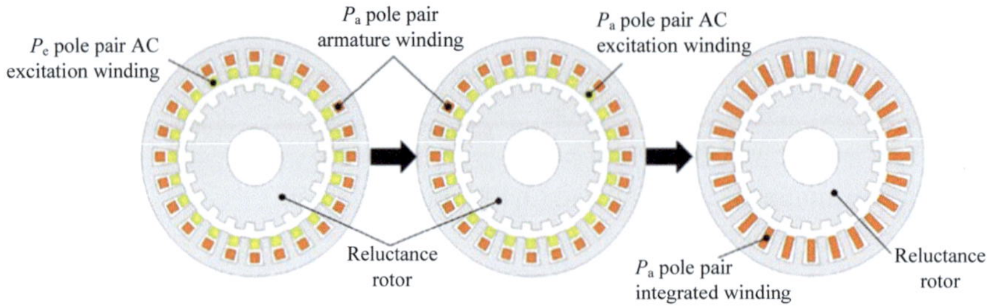

Fig. 3.22 Evolution from brushless doubly-fed reluctance machine to vernier reluctance machine

$$\Omega_r = -\frac{2P_a\Omega_a}{P_r} \tag{3.43}$$

In the vernier reluctance machine, the AC excitation current and the armature current within the integrated winding exhibit the same frequency. Consequently, the resultant current in the integrated winding can be considered as comprising two components, each with identical frequency. In summary, the slot-pole combination of the vernier reluctance machine should adhere to the following relationship:

$$2P_a = |Z_s - Z_r| \tag{3.44}$$

Due to the absence of a distinct excitation unit, this type of machine is typically regarded as a reluctance machine and is hence referred to as the vernier reluctance machine in the literature [26].

The "vernier" effect in the vernier reluctance machine manifests as a speed amplification of its air gap specific permeance. For the doubly-salient topology, Eq. (2.44) provides the expression for the air gap specific permeance. Building on this foundation, the rotational speed of the working specific permeance can be calculated as:

$$\Omega_\lambda = \frac{Z_r}{|Z_s - Z_r|}\Omega_r \tag{3.45}$$

When the number of stator and rotor teeth is closely matched, the rotational speed of the working specific permeance is significantly higher than that of the rotor. In essence, a minor displacement in the rotor's position results in a substantial movement of the working specific permeance, similar to the functionality of a vernier caliper, thus justifying the nomenclature. Furthermore, according to Eqs. (3.44) and (3.45), the vernier reluctance machine can be considered as an integrated system comprising a synchronous reluctance machine with the armature winding's pole pair number equal to $|Z_s - Z_r|/2$ and a gearbox with a reduction ratio of $Z_r/|Z_s - Z_r|$. This configuration highlights that the torque density of the vernier reluctance machine surpasses that of the synchronous reluctance machine, emphasizing its enhanced mechanical efficiency.

3.6.4 Asynchronous Induction Machine

By replacing the AC excitation winding of the brushless doubly-fed reluctance machine with the self-short-circuit winding, the induction FMM is transformed into what is known as the asynchronous induction machine [27], as illustrated in Fig. 3.23. The slot-pole combination of the asynchronous induction machine adheres to the requirements of either Eq. (3.33) or Eq. (3.34), aligning with that of the brushless doubly-fed reluctance machine. It is important to note that the asynchronous induction machine is distinct from the flux modulation induction machine discussed in Sect. 3.3.4. In the flux modulation induction machine, the modulation unit is stationary, and the self-short-circuit winding is located in the rotor, generating a rotary excitation MMF. Conversely, in the asynchronous induction machine, the modulation unit is part of the rotor, and the self-short-circuit winding is housed in the stator, also generating a rotary excitation MMF. This configuration makes the asynchronous induction machine more robust and reliable. Additionally, the starting torque of the asynchronous induction machine can be increased by introducing series resistance into the self-short-circuit winding.

The operating principle of the asynchronous induction machine is as follows: Initially, when the armature winding is energized, it produces a rotating armature MMF with a pole pair number P_a. This MMF is then modulated by the rotating air gap specific permeance with a pole pair number $|Z_s - Z_r|$, resulting in a no-load air gap flux density with a pole pair number $||Z_s - Z_r| \pm P_a|$. According to the pole pair number relationship of the asynchronous induction machine, one of these flux density harmonics corresponds to the pole pair number of the self-short-circuit winding (P_e). This correspondence induces back-EMF and current in the self-short-circuit winding. The current in this winding subsequently produces a rotating MMF, which interacts with the air gap specific permeance to generate a rotating no-load air gap flux density.

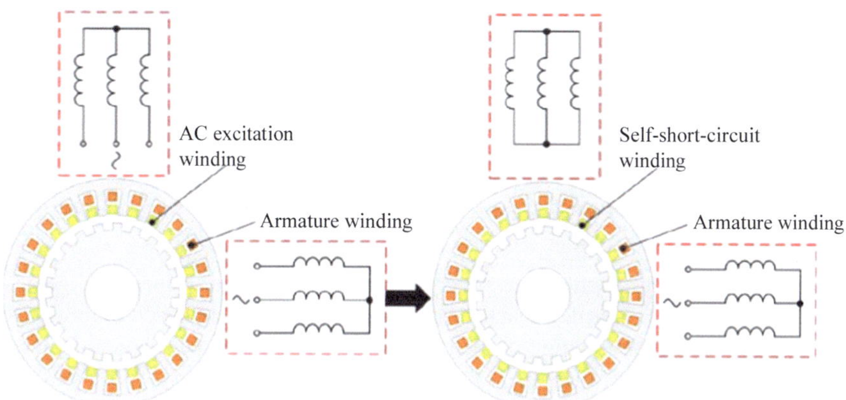

Fig. 3.23 Evolution from brushless doubly-fed reluctance machine to asynchronous induction machine

The interaction of the two air gap flux densities, from the self-short-circuit winding and the armature winding, ultimately generates electromagnetic torque on the rotor. In summary, the output characteristics of the asynchronous induction machine are similar to those of both the brushless doubly-fed reluctance machine and the conventional asynchronous machine, with the synchronous speed and slip ratio defined according to Eqs. (3.39) and (3.40).

3.7 Flux Modulation Transmission Device

In addition to the previously mentioned FMMs, two additional flux modulation transmission devices can be derived from the basic FMM topology. Although these devices serve different functions from FMMs, their operating principles are consistent with those of FMMs. Here, we provide a brief introduction to flux modulation transmission devices.

A critical distinction between a transmission device and an electrical machine is the absence of an armature unit or electrical port in the former. Instead of the armature unit found in electrical machines, transmission devices incorporate an additional excitation unit. As a result, these devices feature two excitation units coupled through a single modulation unit. Considering the four types of excitation units discussed in Sect. 3.1, both DC and AC excitation windings require electrical ports. Therefore, only two types of excitation units remain viable for transmission devices—permanent magnets and self-short-circuit windings. However, the self-short-circuit winding can only passively receive an external rotating magnetic field and generate an excitation MMF through electromagnetic induction. Consequently, it is not feasible for both excitation units to be self-short-circuit windings simultaneously, resulting in only two types of flux modulation transmission devices. The first type incorporates both excitation units composed of permanent magnets, termed the flux modulation magnetic gear [28]. The second type features one excitation unit composed of permanent magnets, while the other utilizes a self-short-circuit winding, referred to as the flux modulation electromagnetic coupler.

3.7.1 Magnetic Gear

The flux modulation magnetic gear, commonly abbreviated as a magnetic gear, is depicted in Fig. 3.24. This device can be conceptualized as a new structure formed by replacing the armature winding of the basic FMM model with permanent magnets. The operating principle functions as follows: initially, the MMF generated by excitation unit 1 is modulated by the ferromagnetic blocks of the modulation unit, producing a magnetic field in the inner air gap. Subsequently, this air gap magnetic field interacts with the magnetic field generated by excitation unit 2, resulting in torque production. The operation mirrors that of the FMM, requiring the pole pair numbers of the magnetic gear to satisfy the following relationship:

$$P_f = P_e \pm P_a \tag{3.46}$$

where P_a represents the pole pair number of excitation unit 1, and P_e denotes the pole pair number of excitation unit 2. Since excitation unit 1 evolves from the armature winding, the corresponding subscript "a" is retained for consistency with previous FMM designs. Excitation unit 2 remains unchanged in the topology evolution, maintaining the subscript "e" as before.

Considering the pole pair combination with $P_f = P_e + P_a$, the torque and power relationships of the magnetic gear are analyzed below. First, the excitation MMF of excitation unit 2 (considering only the fundamental wave) is expressed as:

$$F_e = F_{e1} \cos P_e(\theta - \Omega_e t) \tag{3.47}$$

The specific permeance of the modulation unit can be represented as:

$$\lambda_f(\theta, t) = \lambda_{f0} + \lambda_{f1} \cos P_f(\theta - \Omega_f t) \tag{3.48}$$

Combining Eqs. (3.47) and (3.48), the modulated air gap flux density originating from excitation unit 2 is calculated as:

$$\begin{aligned} B_e(\theta, t) &= \frac{\mu_0}{g} F_e \lambda_f \\ &= \frac{\mu_0}{g} F_{e1} \left\{ \begin{array}{l} \lambda_{f0} \cos P_e(\theta - \Omega_e t) + \frac{1}{2}\lambda_{f1} \cos[P_a\theta - (P_f\Omega_f - P_e\Omega_e)t] \\ +\frac{1}{2}\lambda_{f1} \cos[(P_e + P_f)\theta - (P_f\Omega_f - P_e\Omega_e)t] \end{array} \right\} \end{aligned} \tag{3.49}$$

where M_e denotes the magnetomotive force, and I_e indicates the current in excitation unit 2. It is observed that the pole pair number of the second term in Eq. (3.49) is

Fig. 3.24 Magnetic gear

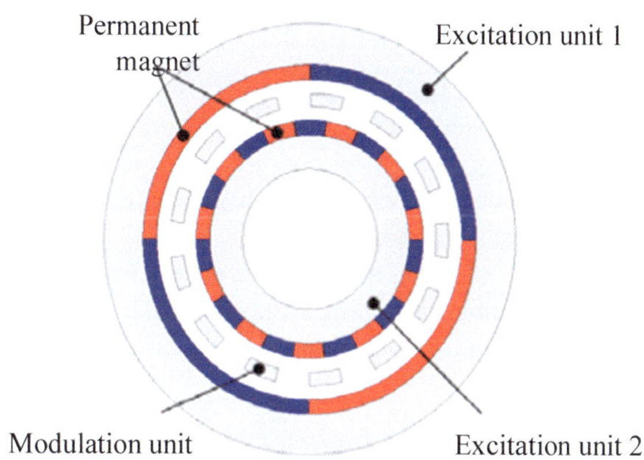

equal to that of excitation unit 1. However, to achieve constant torque transmission, both the pole pair number and the rotation speed must be the same. Consequently, the rotation speed of excitation unit 1 should be determined by:

$$\Omega_{\mathrm{a}} = \frac{P_{\mathrm{f}}\Omega_{\mathrm{f}} - P_{\mathrm{e}}\Omega_{\mathrm{e}}}{P_{\mathrm{a}}} \tag{3.50}$$

According to the Newton's third law:

$$T_{\mathrm{e}} + T_{\mathrm{a}} + T_{\mathrm{f}} = 0 \tag{3.51}$$

The law of conservation of energy also implies:

$$\Omega_{\mathrm{e}}T_{\mathrm{e}} + \Omega_{\mathrm{a}}T_{\mathrm{a}} + \Omega_{\mathrm{f}}T_{\mathrm{f}} = 0 \tag{3.52}$$

By combining Eqs. (3.50)–(3.52):

$$\frac{T_{\mathrm{f}}}{T_{\mathrm{a}}} = -\frac{P_{\mathrm{f}}}{P_{\mathrm{a}}},\ \frac{T_{\mathrm{e}}}{T_{\mathrm{a}}} = \frac{P_{\mathrm{e}}}{P_{\mathrm{a}}} \tag{3.53}$$

Given the three-unit structure of the magnetic gear, selecting two units as input and output ports is necessary, while the remaining unit remains stationary. According to the derived Eq. (3.53), the torque across the three units is proportionate to their respective pole pair numbers. Therefore, excitation unit 1, having fewer pole pairs, must be chosen as one of the ports. Failure to select excitation unit 1 would hinder torque transformation since excitation unit 2 and the modulation unit closely match in pole pair number and torque.

Consequently, the magnetic gear operates in two modes. First, excitation units 1 and 2 function as mechanical ports with the modulation unit stationary. The torque and speed relationships of excitation units 1 and 2 are represented as:

$$\frac{\Omega_{\mathrm{e}}}{\Omega_{\mathrm{a}}} = -\frac{P_{\mathrm{a}}}{P_{\mathrm{e}}},\ \frac{T_{\mathrm{e}}}{T_{\mathrm{a}}} = \frac{P_{\mathrm{e}}}{P_{\mathrm{a}}} \tag{3.54}$$

As observed, the magnetic gear serves a similar function to a mechanical gear, with both devices acting to vary speed and torque. According to the law of conservation of energy, when the output torque increases, the output speed must decrease proportionately, and vice versa. In this operational mode, the rotation directions of the two ports are opposite.

Second, excitation unit 2 can remain stationary, with excitation units 1 and the modulation unit functioning as mechanical ports. The torque and speed relationships are specified in the diagrams and descriptions provided.

$$\frac{\Omega_{\mathrm{f}}}{\Omega_{\mathrm{a}}} = \frac{P_{\mathrm{a}}}{P_{\mathrm{f}}},\ \frac{T_{\mathrm{f}}}{T_{\mathrm{a}}} = -\frac{P_{\mathrm{f}}}{P_{\mathrm{a}}} \tag{3.55}$$

In this operational mode, the rotation directions of the two ports are the same.

Unlike the FMM, where rotor torque can be controlled by adjusting the armature current, the magnetic gear lacks an armature winding, and the excitation MMF from permanent magnets is non-adjustable. Thus, torque control in magnetic gears is achieved solely through the manipulation of the torque angle. As illustrated in Fig. 3.25, the torque angle (γ) represents the angle between the modulated magnetic field generated by excitation unit 2 and the centerline of the permanent magnet in excitation unit 1. With excitation units 1 and 2 serving as input and output ports, when $\gamma = 0$, the electromagnetic torque across the ports is zero, reaching its maximum at $\gamma = 90°$. Combining this with Eq. (3.53), the relationship is derived as:

$$\begin{aligned} T_{\mathrm{e}} &= T_{\max} \sin \gamma \\ T_{\mathrm{a}} &= \frac{P_{\mathrm{a}}}{P_{\mathrm{e}}} T_{\max} \sin \gamma \end{aligned} \tag{3.56}$$

where $T_{\max}$ represents the maximum output torque of the magnetic gear.

The variations of T_{e} and T_{a} with γ are presented in Fig. 3.26. It is evident that when $\gamma < 90°$, an increase in input mechanical torque prompts an automatic adjustment of γ, causing the output torque to rise proportionally. However, when γ exceeds 90°, an increase in input mechanical torque does not proportionally raise the output torque, leading to an imbalance between input mechanical torque and output electromagnetic torque, and potentially resulting in a step-out fault in the magnetic gear. Generally, to ensure safe operation, the operating torque of the magnetic gear should not exceed 70% of the maximum torque to avoid operational failures and ensure reliability.

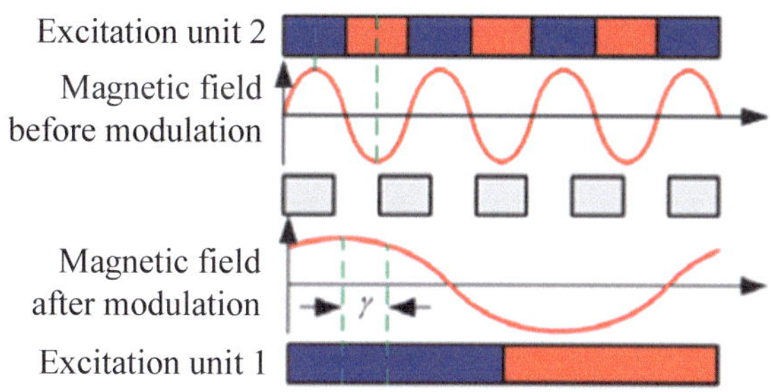

Fig. 3.25 Definition of the torque angle in magnetic gear

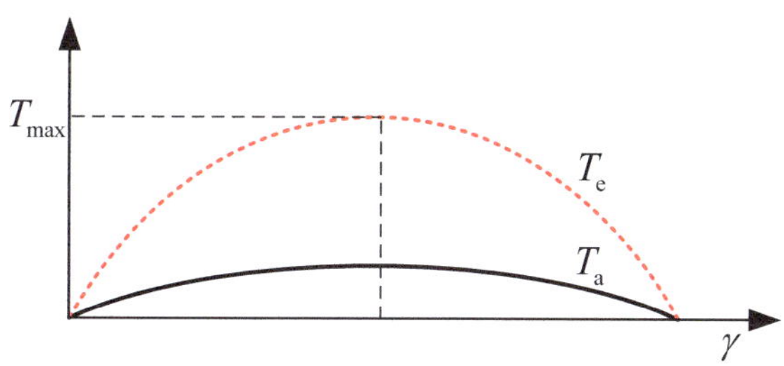

Fig. 3.26 Torque angle characteristics

Unlike mechanical gears, magnetic gears transmit torque through magnetic fields, ensuring non-contact transmission. This characteristic significantly reduces mechanical wear and vibration, lowers noise levels, and diminishes maintenance requirements. Additionally, magnetic gears offer overload protection capabilities. However, their torque density is typically lower than that of mechanical gears, and the extensive use of permanent magnets contributes to higher costs. Consequently, magnetic gears are particularly well-suited for applications with stringent sealing requirements or elevated maintenance costs, such as underwater propulsion, offshore wind power generation, and tidal power generation.

The pole ratio of the FMM can be elucidated by examining the port characteristics of the magnetic gear. A comparison between Figs. 1.5b and 3.24 reveals that the primary distinction lies in the replacement of the armature unit in the FMM with excitation unit 1. Despite this change, both configurations serve the same fundamental purpose—generating a rotating MMF. Consequently, from the perspective of the modulation unit and excitation unit 2, this substitution does not affect their physical properties. Thus, the FMM and the magnetic gear share identical torque relationships, represented by the following equation:

$$\frac{T_{\mathrm{e}}}{T_{\mathrm{a}}} = \frac{P_{\mathrm{e}}}{P_{\mathrm{a}}} = \mathrm{PR} \tag{3.57}$$

where T_{e} indicates the torque of the excitation unit, and T_{a} denotes the torque subjected to the armature unit.

For the FMM, examining the system from the armature unit's perspective reveals a rotating magnetic field with fewer poles and higher speed, similar to what the armature unit of a conventional permanent magnet machine experiences. Therefore, both armature units generate comparable electromagnetic torque. In summary, the FMM amplifies the torque of a conventional permanent magnet machine by a factor of PR, a concept that will be explored further in Chap. 4.

3.7.2 Flux Modulation Electromagnetic Coupler

Turning to the magnetic gear illustrated in Fig. 3.24, replacing the permanent magnets of excitation unit 1 with a self-short-circuit winding results in the topology of the flux modulation electromagnetic coupler, as depicted in Fig. 3.27a. To simplify the structure further, the self-short-circuit winding can be substituted with a copper ring, shown in Fig. 3.27b. The operational principle involves the MMF generated by the permanent magnets of excitation unit 2 being modulated by the ferromagnetic blocks of the modulation unit. This modulation creates a rotary magnetic field in the outer air gap, inducing back-EMF and current in the self-short-circuit winding. The current in this winding then produces an MMF, which interacts with the ferromagnetic blocks to generate a rotary air gap magnetic field. The two air gap magnetic fields from the self-short-circuit winding and the permanent magnets interact to facilitate

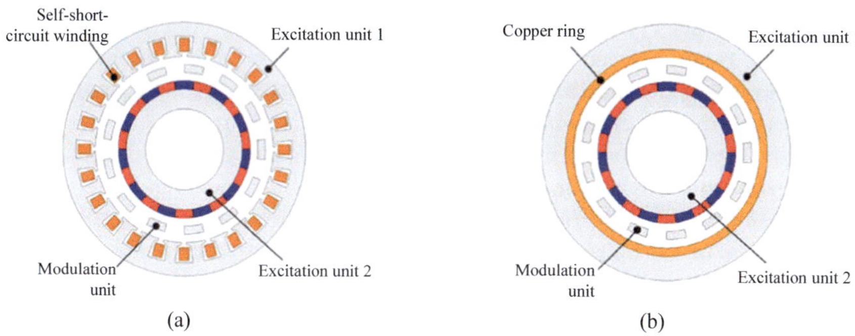

Fig. 3.27 Flux modulation magnetic coupler with different self-short-circuit structures. **a** Winding. **b** Copper ring

torque transmission. As the flux modulation electromagnetic coupler evolves from the magnetic gear, it must adhere to the pole pair numbers specified in Eq. (3.46).

By introducing the torque and speed characteristics of the flux modulation electromagnetic couplers, we establish a definition for the synchronous speed of excitation unit 1:

$$\Omega_{a0} = \frac{P_e \Omega_e}{P_a} \tag{3.58}$$

The slip ratio is defined as:

$$s = \frac{\Omega_{a0} - \Omega_a}{\Omega_{a0}} \tag{3.59}$$

Similar to the induction machine, the electromagnetic torque of excitation unit 1 can be derived as:

$$T_a = \frac{3I_a^2 R_a}{s} = \frac{3\Psi_{a1}^2 P_a^2 \Omega_a R_a}{\frac{R_a^2}{s} + s(\Omega_a P_a L_a)^2} \tag{3.60}$$

where ψ_{a1} denotes the fundamental flux linkage induced in the self-short-circuit winding. Notably, the inductance L_a in this equation refers to winding inductance, which is distinct from the leakage inductance in the torque equation of induction machines. Equation (3.60) demonstrates that when excitation unit 1 operates near the synchronous speed Ω_{a0}, the electromagnetic coupler can still transmit torque even if the rotation speed Ω_a changes. Additionally, similar to the induction machine, the ratio of the output power of excitation unit 1 p_{a_out} to the copper loss p_{a_Cu} is:

$$p_a : p_{a\_out} : p_{a\_Cu} = 1 : (1 - s) : s \tag{3.61}$$

Therefore, enhancing the efficiency of the electromagnetic coupler necessitates aligning the rotation speed of excitation unit 1 (Ω_a) with the synchronous speed (Ω_{a0}), minimizing slip ratio and copper loss. Similar to the magnetic gear, the electromagnetic torque of each unit in the flux modulation electromagnetic coupler is proportional to the corresponding pole pair number. However, due to copper losses in the self-short-circuit winding and copper ring, the input and output powers of the two mechanical ports differ, affecting overall system efficiency.

The results of the finite element simulation for an electromagnetic coupler are shown in Fig. 3.28. Figure 3.28a demonstrates the production of a 2-pole-pair magnetic field under the flux modulation effect of the modulation unit, generated by a 22-pole-pair permanent magnet MMF. Figure 3.28b depicts the torque waveforms of the two rotors when the permanent magnet rotor operates at 300 r/min and the copper rotor at 3080 r/min. While Eq. (3.50) predicts a synchronous speed ratio of 11, the observed speed ratio in Fig. 3.28b is 10.2. This discrepancy is due to the asynchronous operation of the copper rotor, resulting in a slip ratio of 0.07—an important distinction from the magnetic gear.

Finally, the role of the flux modulation electromagnetic gear is clarified in the context of wind power generation systems. Blade speeds in these systems typically remain low, around 10 r/min. Furthermore, there is a need to adjust the blade speed in response to variations in wind speed to capture maximum wind energy. Consequently, a mechanical transmission device that acts as a gear to increase shaft speed is crucial, which in turn reduces the weight and volume of the generator. Simultaneously, the device's output speed should remain stable to ensure the generator operates at a synchronous speed. Traditionally, mechanical gears are used for this purpose, but they introduce challenges such as mechanical wear, significant vibration, noise, and frequent maintenance. As a result, researchers have explored non-contact electromagnetic solutions. Figure 3.29 illustrates one such solution that combines a magnetic gear and an electromagnetic coupler. As shown in Fig. 3.29a, the magnetic gear enhances the rotation speed, while the electromagnetic coupler stabilizes the

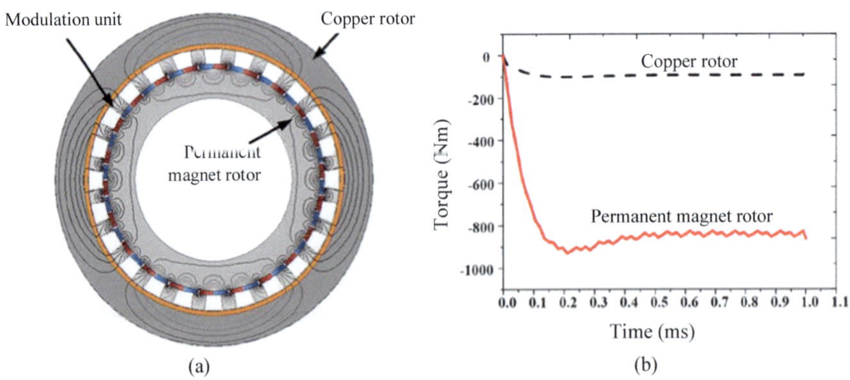

Fig. 3.28 Finite element simulation results of flux modulation electromagnetic coupler: **a** flux lines. **b** Torque

speed. To simplify the system, the permanent magnet rotor of the electromagnetic coupler and the high-speed permanent magnet rotor of the magnetic gear can be integrated, as depicted in Fig. 3.29b. It is noteworthy that the function of the integrated structure in Fig. 3.29b can be achieved by the more compact and practical flux modulation electromagnetic coupler, as illustrated in Fig. 3.30.

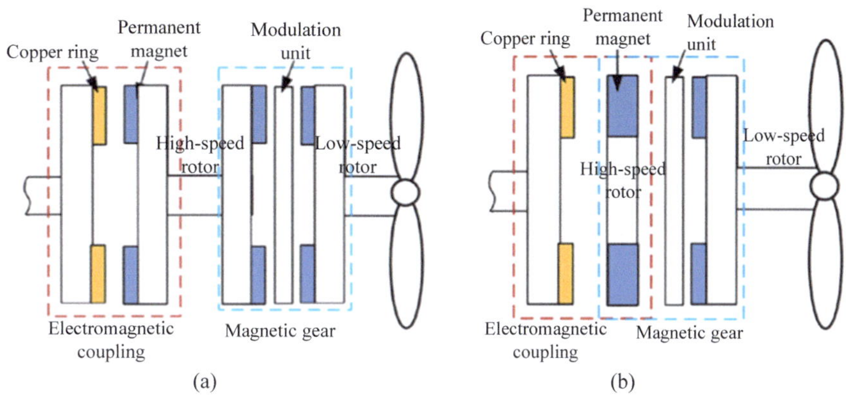

Fig. 3.29 Magnetic gear and electromagnetic coupler system for wind power generation. **a** Cascade structure. **b** Integrated structure

Fig. 3.30 Flux modulation electromagnetic coupler for wind power generation

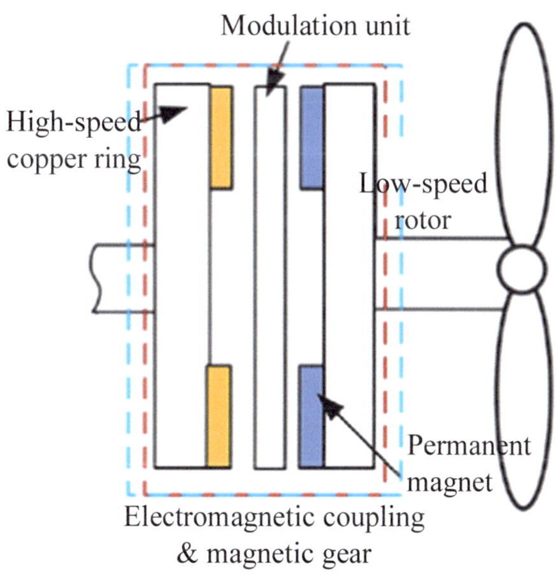

3.8 Summary of This Chapter

This chapter introduces the fundamental FMM model and explores various FMM topologies that have evolved based on the motion of each functional unit and the structure of the excitation unit. This exploration establishes the broad and diverse family of FMMs.

(1) *Motion of Functional Units*: FMMs are categorized into four groups based on the motion of each functional unit: stationary three units:

- Stationary three units: Traditional DC machines and synchronous machines are exemplified by the stationary three units FMM and the stationary modulation unit FMM, respectively.
- Stationary modulation unit: This category typically involves simpler operational dynamics.
- Stationary excitation unit: Features a basic reluctance rotor, focusing on simplified design and operation.
- Non-stationary units: This versatile group may include configurations with either a reluctance rotor or multiple mechanical and electrical ports, adapting to more complex applications.

(2) *Excitation Unit Structures*: The classification extends further into four types based on the structure of the excitation unit:

- Permanent magnet type: This includes rotor permanent magnet, stator permanent magnet, and dual-mechanical-port dual-electrical-port types, leveraging the robust and efficient properties of permanent magnets.
- Electrically excited type: Comprises configurations like rotor electrically excited FMMs with brushes and stator electrically excited FMMs without brushes, offering a balance between traditional designs and modern flux modulation techniques.
- AC excitation type: Characterized by devices such as the brushless doubly-fed reluctance machine and vernier reluctance machine, these FMMs utilize flux modulation to couple excitation and armature MMFs effectively.
- Induction type: Distinguished by the placement of the self-short-circuit winding, either on the stator or rotor side, adapting the induction principle to FMM applications.

(3) *Flux Modulation Electromagnetic Devices*: This category introduces two types of flux modulation electromagnetic devices:

- Magnetic gears: Analogous to mechanical gears, these devices provide constant speed and torque transmission, mimicking the reliability and predictability of mechanical systems but with enhanced durability and reduced maintenance.

- Flux modulation electromagnetic couplers: These offer a fixed torque ratio between input and output ports with an adjustable speed ratio, making them ideal for variable-speed, constant-frequency power generation applications.

References

1. Li DW (2015) Research on flux modulation permanent magnet machine (in Chinese). Huazhong University of Science and Technology, Wuhan
2. Li DW, Zou TJ, Qu RH et al (2018) IEEE Trans Ind Appl 54(2):1320
3. Li DW, Qu RH, Li J et al (2016) IEEE Trans Ind Appl 52(1):125
4. Zou TJ (2018) Research on theory and topologies of high-performance flux modulation permanent magnet machines (in Chinese). Huazhong University of Science and Technology, Wuhan
5. Li J, Chau K, Jiang JZ et al (2010) IEEE Trans Magn 46(6):1475
6. Weh H, May H (1986) Achievable force densities for permanent magnet excited machines in new configurations. In: International conference on electrical machines, Munchen, pp 1107–1111
7. Gao YT, Qu RH, Li DW et al (2016) IEEE Trans Appl Supercond 26(4):1
8. Li DW, Gao YT, Qu RH et al (2018) IEEE Trans Ind Appl 54(1):172
9. Gao YT (2017) Theoretical analysis and topology investigation of flux reversal permanent magnet machines (in Chinese). Huazhong University of Science and Technology, Wuhan
10. Gao YT, Qu RH, Li DW et al (2017) IEEE Trans Ind Appl 53(3):2110
11. Gao YT, Qu RH, Li DW et al (2017) IEEE Trans Ind Appl 53(5):4232
12. Gao YT, Qu RH, Li DW et al (2016) IEEE Trans Ind Appl 52(4):2856
13. Li DW, Qu RH, Li J et al (2016) IEEE Trans Energy Convers 31(1):106
14. Liao YF, Liang F, Lipo T (1992) A novel permanent magnet machine with doubly salient structure. In: IEEE Industry applications society annual meeting, Houston, pp 308–314
15. Jia SF (2017) Research on stator DC-current excited vernier reluctance machines (in Chinese). Huazhong University of Science and Technology, Wuhan
16. Jia SF, Qu RH, Li J et al (2016) IEEE Trans Energy Convers 31(3):935
17. Li ZM, Yu ZX, Kong WB et al (2022) IEEE Trans Ind Electron 69(5):4555
18. Chen JT, Zhu ZQ, Iwasaki S et al (2010) Low-cost flux-switching brushless AC machines. In: IEEE Vehicle power and propulsion conference, Lille, 2010. IEEE, pp 1–6
19. Li HS, Liang F, Zhao Y et al (1995) IEEE Trans Ind Appl 31(1):99
20. Cheng Y, Zhang YZ, Qu RH et al (2021) IEEE Trans Appl Supercond 31(5):1
21. Ren X (2019) Research on multi-mechanical-port multi-electrical-port machines (in Chinese). Huazhong University of Science and Technology, Wuhan
22. Li DW, Qu RH, Ren X et al (2016) Brushless dual-electrical-port, dual mechanical port machines based on the flux modulation principle. In: IEEE energy conversion congress and exposition, Milwaukee. IEEE, pp 1–8
23. Ren X, Li DW, Qu RH et al (2017) IEEE Trans Magn 53(11):1
24. Ren X, Li DW, Qu RH et al (2019) IEEE Trans Ind Electron 66(8):6128
25. Liao YF, Xu LY, Zhen L (1996) IEEE Trans Ind Appl 32(5):1195
26. Lee CH (1963) IEEE Trans Power Appar Syst 82(66):343
27. Li HM, Li QF (1982) Electromagnetic reduction machine (in Chinese). Machinery Industry Press, Beijing
28. Atallah K, Howe D (2001) IEEE Trans Magn 37(4):2844

Chapter 4
Characterization of Flux Modulation Machines

The flux modulation theory and the various types of FMMs have been extensively discussed in Chaps. 2 and 3, respectively. Although FMMs operate on similar principles, their topologies differ due to the unique arrangements, types, and structures of each unit. Consequently, FMMs exhibit both common characteristics and individualized traits due to these topological variations. This chapter begins by deriving the electromagnetic torque and power factor of the basic model, highlighting the shared features of FMMs. Following this, the chapter will analyze the average torque, torque ripple, and power factor of different FMM types, elucidating the reasons behind their unique characteristics.

4.1 Basic Idea

As electromechanical energy conversion devices, the performance of electrical machines is evaluated through both "electrical" and "mechanical" aspects. The primary mechanical performance indicators include average torque and pulsating torque, while the electrical indicators encompass voltage, current, power factor, and voltage distortion. According to electrical machine theory, voltage and current can be regulated by adjusting the number of coil turns. Therefore, the fundamental electrical indicators are the power factor and voltage distortion. Additionally, pulsating torque is significantly influenced by the harmonics of the no-load back-EMF, making voltage distortion a crucial factor in pulsating torque. In summary, this chapter focuses on three key indicators of FMM performance: average torque, pulsating torque, and power factor.

R. Qu et al., *Flux Modulation Machine*, https://doi.org/10.1007/978-981-95-4746-3_4

4.1.1 Average Torque and Pulsating Torque

According to the Maxwell stress tensor method, the instantaneous torque of a machine with a single air gap can be calculated as follows:

$$T = r_g l_{\text{stk}} \int_0^{2\pi} \frac{B_{\text{rad}} B_{\text{tal}}}{2\mu_0} \mathrm{d}\theta \tag{4.1}$$

where B_{rad} and B_{tal} represent the radial and tangential components of the air gap flux density, respectively. Without considering the saturation effect, the magnetic field is the sum of the excitation field and the armature field. Consequently, Eq. (4.1) can be further expanded as:

$$\begin{aligned} T &= r_g l_{\text{stk}} \int_0^{2\pi} \frac{(B_{\text{rad\_e}} + B_{\text{rad\_a}})(B_{\text{tal\_e}} + B_{\text{tal\_a}})}{2\mu_0} \mathrm{d}\theta \\ &= r_g l_{\text{stk}} \int_0^{2\pi} \frac{B_{\text{rad\_e}} B_{\text{tal\_e}}}{2\mu_0} \mathrm{d}\theta + r_g l_{\text{stk}} \int_0^{2\pi} \frac{B_{\text{rad\_e}} B_{\text{tal\_a}} + B_{\text{rad\_a}} B_{\text{tal\_e}}}{2\mu_0} \mathrm{d}\theta \\ &\quad + r_g l_{\text{stk}} \int_0^{2\pi} \frac{B_{\text{rad\_a}} B_{\text{tal\_a}}}{2\mu_0} \mathrm{d}\theta \\ &= T_{\text{cog}} + T_{\text{exc}} + T_{\text{rel}} \end{aligned} \tag{4.2}$$

Equation (4.2) breaks down into three torque components. The first term, containing only the excitation field, represents the cogging torque (T_{cog}), which produces pulsating torque when the armature unit is not energized. The second term, expressed as the product of the excitation and armature fields, constitutes the excitation torque (T_{exc}), reflecting the interaction between the two fields. The third term, exclusively involving the armature field, results in the reluctance torque (T_{rel}).

The nature of these three torque components is then investigated. The cogging torque (T_{cog}) always has an average value of zero, generating only pulsating torque. The excitation torque T_{exc} serves as the primary source of output torque in permanent magnet and electrically excited machines, excluding synchronous reluctance machines and vernier reluctance machines, which lack this torque component. The reluctance torque (T_{rel}) is produced solely by the armature field, with the average value potentially being zero, depending on the machine type. However, as long as reluctance torque exists, it introduces pulsating torque. The reluctance torque is present only when the rotor exhibits a reluctance effect.

In summary, to analyze the pulsating torque of FMMs, it is imperative to study each of the three torque components mentioned above. For permanent magnet and electrically excited FMMs, the average torque is entirely contributed by the excitation torque, while the reluctance torque minimally contributes to the average torque. For reluctance FMMs, although the excitation torque is absent, the reluctance torque can be equated to the excitation torque. Therefore, when analyzing the average torque, only the excitation torque needs consideration.

For permanent magnet or electrically excited FMMs with a single mechanical port, the excitation torque can be calculated as follows [1]:

$$T_{\mathrm{exc}} = \frac{e_{\mathrm{A}} i_{\mathrm{A}} + e_{\mathrm{B}} i_{\mathrm{B}} + e_{\mathrm{C}} i_{\mathrm{C}}}{\Omega_{\mathrm{r}}} \tag{4.3}$$

where e_{A}, e_{B}, e_{C} denote the instantaneous values of no-load back-EMF of phase A, B, and C respectively, and i_{A}, i_{B}, and i_{C} indicate the instantaneous values of armature currents of phase A, B, and C, respectively.

Most permanent magnet machines operate with inverters generating sinusoidal armature currents, and as a result, pulsating torque primarily arises from no-load back-EMF harmonics. Generally, the three-phase back-EMF and current under $I_{\mathrm{d}} = 0$ control can be formulated as:

$$\begin{aligned} e_{\mathrm{A}}(t) &= \sum_{n=1}^{\infty} E_{an} \cos(n\omega_{\mathrm{a}} t) \\ e_{\mathrm{B}}(t) &= \sum_{n=1}^{\infty} E_{an} \cos\left(n\omega_{\mathrm{a}} t - \frac{2\pi n}{3}\right) \\ e_{\mathrm{C}}(t) &= \sum_{n=1}^{\infty} E_{an} \cos\left(n\omega_{\mathrm{a}} t + \frac{2\pi n}{3}\right) \end{aligned} \tag{4.4}$$

$$\begin{aligned} i_{\mathrm{A}}(t) &= I_{\mathrm{a1}} \cos(\omega_{\mathrm{a}} t) \\ i_{\mathrm{B}}(t) &= I_{\mathrm{a1}} \cos\left(\omega_{\mathrm{a}} t - \frac{2\pi}{3}\right) \\ i_{\mathrm{C}}(t) &= I_{\mathrm{a1}} \cos\left(\omega_{\mathrm{a}} t + \frac{2\pi}{3}\right) \end{aligned} \tag{4.5}$$

According to Eq. (4.3), the excitation torque can be determined as:

$$T_{\mathrm{exc}} = \frac{3}{2}\left[E_{\mathrm{a1}} I_{\mathrm{a1}} + \sum_{n=3m+1} E_{an} I_{\mathrm{a1}} \cos(n-1)\omega_{\mathrm{a}} t + \sum_{n=3m-1} E_{an} I_{\mathrm{a1}} \cos(n+1)\omega_{\mathrm{a}} t\right] \tag{4.6}$$

It shows that the average torque is generated by the fundamental back-EMF. The value of the average torque is:

$$T_{\mathrm{avg}} = \frac{3}{2} E_{\mathrm{a1}} I_{\mathrm{a1}} \tag{4.7}$$

Furthermore, the back-EMF harmonics lead to the pulsating excitation torque, i.e.,

$$T_{\text{exc\_ rip}} = \frac{3}{2}\sum_{n=3m\pm1} E_{\text{a}n} I_{\text{a}1} \cos 3m\omega_{\text{a}} t \tag{4.8}$$

Using Eq. (4.8), the frequency of excitation torque pulsation is $3m$ times the fundamental wave. Assuming that the back-EMF does not contain even harmonics, the frequency of excitation torque pulsation is a multiple of 6, and the magnitude of each torque pulsation is proportional to the corresponding back-EMF harmonics. Therefore, the magnitude of the excitation torque pulsation is determined by the ratio of the harmonics to the fundamental waveform of the no-load back-EMF. Since no-load back-EMF harmonics depend on the no-load air gap magnetic field, the core of the calculation lies in the quantitative determination of the no-load air gap magnetic field.

In addition to the excitation torque, the cogging torque also depends on the no-load air gap magnetic field, while the reluctance torque is determined by the armature field. The pulsation of the cogging torque and reluctance torque will be explored in later sections.

4.1.2 Power Factor

The power factor (PF) of a machine is formally defined as the ratio of real power to apparent power. However, due to the complexity of determining voltage harmonics, the power factor is often approximated as the cosine of the phase difference between the fundamental waves of the phase voltage and the phase current. The phase diagram, as illustrated in Fig. 4.1, can also be employed to calculate the power factor of FMMs. This phase diagram is plotted under Id = 0 control, neglecting the winding resistance, with the dots on the voltage and current symbols representing the time vectors.

Then, the power factor can be stated as:

$$\text{PF} = \cos\gamma = \frac{1}{\sqrt{1+\left(\frac{\omega_{\text{a}} L_a I_{\text{a}1}}{E_{\text{a}1}}\right)^2}} = \frac{1}{\sqrt{1+\left(\frac{\psi_{\text{a}1}}{\psi_{\text{e}1}}\right)^2}} \tag{4.9}$$

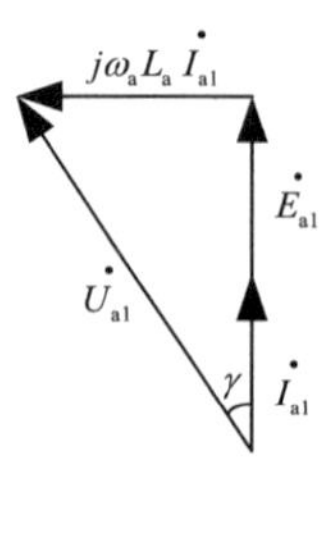

Fig. 4.1 Phase diagram of the FMM ($I_{\text{d}} = 0$ control)

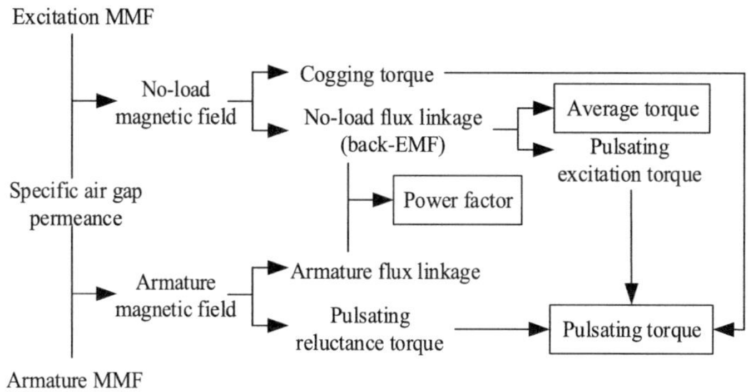

Fig. 4.2 Basic idea of FMM performance analysis

where L_a represents the armature winding inductance, E_{a1} denotes the magnitude of the fundamental phase back-EMF, ψ_{a1} indicates the magnitude of the fundamental flux linkage generated by the armature current, and ψ_{e1} is the magnitude of the fundamental flux linkage generated by the excitation field. Using Eq. (4.9), the power factor of the FMM can be determined by the ratio of the two flux linkages i.e. ψ_{a1} and ψ_{e1}.

The fundamental approach to FMM performance analysis is depicted in Fig. 4.2. The initial step involves calculating the no-load and armature magnetic fields based on the MMF/specific permeance model. Subsequent derivations of torque and power factor performances are then provided.

4.2 Performance Analysis of the Basic Model of Flux Modulation Machine

Figure 1.5 presents the basic model of the FMM, which only includes a single-layer modulation unit. However, as introduced in Sect. 2.2.4, there are also configurations with bilateral slotting, i.e., FMMs with a double-layer modulation unit, such as the flux-switching permanent magnet machine and the electrically excited doubly-salient machine. Therefore, it is necessary to extend the basic model. The extended basic models are illustrated in Fig. 4.3. Figure 4.3a represents a conventional permanent magnet machine, which can be considered a special case of an FMM without the modulation unit. Figure 4.3b, c depict the basic models of FMMs with single-layer and double-layer modulation units, respectively.

In the following analysis of the basic model's performances, we adhere to the flow in Fig. 4.2. First, the following assumptions are made:

(1) Infinite permeability of the stator and rotor cores, i.e., disregarding saturation effects and reluctance of stator and rotor cores.
(2) Neglecting end effects and edge effects.

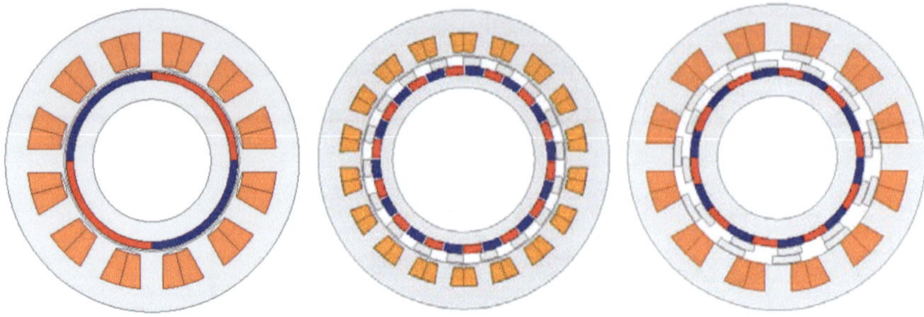

(a) (b) (c)

Fig. 4.3 Basic models of FMMs: **a** conventional permanent magnet machine, **b** single-layer modulation unit, **c** double-layer modulation unit

(3) Only fundamental components of the excitation and armature winding MMFs exist. The specific permeances of both single-layer and double-layer modulation units only have DC and fundamental components.
(4) Disregarding the leakage inductances of the armature winding.
(5) Initially, the peak positions of both the MMF and the permeance are at the origin of the stator coordinate system.

Notably, the third assumption suggests that the air gap magnetic field in the basic model is ideally sinusoidal and will not produce any pulsating torque. Therefore, only fundamental back-EMF, average electromagnetic torque, and power factor characteristics can be analyzed with this ideal assumption. The study of pulsating torque largely varies among various types of FMMs; hence, the corresponding analysis of pulsating torque will be conducted in subsequent sections based on specific FMM topologies.

4.2.1 Air Gap Magnetic Field

The basic model has n layers of modulation units, where $n = 0, 1, 2$ corresponds to the topologies (a), (b), and (c) of Fig. 4.3, respectively. According to the flux modulation theory, the slot-pole combination of the basic model needs to satisfy:

$$P_{\mathrm{a}} = \sum_{m=1}^{n} \mathrm{sgn}_m P_{\mathrm{f}m} \pm P_{\mathrm{e}} \tag{4.10}$$

where P_{e} is the pole pair number of excitation unit, P_{a} is the pole pair number of armature unit, $P_{\mathrm{f}m}$ is the pole pair number of mth layer modulation unit, and sgn_m can be taken as ± 1.

Subsequently, the air gap flux density generated by the excitation unit is:

$$B_{\mathrm{e}}(\theta,t)=\frac{\mu_0}{g}F_{\mathrm{e}1}\cos(P_{\mathrm{e}}\theta-\omega_{\mathrm{e}}t)\prod_{m=0}^{n}\left[\lambda_{\mathrm{f}m0}+\lambda_{\mathrm{f}m1}\cos(P_{\mathrm{f}m}\theta-\omega_{\mathrm{f}m}t)\right] \tag{4.11}$$

where $F_{\mathrm{e}1}$ represents the fundamental magnitude of the excitation MMF, Ω_{e} is the mechanical speed of the excitation unit, and ω_{e} denotes the electric angular velocity of the excitation unit ($\omega_{\mathrm{e}}=P_{\mathrm{e}}\Omega_{\mathrm{e}}$). $\lambda_{\mathrm{f}m0}$ and $\lambda_{\mathrm{f}m1}$ are the DC term and fundamental term of the specific permeance of the mth layer modulation unit, respectively. Moreover, $\omega_{\mathrm{f}m}$ is the electric angular velocity of the mth layer modulation unit, and $\Omega_{\mathrm{f}m}$ is the mechanical speed of the mth layer modulation unit ($\omega_{\mathrm{f}m}=P_{\mathrm{f}m}\Omega_{\mathrm{f}m}$).

According to the rule of pole pair numbers of the basic model, Eq. (4.11) can be further expressed as:

$$\begin{aligned}B_{\mathrm{e}}(\theta,t)&=\frac{\mu_0}{g}F_{\mathrm{e}1}\frac{\prod_{m=0}^{n}\lambda_{\mathrm{f}m1}}{2^n}\cos\left\{P_{\mathrm{a}}\theta\pm\left[\omega_{\mathrm{e}}+\sum_{m=0}^{n}\mathrm{sgn}_m\omega_{\mathrm{f}m}\right]t\right\}\\&=B_{P_{\mathrm{a}}}\cos(P_{\mathrm{a}}\theta-\omega_{\mathrm{a}}t)\end{aligned} \tag{4.12}$$

where $B_{P\mathrm{a}}$ denotes the magnitude of the working flux density and ω_{a} represents the electric angular velocity of the working flux density. $B_{P\mathrm{a}}$ can be deduced as:

$$B_{P_{\mathrm{a}}}=\begin{cases}\frac{\mu_0}{g}F_{\mathrm{e}1}, & n=0\\ \frac{\mu_0}{g}\frac{\lambda_{\mathrm{f}11}}{2}F_{\mathrm{e}1}, & n=1\\ \frac{\mu_0}{g}\frac{\lambda_{\mathrm{f}11}\lambda_{\mathrm{f}21}}{4}F_{\mathrm{e}1}, & n=2\end{cases} \tag{4.13}$$

Using Eq. (4.13), the working flux density of the basic model can be analyzed as follows:

(1) When $n=0$, corresponding to the conventional permanent magnet machine, the modulation unit has a specific permeance of one.

(2) When $n=1$, the excitation MMF undergoes only single-layer modulation. In Sect. 2.2.2, the DC term and the fundamental term of the air gap specific permeance have been quantitatively compared. It shows that when the excitation unit is composed of surface-mounted permanent magnets, the equivalent air gap length is large, and thus the fundamental term of the specific permeance $\lambda_{\mathrm{f}11}$ is only about 0.3. (For reference, the DC term λ_0 can be close to 0.9 for the conventional permanent magnet machine.) When the excitation unit is composed of interior permanent magnets or excitation winding, the equivalent air gap length is small, and $\lambda_{\mathrm{f}11}$ can approach 0.5 at most. By substituting the above data into Eq. (4.13), it is observed that when the excitation unit is a surface-mounted permanent magnet structure, the working flux density $B_{P\mathrm{a}}$ is less than 1/6 of the conventional permanent magnet machine, while in other cases, $B_{P\mathrm{a}}$ is less than 1/4 of the conventional permanent magnet machine.

(3) When $n=2$, the excitation MMF undergoes double-layer modulation. Similarly, it can be deduced that when the excitation unit is a surface-mounted permanent magnet structure, the working density $B_{P\mathrm{a}}$ is less than 1/36 of the conventional

permanent magnet machine, while in other cases, B_{Pa} is less than 1/16 of the conventional permanent magnet machine.

Based on the above analysis, the air gap working magnetic field of the basic model of FMMs is very weak, much lower than the conventional permanent magnet machine. Furthermore, with a increase in the layer of the modulation, the working flux density decreases.

4.2.2 Back-Electromotive Force and Electromagnetic Torque

Assuming that the armature winding is an ideal winding, i.e., only the fundamental harmonic of the armature MMF is considered, the winding function can be represented as:

$$\begin{cases} N_A(\theta) = \frac{2}{\pi}\frac{N_{sa}}{P_a}k_{waP_a}\cos(P_a\theta) \\ N_B(\theta) = \frac{2}{\pi}\frac{N_{sa}}{P_a}k_{waP_a}\cos\left(P_a\theta - \frac{2}{3}\pi\right) \\ N_C(\theta) = \frac{2}{\pi}\frac{N_{sa}}{P_a}k_{waP_a}\cos\left(P_a\theta + \frac{2}{3}\pi\right) \end{cases} \tag{4.14}$$

where $N_A(\theta)$, $N_B(\theta)$, and $N_C(\theta)$ denote the winding functions of phases A, B, and C, respectively. Also, N_{sa} is the number of series turns per phase of the armature winding, and k_{waPa} is the winding factor. For the subscript of the winding factor, "a" represents the armature to distinguish it from the AC excitation winding, and "P_a" is the corresponding pole pair number. This distinction is necessary due to the rich harmonic content in FMMs, requiring the identification of multiple working magnetic fields. Using the winding function, the back-EMF of phase A winding can be calculated as [2]:

$$\begin{aligned} e_A(t) &= -\frac{d\psi_A}{dt} = -\frac{d}{dt}\left[r_g l_{stk}\int_0^{2\pi} B_e(\theta, t)N_A(\theta)d\theta\right] \\ &= 2\frac{N_{sa}}{P_a}r_g l_{stk}\omega_a k_{waP_a}B_{P_a}\sin\omega_a t = E_{a1}\sin\omega_a t \end{aligned} \tag{4.15}$$

where r_g represents the air gap radius, l_{stk} denotes the effective stack length, E_{a1} represents the magnitude of the fundamental back-EMF, and ψ_A signifies the no-load phase flux linkage.

(1) When $n = 0$, the excitation unit of the conventional permanent magnet machine rotates. Hence, according to Eq. (4.12), one can have $\omega_a = P_e\Omega_e = P_a\Omega_e$. Then, the back-EMF is acquired as:

$$E_{a1} = 2r_g l_{stk}N_{sa}\Omega_e k_{waP_a}B_{P_a} \tag{4.16}$$

(2) When $n > 0$, there are two possibilities of the rotating unit in FMMs, i.e., the excitation unit rotates or the modulation unit rotates, and the pole pair numbers of both units are not equal to that of the armature unit. When the excitation unit rotates:

$$E_{\mathrm{a1}} = 2r_{\mathrm{g}}l_{\mathrm{stk}}N_{\mathrm{sa}}\Omega_{\mathrm{e}}k_{\mathrm{wa}P_{\mathrm{a}}}B_{P_{\mathrm{a}}}\frac{P_{\mathrm{e}}}{P_{\mathrm{a}}} \tag{4.17}$$

When the mth modulation unit rotates, $\omega_a = P_{\mathrm{f}m}\Omega_{\mathrm{f}m}$, which yields:

$$E_{\mathrm{a1}} = 2r_{\mathrm{g}}l_{\mathrm{stk}}N_{\mathrm{sa}}\Omega_{\mathrm{f}m}k_{\mathrm{wa}P_{\mathrm{a}}}B_{P_{\mathrm{a}}}\frac{P_{\mathrm{f}m}}{P_{\mathrm{a}}} \tag{4.18}$$

Comparing Eqs. (4.16)–(4.18), it is observed that since the pole pair number of the rotating unit (either excitation unit or modulation unit) is not equal to that of the armature unit in FMMs, there is an additional term P_{e}/P_a or $P_{\mathrm{f}m}/P_a$ in the back-EMF expression compared to the conventional permanent magnet machine. The additional term is the pole ratio of FMMs, which was introduced in Chap. 3.

Using Eq. (4.7), the average torque of the basic model can be derived as:

(1) When $n = 0$:

$$T_{\mathrm{avg}} = 3r_{\mathrm{g}}l_{\mathrm{stk}}N_{\mathrm{s}}k_{\mathrm{wa}P_{\mathrm{a}}}B_{P_{\mathrm{a}}}I_{\mathrm{a1}} \tag{4.19}$$

(2) When $n > 0$:

$$T_{\mathrm{avg}} = 3r_{\mathrm{g}}l_{\mathrm{stk}}N_{\mathrm{s}}k_{\mathrm{wa}P_{\mathrm{a}}}B_{P_{\mathrm{a}}}I_{\mathrm{a1}}\frac{P_{\mathrm{r}}}{P_{\mathrm{a}}} \tag{4.20}$$

In Eq. (4.20), the numerator of the pole ratio is consistently represented by the rotor pole pair number P_{r} to standardize scenarios involving both the rotating excitation unit and the rotating modulation unit. When comparing Eqs. (4.19) and (4.20), it is evident that the torque of the FMM displays a characteristic known as the pole ratio amplification effect. However, it is critical to acknowledge that the working flux density $B_{P\mathrm{a}}$ of the FMM is significantly lower than that observed in conventional permanent magnet machines. Specifically, in configurations where the FMM includes a surface-mounted permanent magnet excitation unit and a single-layer modulation unit, the working flux density is less than one-sixth of that in conventional permanent magnet machines. Consequently, the torque output of the FMM can only approximate that of conventional designs when the pole ratio of the FMM exceeds 5. In contrast, when the FMM incorporates an interior permanent magnet excitation unit, the working flux density may be less than one-fourth of that in conventional permanent magnet machines. Nevertheless, due to the additional MMF effect provided by the interior permanent magnet excitation unit, as discussed in Chap. 2, the torque of the FMM can reach comparable levels only if a high pole ratio is maintained. Therefore, for the basic model operating with a single working magnetic field, achieving competitive torque relative to conventional permanent magnet machines necessitates a high pole ratio.

4.2.3 Power Factor

According to Eq. (4.9), the power factor is a function of the ratio of the phase flux linkage generated by the excitation unit ψ_{e1} to the phase flux linkage generated by the armature current ψ_{a1}. The greater this ratio, the higher the power factor. In the ideal basic model, the armature winding associated with P_a pole pairs solely interlinks the air gap flux density corresponding to those pole pairs. Therefore, the ratio of these two flux linkages equals the ratio of their respective air gap flux densities for the P_a pole pair. For the basic model of the conventional permanent magnet machine depicted in Fig. 4.3a, which lacks a modulation unit, the air gap flux density is solely influenced by the DC component of specific permeance. As a result, the ratio of the excitation flux linkage ψ_{e1} to the armature flux linkage ψ_{a1} can be obtained as:

$$\frac{\psi_{e1}}{\psi_{a1}} = \frac{\frac{F_{e1}\lambda_{f0}}{P_a}}{\frac{F_{a1}\lambda_{f0}}{P_a}} = \frac{F_{e1}}{F_{a1}} \tag{4.21}$$

Indeed, the ratio of the two flux linkages is equivalent to the ratio of the fundamental amplitudes of the excitation and armature MMFs. For the basic model of the single-layer modulation unit depicted in Fig. 4.3b, the no-load P_a pole pair air gap flux density is generated by the flux modulation effect, while the armature current directly produces the P_a pole pair air gap flux density without this modulation effect. Consequently, ψ_{e1} is associated with the fundamental term of the specific permeance λ_{f1}, while ψ_{a1} correlates with the DC term of the specific permeance λ_{f0}, resulting in:

$$\frac{\psi_{e1}}{\psi_{a1}} = \frac{\frac{F_{e1}\lambda_{f1}}{2P_a}}{\frac{F_{a1}\lambda_{f0}}{P_a}} = \frac{F_{e1}}{F_{a1}}\frac{\lambda_{f1}}{2\lambda_{f0}} \tag{4.22}$$

Similarly, for the basic model of the double-layer modulation unit in Fig. 4.3c, the ratio is:

$$\frac{\psi_{e1}}{\psi_{a1}} = \frac{\frac{F_{e1}\lambda_{s1}\lambda_{r1}}{4P_a}}{\frac{F_{a1}\lambda_{s0}\lambda_{r0}}{P_a}} = \frac{F_{e1}}{F_{a1}}\frac{\lambda_{s1}\lambda_{r1}}{4\lambda_{s0}\lambda_{r0}} \tag{4.23}$$

It can be seen that both the basic models with single- and double-layer modulation units will experience a more than 50% decrease in the ratio of ψ_{e1} to ψ_{a1} compared to the conventional permanent magnet machine, resulting in a significant reduction of their power factors.

Equations (4.21)–(4.23) allow an intuitive comparison of the power factors of the three basic models. Furthermore, if a quantitative calculation of power factor is needed, the excitation MMF and armature MMF should be calculated. The excitation MMF F_e can be considered as a constant value, and the armature MMF F_a is given as:

$$F_a(\theta, t) = N_A i_A + N_B i_B + N_C i_C = \frac{3}{\pi}\frac{N_{sa}}{P_a} k_{waP_a} I_{a1} \cos(P_a\theta - \omega_a t) \tag{4.24}$$

Substituting Eq. (4.24) into Eq. (4.9), and characterizing the phase current I_{a1} in terms of electric loading, i.e.,

$$A = \frac{3N_{sa}I_{a1}}{\sqrt{2}\pi r_g} \tag{4.25}$$

Thus, the power factor of the basic model in Fig. 4.3a can be derived as follows:

$$\mathrm{PF} = \frac{1}{\sqrt{1 + \left(\frac{\sqrt{2}Ar_g k_{waP_a}}{F_{e1}P_a}\right)^2}} \tag{4.26}$$

For the basic model of the single-layer modulation unit in Fig. 4.3b, the power factor is:

$$\mathrm{PF} = \frac{1}{\sqrt{1 + \left(\frac{\sqrt{2}Ar_g k_{waP_a}}{F_{e1}P_a}\frac{2\lambda_{f0}}{\lambda_{f1}}\right)^2}} \tag{4.27}$$

For the basic model of the double-layer modulation unit in Fig. 4.3c, the power factor is:

$$\mathrm{PF} = \frac{1}{\sqrt{1 + \left(\frac{\sqrt{2}Ar_g k_{waP_a}}{F_{e1}P_a}\frac{4\lambda_{s0}\lambda_{r0}}{\lambda_{s1}\lambda_{r1}}\right)^2}} \tag{4.28}$$

Comparing Eqs. (4.26)–(4.28), let us discuss the key parameters affecting the power factors of the basic models below:

(1) An increase in electric loading A leads to a lower power factor.
(2) As machine power increases, it results in the growth of machine size and the air gap radius (r_g). Besides, under the same cooling conditions, electric loading (A) also increases. Hence, both factors contribute to the decrease in power factor.
(3) When the armature winding pole pair (P_a) remains unchanged, and the pole ratio increases, the pole pair of the modulation unit (P_f) increases. Therefore, the fundamental term λ_{f1} of the specific permeance decreases, and the constant term λ_{f0} increases, as analyzed in Sect. 2.2.2, leading to a lower power factor.

(4) When the pole ratio is constant, and the armature winding pole pair (P_a) increases, the power factor of the basic model in Fig. 4.3a improves. For the basic model in Fig. 4.3b and c, the increase in armature winding pole pair will give rise to the modulation unit pole pair (P_f), leading to a decrease in λ_{f1}. Under the joint influences of P_f and λ_{f1}, the power factor will first increase and then decrease afterward.

In summary, parameters (1) and (2) are applicable to both conventional machines and FMMs, while parameters (3) and (4) are specific to FMMs.

4.2.4 Relationship of Actual Flux Modulation Machines with the Basic Model

The analysis indicates that the basic model of FMMs exhibits a weak working flux density. Despite benefiting from the pole ratio amplification effect, which enables output torques that are comparable to those of conventional permanent magnet machines, the inherent weak working flux density results in a low power factor. This suggests that the basic model lacks distinctive performance advantages over traditional designs. However, actual FMMs differ from their basic models in that they represent an integration of multiple basic units. For instance, Table 3.4 in the analysis illustrates the magnetic field harmonics of the flux reversal permanent magnet machine. When $l = 1$, the harmonics listed in the 2nd and 3rd rows manifest the fundamental frequency and are thus classified as working harmonics. These harmonics indicate that different basic models are at play for each harmonic. Taking the example of the harmonics in the 2nd row, the pole pair number of the excitation unit is $hZ_s/2$, that of the modulation unit is Z_r, and that of the armature unit is $hZ_s/2 + Z_r$. Given that the harmonic order h can assume multiple values, and the armature winding function also encompasses multiple harmonics, the actual flux reversal permanent magnet machine effectively integrates multiple basic models. Consequently, the output torque of actual FMMs is the cumulative result of all the basic models it incorporates. This aggregation leads to output torques in actual FMMs that exceed those of any individual basic model. This finding explains why the torque expressions for actual FMMs incorporate multiple harmonics, each with different amplification factors. Although the actual FMMs retain certain characteristics of the basic model—such as the positive correlation between output torque and the pole ratio—these actual machines exhibit enhanced performance due to their composite structure. Regarding power factor, while the excitation flux linkage ψ_e in actual FMMs is a combination of several basic models, the armature flux linkage ψ_a derives solely from the P_a pole pair armature field. This configuration results in a higher power factor for actual FMMs compared to their basic counterparts. Thus, while actual FMMs maintain some power factor characteristics of the basic model, as discussed in Sect. 4.2.3, they also display improved power factor due to the integrated effects of multiple basic models.

This subsection has effectively delineated the foundational principles of the basic model of FMMs and examined its characteristics in terms of torque and power factor. It has been elucidated that while the basic model alone does not offer performance advantages over conventional permanent magnet machines, the aggregation of multiple basic models in actual FMMs, conceptualized as a multi-working harmonic machine, results in a significant enhancement of both torque and power factor. The following analyses are conducted on actual FMM topologies.

4.3 Average Torque

To comprehensively understand the performance characteristics of different FMM types and elucidate their distinctions and connections, this chapter further categorizes FMMs into two groups: permanent magnet excitation and electric excitation, based on the types of excitation. The excitation structures utilizing DC excitation winding, AC excitation winding, and self-short-circuit excitation winding are regarded as electrically excited FMMs.

4.3.1 Permanent Magnet Excited Flux Modulation Machine

First, the excitation MMFs of permanent magnet excited FMMs are analyzed as follows.

(1) Vernier permanent magnet machines and transverse-flux machines employ surface-mounted permanent magnets as the excitation unit, with the excitation MMF given by Eq. (2.38).
(2) Flux reversal permanent magnet machines also utilize surface-mounted permanent magnet excitation, but the arrangement of permanent magnets is non-uniform, leading to relatively larger MMF harmonic contents. The excitation MMF can be acquired through Fourier decomposition, as presented in Eq. (4.29) [3]:

$$\begin{aligned} F_{\mathrm{e}}(\theta) &= F_{\mathrm{eh}} \sin \frac{Z_{\mathrm{s}} h}{2} \theta \\ &= \sum_{h=1,3,5,\ldots} \frac{4}{\pi h} \frac{B_{\mathrm{r}} h_{\mathrm{m}}}{\mu_0 \mu_{\mathrm{r}}} \left[1 + (-1)^{\frac{1+h}{2}} \sin \frac{h \pi b_{\mathrm{s}}}{2 t_{\mathrm{s}}} \right] \sin \frac{Z_{\mathrm{s}} h}{2} \theta \end{aligned} \tag{4.29}$$

where F_{eh} is the amplitude of hth permanent magnet excitation MMF, B_r is the remanence of the permanent magnet, h_m is the permanent magnet thickness, μ_0 is the vacuum permeability, μ_r is the relative permeability of the permanent magnet, P_e is the pole pair number of excitation permanent magnets, Z_s is the

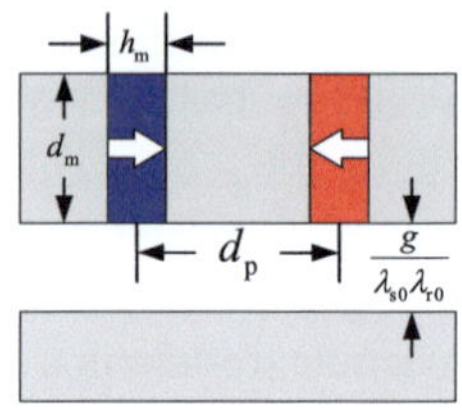

Fig. 4.4 Spoke-magnetization permanent magnet structure

number of stator teeth, b_s is the stator slot opening width, t_s is the stator slot pitch.

(3) Flux switching permanent magnet machines and doubly-salient permanent magnet machines utilize spoke-magnetization permanent magnets as the excitation unit, as illustrated in Fig. 4.4. Here, h_m is the permanent magnet thickness, d_m is the permanent magnet width, d_p is the pole pitch, and g is the air gap length. Considering that both sides of the air gap are slotted, the equivalent air gap length is calculated as $g/\lambda_{s0}\lambda_{r0}$.

Based on the MMF/specific permeance model, the air gap flux density magnitude can be determined as:

$$B_{max} = \frac{B_r h_m d_m}{d_p h_m + \frac{1}{\lambda_{s0}\lambda_{r0}}\mu_r d_m g} \tag{4.30}$$

If the additional MMF caused by slotting is not considered, the magnetic potential generated by the permanent magnets on the surface of the excitation unit forms a square wave with positive and negative symmetry, represented by the following Fourier series.

$$F_e(\theta, t) = \sum_{h=1}^{\infty} \frac{4}{h\pi} \frac{B_{max} g}{\lambda_{s0}\lambda_{r0}\mu_0} \cos hP_e(\theta - \Omega_e t) \tag{4.31}$$

1. **Vernier permanent magnet machine**

The air gap flux density of the vernier permanent magnet machine can be calculated as follows:

$$\begin{aligned} B_e(\theta, t) &= \frac{\mu_0}{g}\lambda_{s0} \sum_{h=1}^{\infty} F_{eh} \cos h(P_e\theta - \omega_e t) \\ &+ \frac{1}{2}\frac{\mu_0}{g} \sum_{h,l=1}^{\infty} F_{eh}\lambda_{sl}\{\cos[(hP_e + lP_f)\theta - h\omega_e t] + \cos[(hP_e - lP_f)\theta - h\omega_e t]\} \end{aligned} \tag{4.32}$$

Here, P_f denotes the pole pair number of the modulation unit. When the vernier machine adopts the straight-tooth stator, $P_f = Z_s$; and when the vernier machine adopts the auxiliary-tooth stator, $P_f = n_f Z_s$, where n_f is the number of auxiliary teeth on each primary tooth.

According to the law of electromagnetic induction, the no-load back-EMF of phase A is [4]:

$$\begin{aligned} e_{\mathrm{A}}(t) &= -\frac{\mathrm{d}\psi_{\mathrm{A}}}{\mathrm{d}t} = -\frac{d}{dt}\left[r_g l_{\mathrm{stk}} \int_0^{2\pi} B(\theta, t) N_{\mathrm{A}}(\theta) \mathrm{d}\theta \right] \\ &= r_g l_{\mathrm{stk}} \frac{\mu_0}{g} \Omega_{\mathrm{e}} N_{\mathrm{s}} \left\{ \sum_{h=1}^{\infty} F_{eh} \left[2\lambda_{\mathrm{s0}} k_{\mathrm{wa}(hP_{\mathrm{e}})} + \sum_{l=1}^{\infty} \lambda_{\mathrm{s}l} \left(\frac{hP_{\mathrm{e}}}{hP_{\mathrm{e}} + lP_{\mathrm{f}}} k_{\mathrm{wa}(hP_{\mathrm{e}}+lP_{\mathrm{f}})} \right. \right. \right. \\ &\quad \left. \left. \left. + \frac{hP_{\mathrm{e}}}{|hP_{\mathrm{e}} - lP_{\mathrm{f}}|} k_{\mathrm{wa}|hP_{\mathrm{e}}-lP_{\mathrm{f}}|} \right) \right] \sin h\omega_{\mathrm{e}} t \right\} \end{aligned} \tag{4.33}$$

Here, $N_{\mathrm{A}}(\theta)$ indicates the winding function of phase A. Different from the ideal winding function in Eq. (4.14), the actual winding function contains a large number of harmonics, which can be expressed as:

$$N_{\mathrm{A}}(\theta) = \sum_{P=1}^{\infty} \frac{2}{\pi} \frac{N_{\mathrm{sa}}}{P} k_{\mathrm{wa}P} \cos(P\theta) \tag{4.34}$$

It can be seen that the harmonic order of the back-EMF is only related to the excitation MMF order h. When $h = 1$, i.e., the fundamental excitation MMF generates the fundamental back-EMF, and when $h > 1$, the excitation MMF harmonics generate the back-EMF harmonics. Only the fundamental back-EMF can contribute to the average torque, where the fundamental back-EMF magnitude can be obtained as:

$$\begin{aligned} E_{\mathrm{a1}} &= r_g l_{\mathrm{stk}} \frac{\mu_0}{g} \Omega_{\mathrm{e}} N_{\mathrm{sa}} F_{\mathrm{e1}} \left[2\lambda_{\mathrm{s0}} k_{\mathrm{wa}(P_{\mathrm{e}})} + \sum_{l=1}^{\infty} \lambda_{\mathrm{s}l} \left(\frac{P_{\mathrm{e}}}{P_{\mathrm{e}} + lP_{\mathrm{f}}} k_{\mathrm{wa}(P_{\mathrm{e}}+lP_{\mathrm{f}})} \right. \right. \\ &\quad \left. \left. + \frac{P_{\mathrm{e}}}{|P_{\mathrm{e}} - lP_{\mathrm{f}}|} k_{\mathrm{wa}|P_{\mathrm{e}}-lP_{\mathrm{f}}|} \right) \right] \end{aligned} \tag{4.35}$$

According to Eq. (4.7), the average torque under $I_{\mathrm{d}} = 0$ control can be calculated as:

$$\begin{aligned} T_{\mathrm{avg}} &= 3 r_g l_{\mathrm{stk}} \frac{\mu_0}{g} I_{\mathrm{a1}} N_{\mathrm{sa}} F_{\mathrm{e1}} \left[\lambda_{\mathrm{s0}} k_{\mathrm{wa}(P_{\mathrm{e}})} + \frac{1}{2} \sum_{l=1}^{\infty} \lambda_{\mathrm{s}l} \left(\frac{P_{\mathrm{e}}}{P_{\mathrm{e}} + lP_{\mathrm{f}}} k_{\mathrm{wa}(P_{\mathrm{e}}+lP_{\mathrm{f}})} \right. \right. \\ &\quad \left. \left. + \frac{P_{\mathrm{e}}}{|P_{\mathrm{e}} - lP_{\mathrm{f}}|} k_{\mathrm{wa}|P_{\mathrm{e}}-lP_{\mathrm{f}}|} \right) \right] \end{aligned} \tag{4.36}$$

The analysis of the torque characteristics of the vernier permanent magnet machine addresses the following aspects:

(1) The comparison of torque expressions for the basic model, as shown in Eqs. (4.19) and (4.20), with those for the vernier permanent magnet machine in

Eq. (4.36) reveals that the average torque of the latter is generated by several basic models and can be categorized into three groups:

(a) A rotating excitation unit with P_e-pole-pairs coupled to a winding with the same number of pole pairs, which forms the basic model depicted in Fig. 3.4a.
(b) A rotating excitation unit with P_e pole pairs, a modulation unit with lP_f-pole-pairs, and an armature winding with $(P_e + lP_f)$-pole pairs, which forms the basic model depicted in Fig. 3.4b.
(c) A rotating excitation unit with P_e pole pairs, a modulation unit with lP_f pole pairs, and an armature winding with $|P_e - lP_f|$-pole pairs, which forms the basic model depicted in Fig. 3.4c.

It is important to recognize that although the number of pole pairs in the armature winding, $P_a = |P_e - P_f|$, the actual armature winding can be regarded as a composite of numerous ideal sinusoidal windings due to the presence of harmonics in the winding function. Therefore, the armature winding can link multiple air gap flux densities across different pole pairs. The working magnetic fields of these three basic models are tooth harmonics to each other; thus, their winding factors are equal in absolute value, and they all contribute to the average torque.

(2) Since the three basic models utilize different numbers of harmonic pole pairs in the armature winding function, they exhibit different pole ratios. The number of pole pairs in the excitation unit, P_e, in the vernier permanent magnet machine approximates that of the modulation unit, P_f. Therefore, the pole ratio of $P_e/|P_e - P_f|$ is significant only when $l = 1$, whereas the pole ratios of the other basic models are approximately one or significantly smaller. Additionally, due to the low working flux density of FMMs, only the two basic models with pole ratios of one and $P_e/|P_e - P_f|$ can yield relatively high torque components, while the contributions of the other basic models are minimal. Consequently, the average torque of the vernier permanent magnet machine can be simplified as follows:

$$T_{\text{avg}} = 3r_g l_{\text{stk}} \frac{\mu_0}{g} N_{\text{sa}} F_{\text{e1}} I_{\text{a1}} \left(\lambda_{\text{s0}} k_{\text{wa}P_e} + \frac{\lambda_{\text{s1}}}{2} \frac{P_e}{P_a} k_{\text{wa}P_a} \right) \tag{4.37}$$

(3) In Eq. (4.37), the direction of torque for two basic models is determined by the sign of the winding factors for their respective armature winding pole pairs. When these winding factors share the same sign, the torques are additive; conversely, they are subtractive. Consequently, ascertaining the sign of the winding factors is essential. Although the distribution factors of tooth harmonics are precisely equal, establishing the positive or negative nature of the pitch factors presents a challenge due to the variable structure and span distance of the winding. To resolve this, a virtual coil method has been introduced. It is assumed that each stator tooth is equipped with one fractional slot concentrated coil, as illustrated in Fig. 4.5 (for the auxiliary-tooth structure, the virtual coil

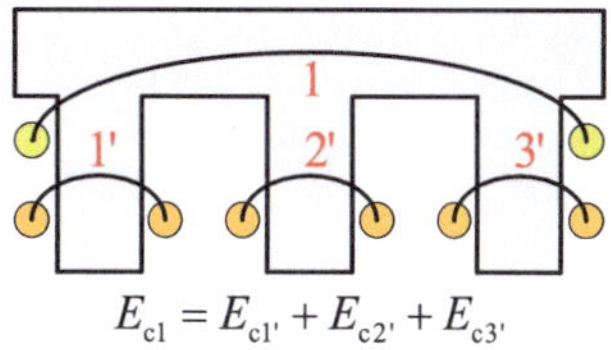

Fig. 4.5 Virtual coil diagram

is wound around each auxiliary tooth). Adhering to the principle of flux continuity, the total flux linkage of the actual coil is equal to the sum of the flux linkages of all virtual coils within the span of the actual coil. This principle also applies to the back-EMF. Hence, the back-EMF of the actual coil, induced by the P-pole-pair magnetic field, can be expressed as:

$$e_{cP}(t) = \tau_c k'_{paP} k'_{daP} E'_{cP1} \sin P\Omega_e t \tag{4.38}$$

where τ_c denotes the actual coil pitch, k'_{paP} represents the P-pole-pair pitch factor of the virtual coil, k'_{daP} signifies the P-pole-pair distribution factor of the virtual coil group within the span of the actual coil, and E'_{cP1} denotes the maximum value of the fundamental back-EMF of the virtual coil. Given that the virtual coil utilizes a concentrated winding structure with a slot number P_f, the two working harmonics of P_e and P_a pole pairs in Eq. (4.37) also qualify as tooth harmonics for the virtual coil. Consequently, their respective distribution factors of the virtual coil are identical. If the corresponding pitch factors of the virtual coil share the same sign, the back-EMFs generated by both are in phase, and the torques can be additive; otherwise, the two torques cancel each other. The pitch factor of the virtual coil can be expressed as follows:

$$k'_{paP} = \sin \frac{P}{P_f}\pi \tag{4.39}$$

For the P_a-pole-pair working magnetic field, since $P_a < P_f$, the k'_{paPa} is positive. When the vernier permanent magnet machine adopts $P_f = P_e + P_a$, thus $P_e < P_f$, the k'_{paPe} is also positive, leading to the superposition of the two torques. Conversely, when the vernier permanent magnet machine adopts $P_f = P_e - P_a$, thus $P_e > P_f$, the k'_{paPe} is negative, resulting in the cancellation of the two torques. Therefore, vernier permanent magnet machines typically adopt $P_f = P_e + P_a$.

(4) Based on the preceding analysis, the torque of the vernier permanent magnet machine primarily consists of two components: a conventional torque component and a modulated torque component. In Eq. (4.37), the conventional torque component is prevalent when the pole ratio is low, while the modulated torque component is predominant when the pole ratio is high. For a clearer understanding of the two torque components, several vernier permanent magnet machines with different pole ratios are analyzed in Fig. 4.6. These machines all utilize $P_f = P_e + P_a$ and are designed within the same dimensions and

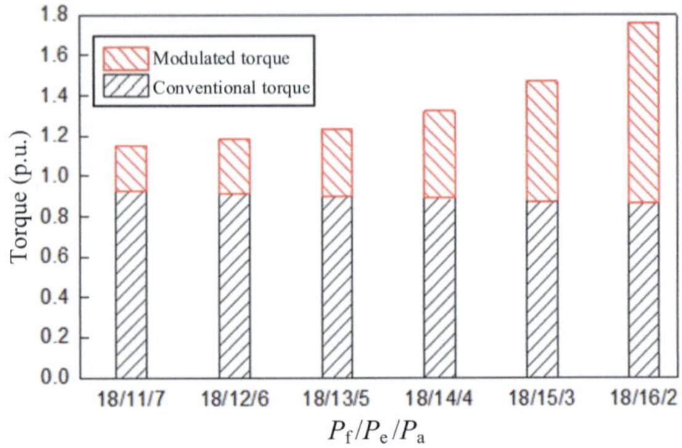

Fig. 4.6 Torque comparison of vernier permanent magnet machines with different pole ratios

electrical loading parameters. As the pole ratio increases, the modulated torque component in the vernier permanent magnet machine augments, significantly enhancing the overall torque capacity of the machine. When the pole ratio is less than two, the conventional torque component is predominant, and the modulated torque component constitutes a relatively minor proportion. At a pole ratio of five, for example, with the stator slot/excitation pole pair/armature pole pair ratio of 18/15/3, both the conventional and modulated torque components are comparable. When the pole ratio exceeds eight, the modulated torque component becomes dominant. Although the conventional torque component in the vernier permanent magnet machine slightly decreases due to stator slotting, the modulated torque component ensures that the vernier permanent magnet machine maintains a significant advantage in torque density compared to conventional permanent magnet machines.

Furthermore, the output torque capabilities of the conventional permanent magnet machine, the vernier permanent magnet machine, and the auxiliary-tooth vernier permanent magnet machine are compared using the same rotor structure but different stator structures, as depicted in Fig. 4.7. The detailed dimensional parameters of these machines are provided in Table 4.1.

Figure 4.8 illustrates the no-load flux lines and flux density distribution across the three types of machines. Compared to the conventional permanent magnet machine, both the vernier permanent magnet machine and the auxiliary-tooth vernier permanent magnet machine exhibit lower flux density in the iron cores. This reduction is attributed to the use of an open-stator-slot structure, which increases the effective air gap length and decreases the non-modulated flux density component, despite the incorporation of the modulated component. In Fig. 4.8b and c, the 10-pole-pair permanent magnet MMFs are modulated to produce a 2-pole-pair magnetic field in the air gap. The no-load back-EMF waveforms of the two vernier machines are similar, with both being approximately 50% higher than that of the conventional permanent magnet machine, as shown in Fig. 4.9. It is noteworthy that the vernier permanent magnet machine features a longer end winding length

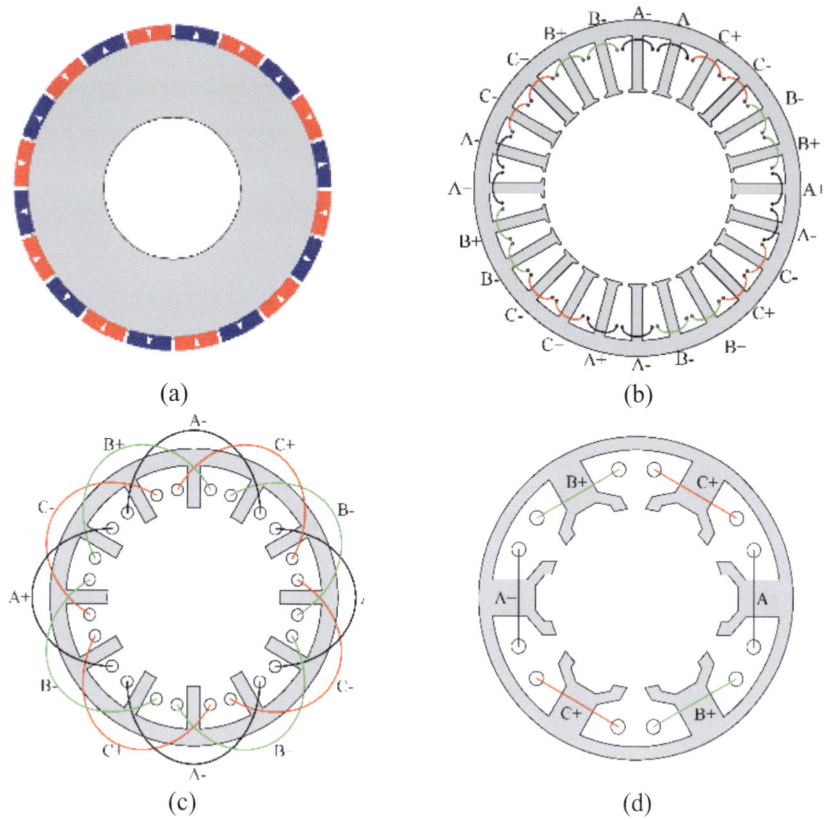

Fig. 4.7 Schematic diagram of the three permanent magnet machines: **a** 10-pole-pair rotor, **b** 24-slot stator of conventional permanent magnet machine, **c** 12-slot stator of vernier permanent magnet machine, and **d** 6-slot stator of auxiliary-tooth vernier permanent magnet machine

due to distributed overlapping winding, whereas the auxiliary-tooth vernier permanent magnet machine, with fewer stator teeth and concentrated non-overlapping winding, reduces winding ends and copper losses. From another perspective, the vernier permanent magnet machine, which employs distributed overlapping winding, displays fewer harmonic contents, primarily the 1st, 5th, and 7th harmonics. In contrast, the auxiliary-tooth vernier permanent magnet machine, utilizing concentrated non-overlapping winding, exhibits richer harmonic contents, including not only the 1st, 5th, and 7th harmonics but also the 2nd and 4th harmonics. The presence of increased harmonics renders the auxiliary-tooth vernier machine more susceptible to saturation, resulting in lower output torque under heavy load conditions. Consequently, the overload torque capability of the auxiliary-tooth vernier permanent magnet machine is inferior to that of the vernier permanent magnet machine with a conventional straight-tooth structure.

Table 4.1 Dimensional parameters of three permanent magnet machines

Parameters	Conventional permanent magnet machine	Vernier permanent magnet machine	Auxiliary-tooth vernier permanent magnet machine
Stator slot number	24	12	6
Pole pair of excitation unit	10	10	10
Pole pair of modulation unit	–	12	12
Pole pair of armature winding	10	2	2
Winding factor	0.933	1	0.866
Rotational speed (rpm)	600		
Stator outer diameter (mm)	130	124	134
Stack length (mm)	70		
Number of series turns per phase	200		

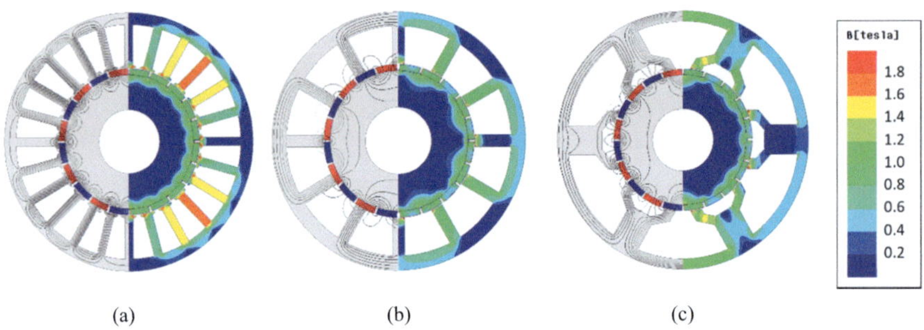

Fig. 4.8 Comparison of no-load flux density and flux lines: **a** conventional permanent magnet machine, **b** vernier permanent magnet machine, and **c** auxiliary-tooth vernier permanent magnet machine

(5) Figure 4.10 examines the output torque of the vernier permanent magnet machine across various pole ratios and winding pole pair numbers P_a [5]. To ensure a fair comparison, the DC copper loss is held constant for these models. In Eq. (4.37), a clear positive correlation exists between the average torque and the pole ratio. However, Fig. 4.10 illustrates that the increase in average torque with the pole ratio slows progressively, and in the latter part of the curve with $P_a = 2$, the torque decreases as the pole ratio increases. This reduction is due to the increase in the stator slot number as the pole ratio increases. Furthermore, as analyzed in Fig. 2.15b, the fundamental term of the specific permeance

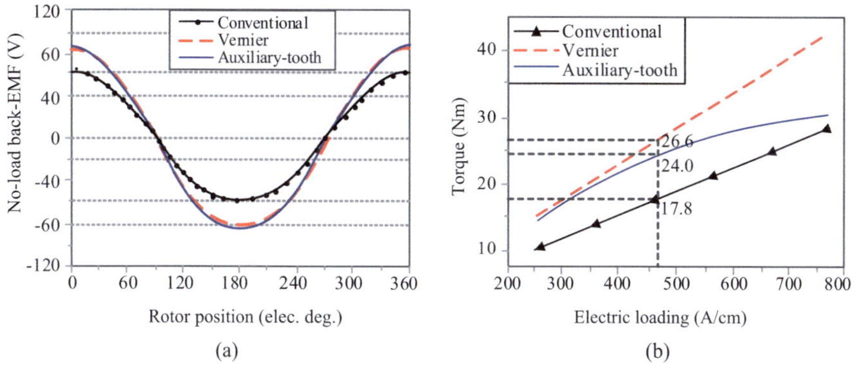

Fig. 4.9 Comparison of no-load back-EMF and torque: **a** no-load back-EMF. **b** Average torque

decreases, which becomes significant when the stator slot number is high. Therefore, there is an optimal pole ratio for achieving maximum torque in the vernier permanent magnet machine. In Fig. 4.10, the machine with $P_a = 2$ displays a lower optimal pole ratio compared to the machine with $P_a = 1$. This lower ratio in the $P_a = 2$ machine is due to a higher stator slot number at the same pole ratio, leading to an earlier inflection point. Moreover, in the low-pole-ratio region, the impact of the fundamental term of the specific permeance is not predominant; instead, the influence of the stator current becomes more significant. Since the winding end of the $P_a = 2$ is shorter than that of the $P_a = 1$, the stator current for the former is higher than the latter under identical copper losses. Consequently, the output torque of the $P_a = 2$ machine exceeds that of the $P_a = 1$ machine in the low-pole-ratio region.

2. Transverse-flux permanent magnet machine

Equation (3.7) specifies the no-load air gap flux density for the transverse-flux permanent magnet machine. Based on this, the no-load phase back-EMF can be derived as

Fig. 4.10 Output torque capability of vernier permanent magnet machine under different pole ratios and winding pole pair numbers

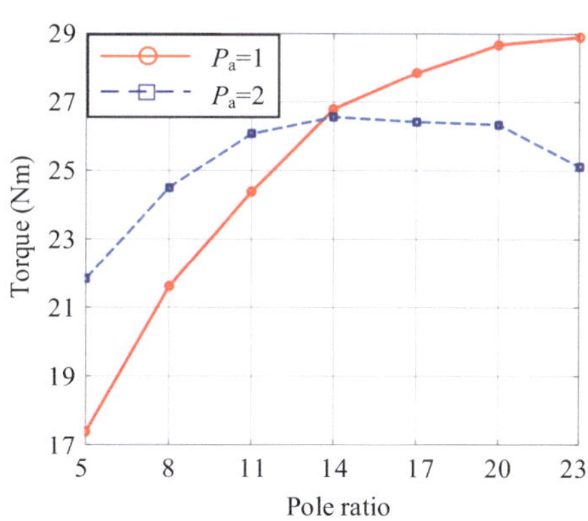

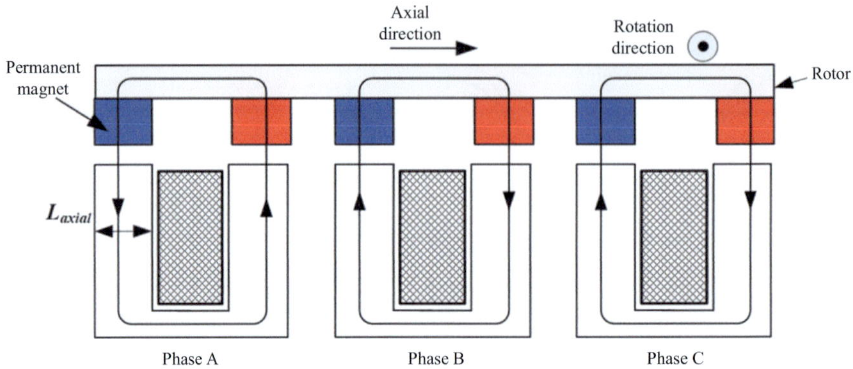

Fig. 4.11 Transverse-flux permanent magnet machine

follows:

$$e_{\mathrm{A}}(t) = -\frac{\mathrm{d}\psi_{\mathrm{A}}}{\mathrm{d}t} = -\frac{\mathrm{d}}{\mathrm{d}t}\left[r_{\mathrm{g}} l_{\mathrm{stk}} N_{\mathrm{sa}} \int_0^{2\pi} B_{\mathrm{e}}(\theta, t)\mathrm{d}\theta\right] = \pi r_{\mathrm{g}} l_{\mathrm{stk}} \frac{\mu_0}{g} \Omega_{\mathrm{e}} N_{\mathrm{sa}} F_{\mathrm{e1}} P_{\mathrm{f}} \lambda_{\mathrm{s1}} \sin\omega_a t \tag{4.40}$$

For the three-phase transverse-flux machine, the phases are physically separated along the axial direction, as depicted in Fig. 4.11. Consequently, the output torque under $I_{\mathrm{d}} = 0$ control can be computed as follows:

$$T_{\mathrm{avg}} = \frac{3}{2}\pi r_{\mathrm{g}} l_{\mathrm{stk}} \frac{\mu_0}{g} I_{\mathrm{a1}} N_{\mathrm{sa}} F_{\mathrm{e1}} P_{\mathrm{f}} \lambda_{\mathrm{s1}} \tag{4.41}$$

Equation (4.41) reveals that the transverse-flux permanent magnet machine differs from the vernier permanent magnet machine in that its torque is derived from a single basic model. This basic model features a distinctive combination of pole pair numbers, where the pole pairs of the excitation unit and modulation unit are equal, and that of the armature unit is zero. Therefore, the concept of a "pole ratio" does not apply in the torque equation, and the torque is directly proportional to the pole pair number of the modulation unit, P_{f}. It is important to note, however, that as P_{f} increases, the pole leakage flux from the permanent magnets also increases, setting an upper limit on the torque density of the transverse-flux permanent magnet machine. In comparison, the torque density of the transverse-flux permanent magnet machine is higher due to the lack of a finite "pole ratio," when contrasted with the vernier permanent magnet machine.

3. **Flux reversal permanent magnet machine**

Table 3.4 outlines the air gap flux density for the flux reversal permanent magnet machine. Based on the winding function, the no-load back-EMF can be formulated as follows [6, 7]:

$$
\begin{aligned}
e_{\mathrm{A}}(t) &= -\frac{\mathrm{d}\psi_{\mathrm{A}}}{\mathrm{d}t} = -\frac{\mathrm{d}}{\mathrm{d}t}\left[r_{\mathrm{g}} l_{\mathrm{stk}} \int_{0}^{2\pi} B_{\mathrm{e}}(\theta, t) N_{\mathrm{A}}(\theta) \mathrm{d}\theta\right] \\
&= r_{\mathrm{g}} l_{\mathrm{stk}} \frac{\mu_0}{g} \Omega_{\mathrm{r}} N_{\mathrm{sa}} \left\{\sum_{h,l=1}^{\infty} F_{\mathrm{e}h} \lambda_{\mathrm{r}l} \left[\frac{lZ_{\mathrm{r}}}{h\frac{Z_{\mathrm{s}}}{2} + lZ_{\mathrm{r}}} k_{\mathrm{wa}\left(h\frac{Z_{\mathrm{s}}}{2} + lZ_{\mathrm{r}}\right)} \right.\right. \\
&\quad \left.\left. - \frac{lZ_{\mathrm{r}}}{\left|h\frac{Z_{\mathrm{s}}}{2} - lZ_{\mathrm{r}}\right|} k_{\mathrm{wa}\left|h\frac{Z_{\mathrm{s}}}{2} - lZ_{\mathrm{r}}\right|}\right] \cos l\omega_{\mathrm{a}} t\right\}
\end{aligned}
\tag{4.42}
$$

It is evident that the magnitude of the back-EMF is exclusively related to the harmonic order of the rotor-specific permeance, l. In analyses focusing on the average torque, the harmonics of the back-EMF are neglected, i.e., setting $l = 1$, yields:

$$
\begin{aligned}
E_{\mathrm{a}1} &= r_{\mathrm{g}} l_{\mathrm{stk}} \frac{\mu_0}{g} \Omega_{\mathrm{r}} N_{\mathrm{sa}} \lambda_{\mathrm{r}1} \left\{\sum_{i=1}^{\infty} F_{\mathrm{e}h} \left[\frac{Z_{\mathrm{r}}}{h\frac{Z_{\mathrm{s}}}{2} + Z_{\mathrm{r}}} k_{\mathrm{wa}\left(h\frac{Z_{\mathrm{s}}}{2} + Z_{\mathrm{r}}\right)} \right.\right. \\
&\quad \left.\left. - \frac{Z_{\mathrm{r}}}{\left|h\frac{Z_{\mathrm{s}}}{2} - Z_{\mathrm{r}}\right|} k_{\mathrm{wa}\left|h\frac{Z_{\mathrm{s}}}{2} - Z_{\mathrm{r}}\right|}\right]\right\} \cos \omega_{\mathrm{a}} t
\end{aligned}
\tag{4.43}
$$

Furthermore, the average torque can be calculated as follows:

$$
T_{\mathrm{avg}} = \frac{3}{2} r_{\mathrm{g}} l_{\mathrm{stk}} \frac{\mu_0}{g} I_{\mathrm{a}1} N_{\mathrm{sa}} \lambda_{\mathrm{r}1} \left\{\sum_{i=1}^{\infty} F_{\mathrm{e}h} \left[\frac{Z_{\mathrm{r}}}{h\frac{Z_{\mathrm{s}}}{2} + Z_{\mathrm{r}}} k_{\mathrm{wa}\left(h\frac{Z_{\mathrm{s}}}{2} + Z_{\mathrm{r}}\right)} - \frac{Z_{\mathrm{r}}}{\left|h\frac{Z_{\mathrm{s}}}{2} - Z_{\mathrm{r}}\right|} k_{\mathrm{wa}\left|h\frac{Z_{\mathrm{s}}}{2} - Z_{\mathrm{r}}\right|}\right]\right\}
\tag{4.44}
$$

In Eq. (4.44), all the harmonics are tooth harmonics of each other; therefore, the absolute values of their winding factors are the same, though their signs may vary. As in the analysis of the vernier permanent magnet machine, the positive or negative sign can be determined using the virtual coil method. Ultimately, Eq. (4.44) can be further simplified:

$$
T_{\mathrm{avg}} = \frac{3}{2} r_{\mathrm{g}} l_{\mathrm{stk}} \frac{\mu_0}{g} I_{\mathrm{a}1} N_{\mathrm{sa}} \lambda_{\mathrm{r}1} k_{\mathrm{waP_a}} \left[\sum_{h=1}^{\infty} (-1)^{\frac{h+1}{2}} F_{\mathrm{e}h} \left(\frac{Z_{\mathrm{r}}}{h\frac{Z_{\mathrm{s}}}{2} + Z_{\mathrm{r}}} + \frac{Z_{\mathrm{r}}}{h\frac{Z_{\mathrm{s}}}{2} - Z_{\mathrm{r}}}\right)\right]
\tag{4.45}
$$

Equation (4.45) shows some similarities with Eq. (4.36) of the vernier permanent magnet machine, such as the coefficients preceding the parentheses. However, significant differences remain. From Eq. (4.45), it is evident that all excitation MMF harmonics and only the fundamental term of the specific permeance contribute to the average torque. In contrast, in Eq. (4.36), only the fundamental term of the excitation MMF and all specific permeance harmonics contribute to the average torque. This divergence stems from the differences in the stationary units of the two machines: for the flux reversal permanent magnet machine, the stationary unit is the excitation MMF, while for the vernier permanent magnet machine, it is the modulation unit. Furthermore, it is important to note that flux reversal permanent magnet machines do not typically have a "dominant" torque component as vernier permanent magnet machines do. The value of the torque components is contingent upon the specific slot-pole combination and the stator slot opening ratio of the machine. For instance, with $Z_s = 12$ and $Z_r = 11$, when harmonic order $h = 1$, it results in the maximal pole ratio of 11/5, and when $h = 3$, the pole ratio of 11/7 is not significantly different. However, for another configuration of $Z_s = 12$ and $Z_r = 16$, when $h = 1$, the pole ratio is 1.6, and when $h = 3$, the maximal pole ratio is 8. Additionally, the value of F_{eh} varies with the stator slot opening ratio. As this ratio increases, F_{e1} decreases while F_{e3} increases. When the stator slot opening ratio is approximately 0.25, F_{e1} and F_{e3} are nearly equivalent. Thus, to enhance the output torque of flux reversal machines, one potential strategy is to maximize the F_{eh} of the dominant working harmonic. For example, when $Z_s = 12$ and $Z_r = 16$, enhancing F_{e3}, corresponding to the maximal pole ratio at harmonic $h = 3$, would significantly improve the output torque [8]. Considering that the specific permeance of the flux reversal permanent magnet machine is less extensively utilized than that of the vernier permanent magnet machine, the former can, at most, match the torque capability of the latter. However, the flux reversal permanent magnet machine has an advantage: its rotor is a simple and robust structure without windings or permanent magnets, which makes it suitable for applications demanding high reliability or reduced rotor costs [8–10].

4. **Flux switching permanent magnet machine**

Table 3.6 enumerates the harmonic components of the air gap flux density in flux switching permanent magnet machines. From this data, the back-EMF can be deduced as [11]:

$$
\begin{aligned}
e_A(t) = r_g l_{stk} \frac{\mu_0}{g} \Omega_e N_s \sum_{l_r=1}^{\infty} \lambda_{rl_r} \Bigg[\sum_{h,l_s=1}^{\infty} \frac{1}{2} F_{eh} \lambda_{sl_s} (-1)^{\frac{h+1}{2}} \Bigg(\frac{l_r Z_r}{\frac{hZ_s}{2} + 2l_s Z_s - l_r Z_r} \\
+ \frac{l_r Z_r}{\frac{hZ_s}{2} - 2l_s Z_s - l_r Z_r} - \frac{l_r Z_r}{\frac{hZ_s}{2} + 2l_s Z_s + l_r Z_r} - \frac{l_r Z_r}{\frac{hZ_s}{2} - 2l_s Z_s + l_r Z_r} \Bigg) k_{wa\left(\frac{hZ_s}{2} - l_r Z_r\right)} \\
+ \sum_{h=1}^{\infty} F_{eh} \lambda_{s0} (-1)^{\frac{h+1}{2}} \Bigg(\frac{l_r Z_r}{\frac{hZ_s}{2} - l_r Z_r} - \frac{l_r Z_r}{\frac{hZ_s}{2} + l_r Z_r} \Bigg) k_{wa\left(\frac{hZ_s}{2} - l_r Z_r\right)} \Bigg] \sin l_r \omega_a t
\end{aligned}
\tag{4.46}
$$

Similarly, the average torque of the flux switching permanent magnet machine can be calculated by setting $l_r = 1$:

$$T_{avg} = \frac{3}{2} r_g l_{stk} \frac{\mu_0}{g} I_{a1} N_{sa} \lambda_{r1} \left[\sum_{h,l_s=1}^{\infty} \frac{1}{2} F_{eh} \lambda_{sl_s} (-1)^{\frac{h+1}{2}} \left(\frac{l_r Z_r}{\frac{hZ_s}{2} + 2l_s Z_s - l_r Z_r} \right. \right.$$
$$\left. + \frac{l_r Z_r}{\frac{hZ_s}{2} - 2l_s Z_s - l_r Z_r} - \frac{l_r Z_r}{\frac{hZ_s}{2} + 2l_s Z_s + l_r Z_r} - \frac{l_r Z_r}{\frac{hZ_s}{2} - 2l_s Z_s + l_r Z_r} \right) k_{wa\left(\frac{hZ_s}{2} - l_r Z_r\right)}$$
$$\left. + \sum_{h=1}^{\infty} F_{eh} \lambda_{s0} (-1)^{\frac{h+1}{2}} \left(\frac{l_r Z_r}{\frac{hZ_s}{2} - l_r Z_r} - \frac{l_r Z_r}{\frac{hZ_s}{2} + l_r Z_r} \right) k_{wa\left(\frac{hZ_s}{2} - l_r Z_r\right)} \right] \tag{4.47}$$

The operational harmonics of the flux switching permanent magnet machine are complex. The fundamental term of the rotor-specific permeance can contribute to the average torque, along with any excitation MMF harmonic and stator-specific permeance harmonic. Therefore, the torque of the flux switching permanent magnet machine does not exhibit an obvious increasing trend with the pole ratio. For example, consider two models with a 12-stator-slot/16-rotor-slot and a 36-stator-slot/16-rotor-slot configuration, both featuring a pole ratio of 8, yet differing in torque capabilities. In the 12-stator-slot/16-rotor-slot configuration, the 3rd excitation MMF harmonic can interact with the DC term of the stator-specific permeance and the fundamental term of the rotor-specific permeance, producing a 2-pole-pair working flux density in the air gap. Additionally, the 2-pole-pair working flux density can be generated by the interaction between the fundamental term of the excitation MMF and the fundamental terms of both the stator and rotor specific permeances. However, both interactions result in relatively low magnitudes of the working flux density. In contrast, in the 36-stator-slot/16-rotor-slot configuration, the 2-pole-pair working flux density in the air gap is produced by the interaction among the fundamental term of the excitation MMF, the DC term of the stator-specific permeance, and the fundamental term of the rotor-specific permeance, resulting in a significantly larger magnitude of the working flux density.

Consequently, the 36-stator-slot/16-rotor-slot model exhibits higher back-EMF and torque than the 12-stator-slot/16-rotor-slot model, even though they share the same pole ratio. This effect becomes more pronounced at lower pole ratios. For instance, analysis and simulations of four flux switching permanent magnet machines with configurations of 12-stator-slot/10-rotor-slot, 12-stator-slot/11-rotor-slot, 12-stator-slot/13-rotor-slot, and 12-stator-slot/14-rotor-slot, with pole ratios of 2.5, 2.2, 2.6, and 3.5 respectively, indicate that the 12-stator-slot/13-rotor-slot machine delivers the highest output torque.

Similar to the flux reversal permanent magnet machine, only the fundamental term of the rotor-specific permeance contributes to the average torque in the flux switching permanent magnet machine. This results in less utilization of the rotor-specific permeance compared to the vernier permanent magnet machine, leading to a lower torque capability. Moreover, due to the spoke-magnetization excitation unit used in the flux switching permanent magnet machine, the additional MMF effect becomes severe at high pole ratios, diminishing the output torque. Only at low pole ratios does the torque density of the flux switching machine surpass that of the conventional permanent magnet machine.

4.3.2 *Electrically Excited Flux Modulation Machine*

In electrically excited machines, the primary distinction from permanent magnet machines lies in the variation of the MMF in the excitation unit. For AC excitation winding and self-short-circuit excitation winding, the analysis considers only the fundamental MMF:

$$F_{\mathrm{e}}(\theta, t) = \frac{3N_{\mathrm{se}}I_{\mathrm{e1}}}{\pi P_{\mathrm{e}}} k_{\mathrm{we}P_{\mathrm{e}}} \cos(P_{\mathrm{e}}\theta - \omega_{\mathrm{e}}t) \tag{4.48}$$

where P_{e} denotes the pole pair number of the electrically excited winding, ω_{e} represents the electrical angular velocity of the excitation MMF, $k_{\mathrm{we}P\mathrm{e}}$ denotes the excitation winding factor for the P_{e} pole pair harmonic, N_{se} signifies the series turns per phase, and I_{e1} refers to the fundamental amplitude of the excitation current.

For DC excitation winding, it is essential to note that the stator slot openings of some DC electrically excited FMMs are often large to maximize the modulation effect of the stator teeth. In such cases, the slotting factor of the stator excitation MMF must be considered and incorporated into the winding factor $k_{\mathrm{we}P\mathrm{e}}$. When the slotting factor is omitted, the MMF of the DC excitation winding is calculated as follows:

$$F_{\mathrm{e}}(\theta, t) = \sum_{h=1}^{\infty} \frac{2N_{\mathrm{se}}I_{\mathrm{e1}}}{\pi P_{\mathrm{e}}h} \cos hP_{\mathrm{e}}(\theta - \Omega_{\mathrm{e}}t) \tag{4.49}$$

When the slotting factor is taken into account, the MMF of the DC excitation winding is:

$$F_{\mathrm{e}}(\theta, t) = \sum_{h=1}^{\infty} \frac{4N_{\mathrm{se}}I_{\mathrm{e1}} \sin\left(\frac{\pi}{2} h \frac{b_{\mathrm{s}}}{t_{\mathrm{s}}}\right)}{\pi^2 P_{\mathrm{e}} h^2 \frac{b_{\mathrm{s}}}{t_{\mathrm{s}}}} \cos hP_{\mathrm{e}}(\theta - \Omega_{\mathrm{e}}t) \tag{4.50}$$

According to Eq. (4.50), when the slot opening is large, for example, where $b_{\mathrm{s}}/t_{\mathrm{s}} = 0.5$, the harmonic of the excitation MMF is inversely proportional to the square of its order. Consequently, the magnitudes of these excitation MMF harmonics are very small and can typically be neglected. Therefore, in the analysis of DC-excited FMMs with large slot openings, the DC excitation MMF can be simplified as follows:

$$F_{\mathrm{e}}(\theta, t) \approx \frac{4N_{\mathrm{se}}I_{\mathrm{e1}} \sin\left(\frac{\pi}{2} \frac{b_{\mathrm{s}}}{t_{\mathrm{s}}}\right)}{\pi^2 P_{\mathrm{e}} \frac{b_{\mathrm{s}}}{t_{\mathrm{s}}}} \cos P_{\mathrm{e}}(\theta - \Omega_{\mathrm{e}}t) \tag{4.51}$$

1. **Electrically excited vernier machine**

An electrically excited vernier machine can be conceptualized as replacing the rotor permanent magnet of a vernier permanent magnet machine with a DC electrically excited winding. Therefore, they adhere to the same torque generation principle. By integrating the torque expression of the vernier permanent magnet machine from Eq. (4.37) and substituting the excitation MMF with that derived in Eq. (4.49), the average torque of the electrically excited vernier machine can be determined as follows:

$$T_{\text{avg}} = \frac{12}{\pi} r_g l_{stk} \frac{\mu_0}{g} N_{\text{sa}} N_{\text{se}} I_{\text{a1}} I_{\text{e1}} \left(\frac{\lambda_{\text{s0}}}{P_{\text{e}}} + \frac{\lambda_{\text{s1}}}{2P_{\text{a}}} \right) k_{\text{wa}P_{\text{a}}} \tag{4.52}$$

In Eq. (4.37), the torque of the vernier permanent magnet machine is divided into two components. The torque generated by the DC term of the specific permeance, λ_{s0}, is independent of the pole ratio, whereas the torque generated by the fundamental term of the specific permeance, λ_{s1}, is proportional to the pole ratio. However, in electrically excited machines, the excitation capacity of the DC excitation winding is inversely proportional to its pole pair number, P_e. Therefore, in Eq. (4.52), the torque produced by the fundamental term of the specific permeance is also inversely proportional to the pole pair number of the excitation unit. Additionally, the torque generated by the fundamental term of the specific permeance is similarly inversely proportional to the pole pair number of the armature unit P_a, which is a characteristic common to conventional electrically excited synchronous machines.

Next, when comparing the output torque of the electrically excited vernier machine to that of a conventional electrically excited synchronous machine with the same armature pole pair number P_a, differences become apparent. The output torque of the conventional electrically excited synchronous machine corresponds to the first term of Eq. (4.52), with P_e replaced by P_a. Since $P_e > P_a$, the torque generated by the conventional electrically excited synchronous machine is significantly greater than the first torque component of the electrically excited vernier machine. Additionally, the stator slot opening in the electrically excited vernier machine is typically large. According to Fig. 2.16, its λ_{s0} may be only about 0.6, considerably lower than that of the conventional electrically excited synchronous machine, which generally features mostly semi-closed slots. Moreover, for the second torque component of the electrically excited vernier machine, there is a coefficient of 1/2 due to the flux modulation process. Furthermore, λ_{s1} is also much smaller than λ_{s0}. Consequently, the second torque component in Eq. (4.45) is only about one-third of the torque of the conventional electrically excited synchronous machine. Therefore, the overall torque of the electrically excited vernier machine is less than that of the conventional electrically excited synchronous machine with the same armature pole pair number.

2. **Flux modulation induction machine**

The operational principles of the flux modulation induction machine are complex. To investigate its operating characteristics, the flux modulation induction machine is assumed to have ideal winding structures for both the stator and rotor. Under this assumption, the armature MMF can be expressed as follows [12]:

$$F_a(\theta, t) = F_{a1} \cos(P_a\theta + \omega_a t) \tag{4.53}$$

Additionally, the modulation unit only considers the DC term and fundamental term of the specific permeance. In this context, the air gap specific permeance can be represented as:

$$\lambda_f(\theta, t) = \frac{\mu_0}{g}(\lambda_{f0} + \lambda_{f1} \cos P_f\theta) \tag{4.54}$$

Then, the air gap flux density can be derived as:

$$\begin{aligned} B_a(\theta, t) = \frac{\mu_0}{g} F_{a1} \bigg\{ \lambda_{f0} \cos(P_a\theta + \omega_a t) + \frac{\lambda_{f1}}{2} \{\cos[(P_f - P_a)\theta - \omega_a t] \\ + \cos[(P_f + P_a)\theta + \omega_a t]\}\} \end{aligned} \tag{4.55}$$

If the slot-pole combination $P_f = P_e + P_a$ is adopted, the pole pair number of the self-short-circuit excitation winding can be determined as $P_e = P_f - P_a$. Consequently, the synchronous speed of the corresponding working magnetic field is $\Omega_{e0} = \omega_a / P_e$. Assuming a slip ratio s, the rotor speed (i.e., excitation unit) is:

$$\Omega_e = (1 - s)\frac{\omega_a}{P_e} \tag{4.56}$$

Owing to the differential speed between the rotor and the armature magnetic field, a back-EMF will be induced in the rotor self-short-circuit winding. The fundamental magnitude of the back-EMF can be calculated as:

$$E_{e1} = s\frac{\mu_0}{g} F_{a1}\lambda_{f1} r_g L_{stk} N_{se} k_{weP_e} \frac{\omega_a}{P_e} = \frac{3}{\pi}\frac{\mu_0}{g} I_{a1}\lambda_{f1} r_g L_{stk} (N_{sa}N_{se}) \left(k_{weP_e} k_{waP_a}\right) s\omega_a \frac{1}{P_e P_a} \tag{4.57}$$

Subsequently, the magnitude of the excitation current generated by the back-EMF can be determined as:

$$I_{e1} = \frac{E_{e1}}{\sqrt{R_e^2 + (s\omega_a L_e)^2}} \tag{4.58}$$

The corresponding phase angle is:

$$\gamma = \text{atan}\left(\frac{s\omega_a L_e}{R_e}\right) \tag{4.59}$$

According to the law of electromagnetic induction, the magnitude of the no-load back-EMF generated in the armature winding can be deduced as:

$$E_{\mathrm{a1}} = \frac{3}{\pi}\frac{\mu_0}{gP_{\mathrm{a}}} r_{\mathrm{g}} L_{\mathrm{stk}} \lambda_{\mathrm{f1}} (N_{\mathrm{sa}} N_{\mathrm{se}}) \left(k_{\mathrm{we}P_{\mathrm{e}}} k_{\mathrm{wa}P_{\mathrm{a}}}\right) \Omega_{\mathrm{e0}} I_{\mathrm{e1}} \cos\gamma \tag{4.60}$$

where Ω_{e0} represents the rotor synchronous speed. In induction machines, the electromagnetic power of the stator cannot be fully converted into mechanical power of the rotor due to the presence of rotor copper loss, which amounts to s times the electromagnetic power. Therefore, the rotor torque is calculated as:

$$T_{\mathrm{e}} = (1 - s)\frac{e_{\mathrm{A}} i_{\mathrm{A}} + e_{\mathrm{B}} i_{\mathrm{B}} + e_{\mathrm{C}} i_{\mathrm{C}}}{\Omega_{\mathrm{e}}} = \frac{e_{\mathrm{A}} i_{\mathrm{A}} + e_{\mathrm{B}} i_{\mathrm{B}} + e_{\mathrm{C}} i_{\mathrm{C}}}{\Omega_{\mathrm{e0}}} \tag{4.61}$$

According to Eq. (4.61), the average torque of the flux modulation induction machine is:

$$T_{\mathrm{avg}} = \frac{9}{2\pi}\frac{\mu_0}{gP_{\mathrm{a}}} r_{\mathrm{g}} L_{\mathrm{stk}} \lambda_{\mathrm{f1}} (N_{\mathrm{sa}} N_{\mathrm{se}}) \left(k_{\mathrm{we}P_{\mathrm{e}}} k_{\mathrm{wa}P_{\mathrm{a}}}\right) I_{\mathrm{a1}} I_{\mathrm{e1}} \cos\gamma \tag{4.62}$$

The flux modulation induction machine operates on a principle similar to that of the conventional induction machine. However, due to the presence of the flux modulation effect, it exhibits distinct torque characteristics compared to its conventional counterpart. Therefore, it is crucial to compare and analyze the torque characteristics of both machines. Since the torque generation process in flux modulation induction machines can be divided into two stages—first, the armature current induces excitation current in the rotor excitation winding, and second, the excitation current interacts with the armature field to generate torque—the electromagnetic performance of these two stages is compared individually.

(1) The armature MMF is modulated by the modulation unit, resulting in the induction of back-EMF in the rotor's self-short-circuit excitation winding. Consequently, the back-EMF in the rotor winding is influenced by the magnitude of the modulated armature field. When compared to the non-modulated armature field, the modulated field is significantly reduced, typically to at most half of its original strength. Additionally, due to the large number of rotor pole pairs, the induced back-EMF is low, being inversely proportional to the pole ratio. As a result, the back-EMF in the rotor excitation winding is considerably small. For example, in a flux modulation induction machine with a pole ratio of 5, the back-EMF induced in the rotor winding might be less than one-tenth of that induced in a conventional induction machine with the same armature pole pair number.

(2) The rotor current in the flux modulation induction machine comprises two components: the d-axis and q-axis components. The magnetic field generated by the d-axis component and the armature magnetic field are 180° out of phase, thus cancelling each other. Consequently, the d-axis component does not contribute

to electromagnetic torque generation, and only the q-axis component functions as the excitation current, with its magnitude given by Eq. (4.62):

$$I_{\mathrm{e1q}} = I_{\mathrm{e1}} \cos \gamma = \frac{E_{\mathrm{e}} R_{\mathrm{e}}}{R_{\mathrm{e}}^2 + (s\omega_{\mathrm{a}} L_{\mathrm{e}})^2} \tag{4.63}$$

In a conventional induction machine, the rotor winding inductance, $s\omega_{\mathrm{a}} L_{\mathrm{e}}$, is significantly larger than the resistance, R_{e}, often with a ratio exceeding 20. In contrast, the flux modulation induction machine, due to its high number of rotor pole pairs, exhibits a substantially smaller inductance. For example, with a pole ratio of 5 and the same armature pole pair number, the rotor magnetizing inductance of the flux modulation induction machine is reduced to 1/25th of that in a conventional induction machine. However, the slot leakage inductance remains nearly the same. Overall, the total rotor winding inductance in the flux modulation induction machine is at most one-fifth of that in the conventional induction machine. When this data is substituted into Eq. (4.63), along with the back-EMF data, it can be deduced that the effective rotor excitation current of the flux modulation induction machine is approximately twice that of the conventional induction machine.

(3) In the flux modulation induction machine, the rotor excitation current induces back-EMF in the armature winding, a phenomenon that aligns with the operation of the electrically excited vernier machine. Despite the presence of a pole ratio amplification factor, the magnitude of the excitation MMF is inversely proportional to the rotor pole pair number. Consequently, the induced armature back-EMF is not influenced by the rotor pole pair number but is inversely proportional only to the armature pole pair number. Therefore, unlike conventional induction machines, the flux modulation induction machine does not demonstrate the back-EMF and torque amplification effects typically observed.

Given that induction machines are frequently analyzed using equivalent circuits, this book also presents the derivation of the equivalent circuit for the flux modulation induction machine and compares it to that of the conventional induction machine. Initially, Eq. (4.58) is expressed in vector form as:

$$\dot{I}_{\mathrm{e1}} = \frac{\dot{E}_{\mathrm{e1}}}{R_{\mathrm{e}} + \mathrm{j} s\omega_{\mathrm{a}} L_{\mathrm{e}}} \tag{4.64}$$

where j is the imaginary unit. Since the frequency of the induced back-EMF in the excitation winding differs from that of the armature winding, the frequency conversion method commonly employed in conventional induction machines is utilized. This method involves substituting the actual rotating excitation unit with a virtual stationary excitation unit while ensuring that the magnitude and phase of the excitation MMFs produced by both units are equivalent. Equation (4.64) can then be transformed accordingly:

$$\dot{I}_{e1} = \frac{\dot{E}_{e1}}{\frac{R_e}{s} + j\omega_a L_e} = \frac{\dot{E}_{e10}}{\frac{R_e}{s} + j\omega_a L_e} \tag{4.65}$$

This transformation implies that the resistance of the virtual excitation unit becomes $1/s$ times that of the actual excitation unit, while the inductance remains unchanged. For electrically excited machines, the no-load back-EMF of the armature winding can be calculated using the mutual inductance, as stated in Eq. (4.65):

$$\dot{E}_{e10} = -j\omega_a M_{ae} \dot{I}_{a1} \tag{4.66}$$

Upon substituting Eq. (4.66) into Eq. (4.65):

$$\dot{I}_{e1} = \frac{-j\omega_a M_{ae} \dot{I}_{a1}}{\frac{R_e}{s} + j\omega_a L_e} \tag{4.67}$$

where M_{ae} represents the maximum value of the mutual inductance between the excitation winding and the armature winding. According to Eq. (4.67), the relationship between the excitation current and the armature current can be characterized by a parallel circuit configuration. In this circuit, I_{a1} is the armature current, I_{e1} is the excitation current, and the impedances in the two branches are $j\omega_a M_{ae}$ and $R_e/s + j\omega_a(L_e - M_{ae})$, respectively, as illustrated in Fig. 4.12a. Notably, the electrical parameters in the two branches are denoted with an apostrophe, indicating the electrical quantities after converting the number of turns. However, when using Fig. 4.12a for analysis, the armature current must be known. Since most induction machines are directly connected to the grid, only the armature voltage is known, and the armature current remains unknown. Therefore, the equivalent circuit requires further refinement.

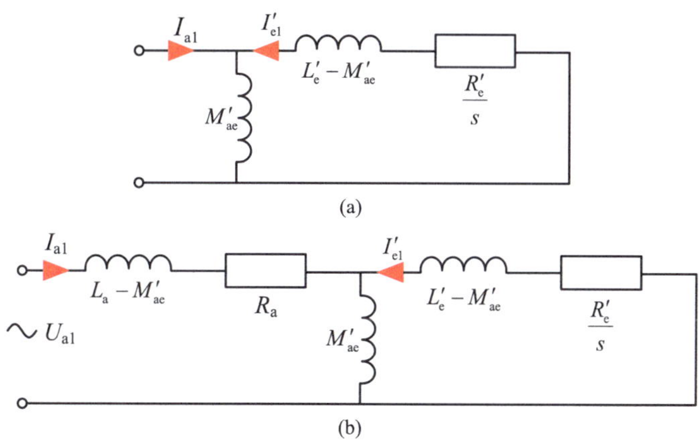

Fig. 4.12 Equivalent circuit of flux modulation induction machine: **a** constant current source and **b** constant voltage source

According to the law of electromagnetic induction, the armature voltage U_{a1} can be formulated as:

$$\begin{aligned}\dot{U}_{a1} &= \dot{I}_{a1}R_a + j\omega_a L_a \dot{I}_{a1} + j\omega_a M'_{ae}\dot{I}'_{e1} \\ &= \dot{I}_{a1}R_a + j\omega_a\left(L_a - M'_{ae}\right)\dot{I}_{a1} + j\omega_a M'_{ae}\left(\dot{I}'_{e1} + \dot{I}_{a1}\right)\end{aligned} \tag{4.68}$$

Based on Eq. (4.68) and Fig. 4.12a, the equivalent circuit of the constant voltage source is depicted in Fig. 4.12b. As observed, the equivalent circuit of the flux modulation induction machine resembles the "T-shaped" equivalent circuit of the conventional induction machine. However, are there any differences between the two in terms of parameters? In the equivalent circuit depicted in Fig. 4.12, the excitation unit undergoes a conversion of turns, meaning the rotor winding is converted to have equal turns with the stator winding, to align the mismatch of stator and rotor inductance parameters. This conversion is similarly performed in conventional induction machines. Therefore, in the conventional induction machine, the armature magnetic field induces the same flux linkage in both the armature and excitation windings. In other words, the armature magnetizing inductance L_{am}, is equal to the converted mutual inductance M'_{ae}. Thus, for the conventional induction machine, there is:

$$L_a - M'_{ae} = L_a - L_{am} = L_{a\sigma} \tag{4.69}$$

Here, the term $L_a - M'_{ae}$ represents the leakage inductance $L_{a\sigma}$. However, for the flux modulation induction machine, the armature magnetic field does not induce identical flux linkage in both the armature winding and the excitation winding. This disparity arises because the armature magnetic field directly induces flux linkage in the armature winding, but only indirectly induces flux linkage in the excitation winding after the flux modulation effect. Consequently, the armature magnetizing inductance M'_{ae} does not correspond to a physical entity in the flux modulation induction machine; it is merely a mathematical expression, and thus Eq. (4.69) cannot be satisfied.

Regarding the calculation of inductances, both the excitation winding inductance L_e and the armature winding inductance L_a include magnetizing inductance and leakage inductance. Since leakage inductance is complex and influenced by various dimensional parameters, only the magnetizing inductances are calculated:

$$\begin{aligned}L_{am} &= \frac{6\sqrt{2}\mu_0 r_g l_{stk}(N_{sa}k_{wa})^2\lambda_{f0}}{\pi g P_a^2} \\ L_{em} &= \frac{6\sqrt{2}\mu_0 r_g l_{stk}(N_{se}k_{we})^2\lambda_{f0}}{\pi g P_e^2} \\ M_{ae} &= \frac{3\sqrt{2}\mu_0 r_g l_{stk}N_{sa}k_{wa}N_{se}k_{we}\lambda_{f1}}{\pi g P_a P_e}\end{aligned} \tag{4.70}$$

It can be derived that:

$$\frac{L_{\mathrm{am}}L_{\mathrm{em}}}{M_{\mathrm{ae}}^2} = 4\left(\frac{\lambda_{\mathrm{f0}}}{\lambda_{\mathrm{f1}}}\right)^2 \gg 1 \tag{4.71}$$

As previously mentioned, in the flux modulation induction machine, the conversion of rotor turns is merely a mathematical operation without physical meaning. Therefore, it is only necessary to ensure that the inductances ($L_{\mathrm{a}} - M'_{\mathrm{ae}}$) and ($L'_{\mathrm{e}} - M'_{\mathrm{ae}}$) are positive. According to Eq. (4.71), both L_{am} and L'_{em} are significantly larger than M'_{ae}, resulting in high values of ($L_{\mathrm{a}} - M'_{\mathrm{ae}}$) and ($L'_{\mathrm{e}} - M'_{\mathrm{ae}}$). Consequently, the inductances on the two arms of the "T-shape" equivalent circuit ($L_{\mathrm{a}} - M'_{\mathrm{ae}}$) and ($L'_{\mathrm{e}} - M'_{\mathrm{ae}}$) are much larger than the leakage inductance $L_{\mathrm{a}\sigma}$ and $L_{\mathrm{e}\sigma}$, respectively. Since the electromagnetic torque is given by:

$$T_{\mathrm{e}} = \frac{P_{\mathrm{e}}}{\Omega_{\mathrm{e0}}} = \frac{P_{\mathrm{e}}I'^2_{\mathrm{e1}}R'_{\mathrm{e}}}{\omega_{\mathrm{a}}s} \tag{4.72}$$

As the inductance of ($L'_{\mathrm{e}} - M'_{\mathrm{ae}}$) is high, there is a substantial current in the M'_{ae} branch, leading to a much smaller I'_{e1} than I_{a1}. Consequently, the torque performance of the flux modulation induction machine is inferior to that of the conventional induction machine, consistent with the previous analysis results based on the torque expression. Additionally, due to the large value of ($L'_{\mathrm{e}} - M'_{\mathrm{ae}}$), it is not possible to neglect the magnetizing inductance and establish a simplified equivalent circuit like that of the conventional induction machine.

3. **Electrically excited vernier reluctance machine**

Stator electrically excited FMMs include the electrically excited vernier reluctance machine, the electrically excited flux switching machine, and the electrically excited doubly-salient machine [13]. As these topologies all feature doubly salient iron cores with stators that contain both excitation and armature windings, their operating principles and performance characteristics are analogous. For clarity, the electrically excited vernier reluctance machine is chosen as the representative of stator electrically excited FMMs. According to the flux modulation theory, the average torque of the electrically excited vernier reluctance machine can be derived as:

$$
\begin{aligned}
T_{\text{avg}} &= \sum_{h=-\infty}^{\infty} T_h + \sum_{h,l_s=-\infty}^{\infty} T_{h,l_s} \\
T_h &= \frac{3}{2} r_g l_{\text{stk}} \frac{\mu_0}{g} N_{\text{sa}} F_{eh} \lambda_{s0} \lambda_{r1} I_{a1} \frac{Z_r}{\frac{hZ_s}{2} + Z_r} k_{\text{wa}\left(\frac{hZ_s}{2}+Z_r\right)} \\
&= \frac{6}{\pi^2} r_g l_{\text{stk}} \frac{\mu_0}{g} (N_{\text{sa}} N_{\text{se}})(I_{a1} I_{e1}) \lambda_{s0} \lambda_{r1} \frac{Z_r}{h^2 \frac{Z_s}{2}\left(\frac{hZ_s}{2} + Z_r\right)} \frac{\sin\left(\frac{\pi}{2} h \frac{b_s}{t_s}\right)}{\frac{b_s}{t_s}} k_{\text{wa}\left(\frac{hZ_s}{2}+Z_r\right)} \\
T_{h,l_s} &= \frac{3}{\pi^2} r_g l_{\text{stk}} \frac{\mu_0}{g} (N_{\text{sa}} N_{\text{se}})(I_{a1} I_{e1}) \lambda_{s|l_s|} \lambda_{r1} \frac{Z_r}{h^2 \frac{Z_s}{2}\left(\frac{hZ_s}{2} + l_s Z_s + Z_r\right)} \\
&\quad \frac{\sin\left(\frac{\pi}{2} h \frac{b_s}{t_s}\right)}{\frac{b_s}{t_s}} k_{\text{wa}\left(\frac{hZ_s}{2}+l_s Z_s+Z_r\right)}
\end{aligned}
\tag{4.73}
$$

In Eq. (4.73), the excitation MMF order h and the stator specific permeance order l_s can assume negative values, simplifying the expression. The negative order represents the slot-pole combination wherein the armature winding pole pair is derived from the difference between half the stator slot number $Z_s/2$ and the rotor slot number Z_r.

In Eq. (4.73), even when considering only the fundamental term of the excitation MMF for a $Z_s/2$ pole pair, the output torque components of the electrically excited vernier reluctance machine remain complex. For simplicity, the output torque primarily comprises two basic models. The first basic model features a single-layer modulation unit structure, where the reluctance rotor serves as the modulation unit with Z_r pole pairs, and the corresponding armature winding pole pair number is $|Z_s/2 \pm Z_r|$. The second basic model involves a double-layer modulation unit structure, where both the salient stator and rotor function as modulation units, and the corresponding armature winding pole pair number is $|Z_s/2 \pm l_s Z_s \pm Z_r|$. All these armature winding pole pairs are tooth harmonics, and thus the absolute values of their winding factors are equal to the fundamental winding factor.

Additionally, substituting the expression of the excitation MMF F_{eh} into Eq. (4.73), the pole ratio of the electrically excited vernier reluctance machine is calculated as $2Z_r/(Z_s P_a)$, which differs from the vernier permanent magnet machine with a pole ratio of P_r/P_a and the electrically excited vernier machine with a pole ratio of $1/P_a$. This difference arises because the excitation MMF of the electrically excited vernier reluctance machine has a pole pair number of $Z_s/2$, and the magnitude of the excitation MMF F_{eh} is inversely proportional to its pole pair number $Z_s/2$. Consequently, the original pole ratio Z_r/P_a is modified to $2Z_r/(Z_s P_a)$.

Next, the torque density of the electrically excited vernier reluctance machine is examined. Consider a model with a stator slot number $Z_s = 6$, rotor slot number $Z_r = 5$, excitation winding pole pair number $P_e = 3$, and armature winding pole pair number $P_a = 2$ as an example. For comparative analysis, the synchronous reluctance machine is chosen as the benchmark due to its prevalent use and its classification as a specific type of stator electrically excited FMM. To ensure a fair comparison, a combination of armature winding pole pair number $P_a = 2$ and rotor pole pair number $Z_r = 4$ is selected for the synchronous reluctance machine. According to the principle of flux modulation, the excitation MMF and armature MMF of the synchronous reluctance machine are considered to be generated by the d-axis and q-axis currents, respectively. Thus, the average torque of the synchronous reluctance machine is calculated as follows:

$$T_{avg} = \frac{3}{\pi} r_g l_{stk} \frac{\mu_0}{g} N_{sa}^2 \left(I_{a1d} I_{a1q}\right) \lambda_{r1} \frac{Z_r}{P_a^2} k_{waP_a}^2 \tag{4.74}$$

Compared to the electrically excited vernier reluctance machine, the stator of the synchronous reluctance machine features semi-closed slots. Therefore, the slotting factor can be neglected, and the stator specific permeance is considered to be one. Moreover, the pole ratio of the synchronous reluctance machine is Z_r/P_a^2. These distinctions underpin the discussion on the average torques of the two machines below:

(1) The synchronous reluctance machine has a pole ratio of $4/2^2 = 1$. Disregarding other coefficients in the torque expression, the per-unit value of the average torque is one.

(2) Electrically excited vernier reluctance machines exhibit various pole pair numbers for the working armature MMFs, with the minimum pole pair number being 2. The corresponding torque components are T_{-1} and $T_{1,-1}$. The T_{-1} component has a pole ratio of 5/6, and it is generated by the DC term of the stator specific permeance λ_{s0}, which is approximately 0.6 under open stator slots. The $T_{1,-1}$ component also has a pole ratio of 5/6, and it is generated by the fundamental term of the stator specific permeance λ_{s1}, which is about 0.4 under open stator slots and is further reduced by a coefficient of 1/2 in the torque expression. Moreover, due to the open stator slots, the slotting factor is around 0.9, which cannot be ignored. Therefore, the resultant per-unit values of T_{-1} and $T_{1,-1}$ are 0.45 and 0.15, respectively. In addition to the aforementioned torque components, the T_1 and $T_{-1,1}$ also have relatively significant per-unit values of 0.11 and 0.04, respectively. Although all torque components are positive, the sum of the aforementioned components totals only about 0.75, which is less than that of the synchronous reluctance machine. Moreover, this analysis assumes that the electric loadings of the excitation and armature windings of the electrically excited vernier reluctance machine are equivalent to the d-axis and q-axis electric loadings of the synchronous reluctance machine, respectively. However, the total RMS value of the former is the algebraic sum, whereas that of the latter is the vector sum. Thus, the current in the synchronous reluctance machine is approximately 40% higher than that of the electrically excited vernier reluctance machine under the same DC copper loss, further accentuating the torque disparity between the two machines.

Therefore, it can be concluded that the electrically excited vernier reluctance machine has a variety of torque components, though they are of low values, resulting in significantly lower torque density compared to the synchronous reluctance machine. A distinctive feature of the electrically excited vernier reluctance machine is its unique structure. Unlike the electrically excited synchronous machine, where both the excitation and armature windings are typically located in the rotor, the electrically excited vernier reluctance machine places both windings in the stator. This design helps avoid issues associated with rotational excitation. Moreover, compared to the synchronous reluctance machine, the electrically excited vernier reluctance machine features self-excitation capability, eliminating the need for external armature controllers and making it a promising option for applications in aviation starting and power generation systems.

4. **Brushless doubly-fed reluctance machine**

There are two principal differences between the brushless doubly-fed reluctance machine and the electrically excited vernier reluctance machine. First, the brushless doubly-fed reluctance machine is excited by AC windings, while the electrically excited vernier reluctance machine utilizes DC windings. Second, the former typically features overlapping excitation and armature windings with pole pair numbers much smaller than the stator slot number Z_s; while the latter generally uses fractional slot concentrated armature windings, and the pole pair number of the excitation winding is $Z_s/2$. Despite these differences, the torque generation mechanisms of the two machines are similar. Consequently, the torque expression for the brushless doubly-fed reluctance machine can be adapted by modifying the pole pair number in Eq. (4.73):

$$\begin{aligned} T_{\text{avg}} = & \frac{9}{2\pi} r_g l_{\text{stk}} \frac{\mu_0}{g} (N_{\text{sa}} N_{\text{se}})(I_{\text{a1}} I_{\text{e1}}) \lambda_{\text{r1}} \left\{ \lambda_{\text{s0}} \frac{Z_r}{P_e(Z_s - P_a)} + \sum_{l_s=1}^{\infty} \frac{(-1)^{l_s+1}}{2} \lambda_{sl_s} \right. \\ & \left. \times \left\{ \frac{Z_r}{P_e[(l_s - 1)Z_s + P_a]} - \frac{Z_r}{P_e[(l_s + 1)Z_s - P_a]} \right\} \right\} k_{\text{wa}P_a} k_{\text{we}P_e} \end{aligned} \tag{4.75}$$

In Eq. (4.75), the omission of excitation MMF harmonics should be noted. This simplification is justified by the significant difference between Z_s and P_e, causing the lowest tooth harmonic order being much larger than P_e and consequently, the magnitude of these harmonics being considerably smaller than the fundamental term. For the analysis of pole ratios in Eq. (4.75), a specific example with $Z_s = 72$, $P_a = 2$, $P_e = 3$, and $Z_r = 67$ is considered. The three pole ratios in Eq. (4.75) each incorporate the component Z_r/P_e, differing in their denominators: $(Z_s - P_a)$, $[(l_s - 1)Z_s + P_a]$, and $[(l_s + 1)Z_s - P_a]$ respectively. When $l_s = 1$, the minimum denominator $[(l_s - 1)Z_s + P_a] = P_a = 2$, represents the basic model of double-layer modulation units. Excluding this minimum denominator of two, the next smallest denominator is $(Z_s - P_a) = 70$, which is significantly higher, resulting in a very low corresponding torque component. Consequently, the brushless doubly-fed reluctance machine predominantly exhibits one torque component, leading to a simplified Eq. (4.75):

$$T_{avg} = \frac{9}{4\pi} r_g l_{stk} \frac{\mu_0}{g} (N_{sa} N_{se})(I_{a1} I_{e1}) \lambda_{s1} \lambda_{r1} \frac{Z_r}{P_e P_a} k_{waP_a} k_{weP_e} \tag{4.76}$$

Therefore, the torque amplification factor for the brushless doubly-fed reluctance machine is $Z_r/(P_e P_a)$. Compared with the torque amplification factor for the synchronous reluctance machine in Eq. (4.74), which is Z_r/P_a^2, it can be observed that the torque amplification factor of the synchronous reluctance machine is significantly lower. In the brushless doubly-fed reluctance machine, the rotor slot number Z_r is substantially higher than both P_e and P_a, whereas in the synchronous reluctance machine, $Z_r = 2P_a$. As a result, the torque amplification factor of the brushless doubly-fed reluctance machine is considerably higher than that of the synchronous reluctance machine, leading to a higher torque density for the former. It should also be noted that for the synchronous reluctance machine, the torque amplification factor $Z_r/P_a^2 = 4/Z$, makes it suitable for high-speed applications with low Z_r. In contrast, the brushless doubly-fed reluctance machine, with its higher Z_r, is better suited for low-speed applications.

5. **Vernier reluctance machine**

Referring to Fig. 3.22, the vernier reluctance machine can be considered a specialized case of the brushless doubly-fed reluctance machine, integrating excitation and armature windings. Using Eq. (4.76), the torque expression for the vernier reluctance machine is derived:

$$T_{avg} = \frac{9}{4\pi} r_g l_{stk} \frac{\mu_0}{g} N_{as}^2 (I_{a1d} I_{a1q}) \lambda_{s1} \lambda_{r1} \frac{Z_r}{P_a^2} k_{waP_a}^2 \tag{4.77}$$

Similar to the brushless doubly-fed reluctance machine, the vernier reluctance machine features a high pole ratio, for example, reaching 70 with $Z_s = 72$, $P_a = 1$, and $Z_r = 70$. Theoretically, this suggests a higher output torque compared to the synchronous reluctance machine. However, it is essential to understand that an increase in the pole ratio does not necessarily translate into a proportional improvement in torque density. As the pole ratio increases, the fundamental specific permeance at the optimal slot opening tends to decrease. Therefore, both the pole ratio and specific permeance effects must be considered when aiming to achieve higher torque density in the vernier reluctance machine.

6. **Asynchronous induction machine**

As previously noted, the asynchronous induction machine can be regarded as a brushless doubly-fed reluctance machine with the excitation winding replaced by a self-short-circuit winding [14, 15]. Although a large number of tooth harmonics in the asynchronous induction machine can theoretically contribute to the average torque, the torque is primarily generated by the fundamental harmonic of MMF. Consequently, the characteristics of the asynchronous induction machine are similar to those of the flux modulation induction machine. This section outlines the differences between the asynchronous induction machine and the flux modulation induction machine. First, in the asynchronous induction machine, the modulation unit rotates with the pole pair number of Z_r, and the armature magnetic field is calculated based on flux modulation theory. Here, only the $|Z_s - Z_r| - P_a$ pole pair harmonic is considered:

$$B_a(\theta, t) = \frac{\mu_0}{g}\frac{\lambda_{s1}\lambda_{r1}}{4}F_{a1}\cos[(Z_s - Z_r - P_a)\theta + (\omega_a + Z_r\Omega_r)t] \tag{4.78}$$

Generally, the synchronous speed of an induction machine is defined as the rotor speed at which the rotation of the armature magnetic field synchronizes with that of the self-short-circuit winding, meaning there is no coupling between the two. Since the self-short-circuit winding of the asynchronous induction machine is stationary, the synchronous speed should be defined as follows:

$$\Omega_{r0} = -\frac{\omega_a}{Z_r} \tag{4.79}$$

This definition aligns with the synchronous speed of the brushless doubly-fed reluctance machine as stated in Eq. (3.39). When the rotor speed deviates from the synchronous speed, the two windings can interact to produce electromagnetic torque. Considering that the asynchronous induction machine differs primarily in the method of excitation current generation from the brushless doubly-fed reluctance machine, their torque expressions are similar. However, a significant difference lies in the uncontrollable current phase of the self-short-circuit winding \gamma, as indicated below:

$$T_{avg} = \frac{9}{4\pi}r_g l_{stk}\frac{\mu_0}{g}(N_{sa}N_{se})(I_{a1}I_{e1})\lambda_{s1}\lambda_{r1}\frac{Z_r}{P_eP_a}k_{waP_a}k_{weP_e}\cos\gamma \tag{4.80}$$

Compared to the flux modulation induction machine with a single-layer modulation unit, the asynchronous induction machine employs a double-layer modulation unit. Consequently, the first denominator in the torque expression changes from 2π to 4π in Eq. (4.80). Additionally, the asynchronous induction machine shares the pole ratio of $Z_r/(P_aP_e)$, identical to other FMMs with stator excitation winding. The asynchronous induction machine can also be analyzed using the "T-shape" equivalent circuit. Due to the effects of bilateral slotting, the mutual inductance M_{ae} between the armature winding and self-short-circuit winding of the asynchronous induction machine is distinct from that in the flux modulation induction machine:

$$M_{ae} = \frac{3\mu_0 r_g l_{stk}N_{sa}N_{se}k_{waP_a}k_{weP_e}\lambda_{s1}\lambda_{r1}}{2\pi g P_aP_e} \tag{4.81}$$

It can be observed that in Eq. (4.81), the denominator changes from π to 2π, and the stator-specific permeance λ_{s1} appears due to the double-layer modulation unit. Furthermore, as the stator slot number and rotor slot number increase to enhance the pole ratio, λ_{s1} and λ_{r1} decrease. As a result, the mutual inductance M_{ae} is much smaller than in the flux modulation induction machine. Therefore, the terms $(L_a - M'_{ae})$ and $(L'_e - M'_{ae})$ in the equivalent circuit, as shown in Fig. 4.12, increase, thus reducing the torque density. To mitigate this reduction, the asynchronous induction machine often incorporates series capacitors in the rotor winding to compensate for the impedance and enhance the machine's torque, making it comparable to that of the conventional induction machine.

4.4 Pulsating Torque

Pulsating torque is a crucial performance metric for electrical machines, affecting speed stability, positioning accuracy, dynamic control performance, and more. Currently, permanent magnet machines are extensively utilized in high-demand sectors such as aerospace, servo processing, automotive drive, and rail transportation, which require high torque density and minimal pulsating torque. In contrast, electrically excited machines are predominantly used in large power generation and home appliances, where the requirements for pulsating torque are relatively lower. This section concentrates on analyzing the pulsating torque characteristics of three types of permanent magnet FMMs: the vernier permanent magnet machine, the flux switching permanent magnet machine, and the flux reversal permanent magnet machine. Other FMMs are not discussed in this section as their pulsating torque can be similarly analyzed. Section 4.1.1 introduces three torque components of FMMs, each capable of producing pulsating torque, and considers them separately as follows.

4.4.1 Excitation Torque Pulsation

According to Eq. (4.8), the excitation torque pulsation can be calculated from the no-load back-EMF.

1. **Vernier permanent magnet machine**

Equation (4.33) describes the back-EMF of the vernier permanent magnet machine, from which the hth back-EMF magnitude can be determined:

$$E_{ah} = r_g l_{stk} \frac{\mu_0}{g} \Omega_e N_{sa} F_{eh} \left[2\lambda_{s0} k_{wa(hP_e)} + \sum_{l=1}^{\infty} \lambda_{sl} \left(\frac{hP_e}{hP_e + lP_f} k_{wa(hP_e+lP_f)} + \frac{hP_e}{|hP_e - lP_f|} k_{wa|hP_e-lP_f|} \right) \right] \tag{4.82}$$

It can be observed that the hth back-EMF harmonic is produced by the hth MMF harmonic, so h can only be odd. Additionally, since multiples of the 3rd harmonic do not contribute, only the $(6m \pm 1)$th harmonic needs consideration. As the pole pair number of the spatial magnetic field produced by the hth back-EMF harmonic is $|hP_e \pm lP_f|$, all these harmonics are tooth harmonics. Consequently, the winding factors have the same absolute values, and the magnitude of the back-EMF harmonic in Eq. (4.82) depends solely on the specific permeance and pole ratio. When $h = 1$, the fundamental back-EMF E_{a1} is primarily composed of the non-modulated component and the modulated component with $l = 1$, featuring pole ratios $P_e/|P_e - P_f|$ and $P_e/(P_e + P_f)$ respectively. For higher harmonic orders ($h \geq 5$), the corresponding pole ratios of the modulated components are $hP_e/|hP_e - P_f|$ and $hP_e/(hP_e + P_f)$ respectively. Since h is at least five, the values of these pole ratios approach one, significantly lower than that of the fundamental non-modulated component. Therefore, with increasing harmonic order h, the modulated back-EMF component constitutes a much smaller portion of the resultant back-EMF harmonic. To maximize the pole ratio, $h = l$ can be selected, with its pole ratio equal to that of the fundamental back-EMF, i.e., P_e/P_a. However, due to the high order of specific permeance, λ_{sl} and the corresponding modulated back-EMF component are negligibly small. Consequently, the components in the back-EMF harmonic differ significantly from those of the fundamental back-EMF. In other words, for back-EMF harmonics, the non-modulated component dominates, and the modulated component is minimal.

The torque ripple index (TRI) is employed to swiftly predict the excitation torque pulsation of the vernier permanent magnet machine with different pole ratios. The TRI represents the ratio of the back-EMF harmonic to the fundamental back-EMF, excluding the impact of the permanent magnet MMF. According to Eq. (4.82), assuming that the phases of different back-EMF components are identical, the TRI expression for the vernier permanent magnet machine can be deduced as [16–18]:

$$\mathrm{TRI} = \frac{\left[\lambda_{s0} + \frac{\lambda_{sl}}{2}\left(\frac{hP_e}{hP_e + P_f} + \frac{hP_e}{|hP_e - P_f|}\right)\right]k_{wa(hP_e)}}{\left[\lambda_{s0} + \frac{\lambda_{s1}}{2}\left(\frac{P_e}{P_e + P_f} + \frac{P_e}{P_a}\right)\right]k_{waP_a}} \tag{4.83}$$

Using Eq. (4.8), the excitation torque pulsation of the vernier permanent magnet machine can be reformulated as:

$$T_{\text{exc-rip}} = \sum_{h=6m\pm1}^{\infty} \mathrm{TRI}\frac{F_{eh}}{F_{e1}} \cos(h \mp 1)\omega_e t \times 100\% \tag{4.84}$$

According to the previous analysis, the fundamental back-EMF E_{a1} of the vernier permanent magnet machine comprises two main components: the non-modulated and modulated components. In contrast, the harmonic back-EMF E_{ah} primarily consists of the non-modulated component. As a result, the torque ripple index (TRI) of the vernier permanent magnet machine is significantly lower than that of conventional permanent magnet machines. With an increase in the pole ratio, the modulated component of the fundamental back-EMF enhances, leading to a decrease in TRI. Figure 4.13 illustrates the variations of TRI under different pole ratios, showing that the larger the pole ratio P_e/P_a, the smaller the TRI of the vernier permanent magnet machine, which translates to reduced excitation torque pulsation.

Furthermore, based on Eq. (4.82), the excitation torque pulsation is composed of multiple components, each characterized by a basic model. The hth permanent magnet MMF harmonic interacts with the lth specific permeance to produce spatial magnetic field harmonics with $|hP_e \pm lP_f|$ pole pairs. This interaction corresponds to the basic model illustrated in Fig. 4.3b, where the excitation pole pair number is hP_e, the modulation unit pole pair number is lP_f, and the armature pole pair number is $|hP_e \pm lP_f|$. When the excitation unit rotates, the corresponding electrical frequency is $h\omega_a$. If the armature is fed with currents of the same frequency, average torque can be generated. However, since the actual frequency of the armature current is ω_a,

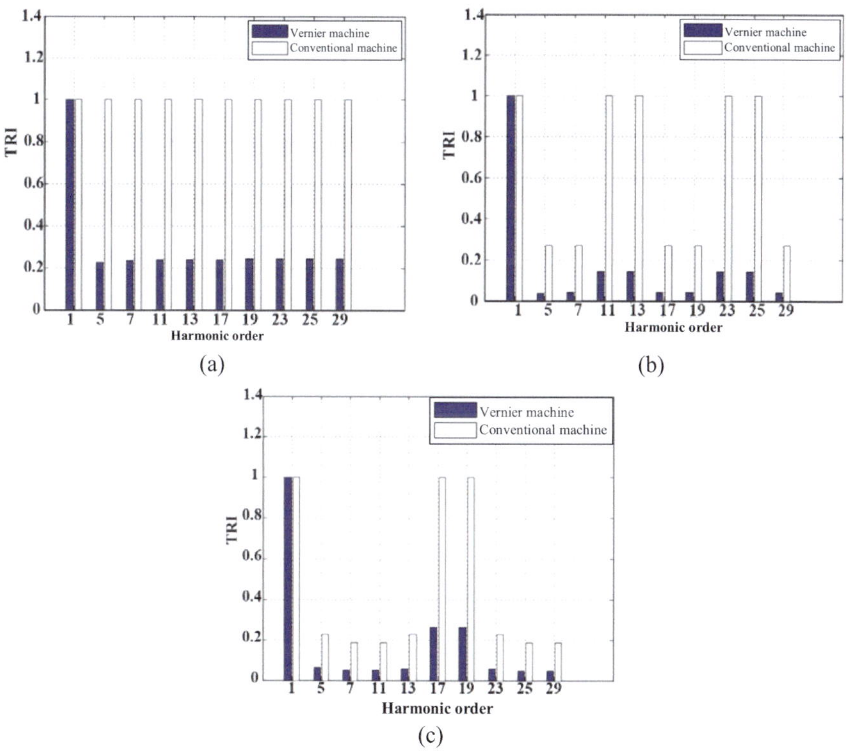

Fig. 4.13 Comparisons of TRI: **a** vernier permanent magnet machine with a pole ratio of 5 and a conventional 12-slot 4-pole permanent magnet machine, **b** vernier permanent magnet machine with a pole ratio of 11 and a conventional 12-slot 2-pole permanent magnet machine, and **c** vernier permanent magnet machine with a pole ratio of 17 and a conventional 18-slot 2-pole permanent magnet machine

these spatial magnetic field harmonics with a frequency of $h\omega_a$ can only produce pulsating torque.

2. **Flux reversal permanent magnet machine**

Equation (4.40) provides the expression for the back-EMF of the flux reversal permanent magnet machine, which is used to derive the lth harmonic magnitude:

$$E_{al} = r_g l_{stk} \frac{\mu_0}{g} \Omega_r N_{sa} \lambda_{rl} \left\{ \sum_{h=1}^{\infty} F_{eh} \left[\frac{lZ_r}{h\frac{Z_s}{2} + lZ_r} k_{wa(h\frac{Z_s}{2}+lZ_r)} - \frac{lZ_r}{\left|h\frac{Z_s}{2} - lZ_r\right|} k_{wa\left|h\frac{Z_s}{2}-lZ_r\right|} \right] \right\} \tag{4.85}$$

In contrast to the vernier permanent magnet machine, the flux reversal permanent magnet machine exhibits a richer array of excitation MMF harmonics F_{eh} due to the uneven arrangement of permanent magnets. Consequently, Eq. (4.85) displays more back-EMF harmonics characterized by both large pole ratios and high excitation MMF magnitudes. For instance, with a slot-pole combination of $Z_s = 12$, $P_a = 2$, and $Z_r = 16$, the pole ratio of the fundamental magnetic field is eight. Additionally, the excitation MMF includes a significant 5th harmonic, capable of interacting with the 2nd harmonic of the rotor-specific permeance, generating a 2-pole-pair air gap magnetic field. This harmonic corresponds to a pole ratio of 16, twice that of the fundamental magnetic field, resulting in pronounced back-EMF harmonics and torque pulsations.

3. **Flux switching permanent magnet machine**

In Eq. (4.46), rotor specific permeance harmonics with order $l_r > 1$ generate back-EMF harmonics and contribute to excitation torque pulsations. Given the similarity between the flux switching permanent magnet machine and the flux reversal permanent magnet machine, where excitation MMF harmonics are richer compared to those in the vernier permanent magnet machine, the flux switching permanent magnet machine also experiences substantial excitation torque pulsations.

4.4.2 *Cogging Torque*

The cogging torque can be considered as the change in magnetic field energy with respect to rotor rotation under no-load conditions. According to the law of energy conservation, the expression can be written as:

$$T_{cog} = -\frac{\partial W_m}{\partial \theta_m} \tag{4.86}$$

where θ_m represents the relative position of the stator and rotor, and W_m is the magnetic field energy. It can be expressed as:

$$W_{\mathrm{m}} = \int\limits_{V} \frac{B_{\mathrm{e}}^2}{2\mu_0}\mathrm{d}V = \frac{\mu_0}{4g^2}\left(R_{\mathrm{og}}^2 - R_{\mathrm{ig}}^2\right)l_{\mathrm{stk}} \int\limits_{0}^{2\pi} F_{\mathrm{e}}^2(\theta, \theta_{\mathrm{m}})\lambda_{\mathrm{f}}^2(\theta, \theta_{\mathrm{m}})\mathrm{d}\theta \tag{4.87}$$

The cogging torque for each FMM can be obtained by substituting their excitation MMF and specific permeance into Eq. (4.87).

1. **Surface-mounted vernier permanent magnet machine**

The surface-mounted vernier permanent magnet machine has a structure similar to the conventional permanent magnet machine, except for the slot-pole combination [19, 20]. Therefore, they almost have the same expression for cogging torque. By using $\mathcal{F}_e = F_e^2$, the $\mathcal{F}_e$ can be expressed as a function with a period of $2P_{\mathrm{e}}$, i.e.,

$$\mathcal{F}_{\mathrm{e}}(\theta, \theta_{\mathrm{m}}) = \mathcal{F}_{\mathrm{e}0} + \sum_{h=1}^{\infty} \mathcal{F}_{\mathrm{e}h} \cos 2hP_{\mathrm{e}}(\theta - \theta_{\mathrm{m}}) \tag{4.88}$$

where the coefficients can be expressed as:

$$\mathcal{F}_{\mathrm{e}0} = \frac{P_{\mathrm{e}}}{\pi}\int\limits_{0}^{\frac{\pi}{P_{\mathrm{e}}}} \mathcal{F}_{\mathrm{e}}(\theta, 0)\mathrm{d}\theta = \frac{\alpha_{\mathrm{m}}}{\pi}\left(\frac{B_{\mathrm{r}}h_{\mathrm{m}}}{\mu_0\mu_{\mathrm{r}}}\right)^2$$

$$\mathcal{F}_{\mathrm{e}h} = \frac{2P_{\mathrm{e}}}{\pi}\int\limits_{0}^{\frac{\pi}{P_{\mathrm{e}}}} \mathcal{F}_{\mathrm{e}}(\theta, 0)\cos(2P_{\mathrm{e}}h\theta)\mathrm{d}\theta = \frac{2}{h\pi}\left(\frac{B_{\mathrm{r}}h_{\mathrm{m}}}{\mu_0\mu_{\mathrm{r}}}\right)^2 \sin h\alpha_{\mathrm{m}}\pi \tag{4.89}$$

Similarly, setting $\Lambda_{\mathrm{s}} = {\lambda_{\mathrm{s}}}^2$, Λ_{s} can be written as:

$$\Lambda_{\mathrm{s}}(\theta) = \Lambda_{\mathrm{s}0} + \sum_{l=1}^{\infty} \Lambda_{\mathrm{s}l} \cos lZ_{\mathrm{s}}\theta$$

$$\Lambda_{\mathrm{s}0} = \lambda_{\mathrm{s}0}^2 + \frac{1}{2}\sum_{m=1}^{\infty} \lambda_{\mathrm{s}m}^2 \quad \Lambda_{\mathrm{s}l} = 2\lambda_{\mathrm{s}0}\lambda_{\mathrm{s}l} + \sum_{m\pm n=l} \lambda_{\mathrm{s}m}\lambda_{\mathrm{s}n} \tag{4.90}$$

In Eq. (4.87), $\mathcal{F}_{\mathrm{e}h}$ and $\Lambda_{\mathrm{s}l}$ need to share the same pole pair number to ensure that the integration result is not zero. Additionally, since the differentiation of θ_{m} is performed in Eq. (4.86), the constant component of W_{m} does not contribute to the cogging torque expression. By taking these considerations into account, the cogging torque expression can be formulated as follows:

$$T_{\mathrm{cog}}(\theta_{\mathrm{m}}) = \frac{\pi\mu_0 n_{\mathrm{c}}}{4g_{\mathrm{ef}}^2}\left(R_{\mathrm{og}}^2 - R_{\mathrm{ig}}^2\right)l_{\mathrm{stk}} \sum_{h=1}^{\infty} h\mathcal{F}_{\mathrm{e}\frac{n_{\mathrm{c}}}{2P_{\mathrm{e}}}h}\Lambda_{\mathrm{s}\frac{n_{\mathrm{c}}}{Z_{\mathrm{s}}}h} \sin hn_{\mathrm{c}}\theta_{\mathrm{m}} \tag{4.91}$$

where n_c represents the least common multiple of Z_s and $2P_e$. According to Eq. (4.91), the cogging torque for the vernier permanent magnet machine consists of an infinite number of components. However, since both the coefficients $\mathcal{F}_{e[(n_c/2P_e)]h}$ and $\Lambda_{s(nc/Zs)h}$ decrease sharply with the increase of order h, the cogging torque is primarily generated by the fundamental components of $\mathcal{F}_{e[(n_c/2P_e)]h}$ and $\Lambda_{s(nc/Zs)h}$, when $h = 1$. The influential parameters on cogging torque are further discussed below.

(1) In terms of expression, the cogging torque is proportional to n_c.
(2) The cogging torque is proportional to the coefficients $\mathcal{F}_{e[(n_c/2P_e)]h}$ and $\Lambda_{s(nc/Zs)h}$, while the magnitude of these coefficients is inversely proportional to their orders $n_c/2P_e$ and n_c/Z_s, respectively.
(3) The cogging torque is also related to the pole arc of permanent magnets α_m. If $\alpha_m = 1$, $\mathcal{F}_{eh} = 0$ in Eq. (4.89). However, due to the pole leakage flux of permanent magnets, the magnetic field at the edge of the permanent magnets is weakened, resulting in a lower calculated pole arc than the actual pole arc and an inability to completely eliminate the cogging torque.
(4) The cogging torque is also influenced by dimensional parameters, among which the equivalent air gap length g_{ef} is a key parameter. The g_{ef} not only appears in the coefficient of Eq. (4.91) but also affects the value of specific permeance Λ_{sl}. Generally speaking, the larger the parameter t_s/g_{ef}, the higher the value of λ_{sl} will be, thus introducing a higher cogging torque.

Based on the above analysis, for vernier permanent magnet machines with similar size and air gap, the goodness factor k_{cog} is defined to qualitatively predict the magnitude of the cogging torque under different combinations of stator slots and rotor poles. The goodness factor is the ratio of the product of the stator slot number (Z_s) and the rotor pole number ($2P_e$) to the least common multiple of the two, i.e.,

$$k_{cog} = n_c \times \frac{2P_e}{n_c} \times \frac{Z_s}{n_c} = \frac{2P_e Z_s}{n_c} \tag{4.92}$$

A smaller goodness factor correlates with a lower cogging torque. Higher pole ratios typically result in smaller goodness factors, indicating reduced cogging torque. For example, consider a 12-stator-slot 22-rotor-pole vernier permanent magnet machine, a 24-stator-slot 20-rotor-pole fractional slot concentrated winding permanent magnet synchronous machine, and a 60-stator-slot 20-rotor-pole integer-slot distributed-winding permanent magnet synchronous machine. Their goodness factors are 2, 4, and 20, respectively. Consequently, the 12-stator-slot 22-rotor-pole vernier permanent magnet machine exhibits the lowest goodness factor and cogging torque. Table 4.2 compares the finite element simulation results of cogging torque and torque ripple under different pole ratios. The model with a pole ratio of 11 demonstrates significantly smaller cogging torque and torque ripple than the model with a pole ratio of 5, aligning with the previous theoretical analysis.

2. **Flux reversal permanent magnet machine**

Table 4.2 Comparison of vernier permanent magnet machines with different pole ratios

Parameters	12-stator-slot 20-rotor-pole model	12-stator-slot 22-rotor-pole model
Pole ratio	5	11
Stator outer diameter (mm)	124	
Axial stack length (mm)	80	70
Phase current (A)	8.48	8
Number of series turns per phase	150	112
Average torque (N m)	20.3	20.8
Torque ripple (%)	2.8	0.4
Cogging torque (mN m)	60	22

The flux reversal permanent magnet machine differs from the vernier permanent magnet machine in that its modulation unit rotates, while the excitation unit remains stationary. However, this rotational distinction does not significantly impact the cogging torque expression. By substituting $P_e = Z_s/2$ and $P_f = Z_r$ into Eq. (4.91), the cogging torque of the flux reversal permanent magnet machine can be determined as:

$$T_{\text{cog}}(\theta_m) = \frac{\pi \mu_0 n_c}{4 g_{\text{ef}}^2}\left(R_{\text{og}}^2 - R_{\text{ig}}^2\right) l_{\text{stk}} \sum_{h=1}^{\infty} h \mathcal{F}_{\text{e}\frac{N_c}{Z_s}h} \Lambda_{\text{r}\frac{N_c}{Z_r}h} \sin h n_c \theta_m \tag{4.93}$$

where n_c denotes the least common multiple of Z_s and Z_r. Compared to the vernier permanent magnet machine, the flux reversal permanent magnet machine features uneven permanent magnet arrangements, resulting in a richer harmonic excitation MMF. The $\mathcal{F}_{eh}$ can be obtained as:

$$\begin{cases} \mathcal{F}_{\text{e}0} = \left(1 - \frac{b_s}{t_s}\right)\left(\frac{B_r h_m}{\mu_0 \mu_r}\right)^2 \\ \mathcal{F}_{\text{e}h} = \frac{2}{h\pi}\left(\frac{B_r h_m}{\mu_0 \mu_r}\right)^2 \sin h\left(1 - \frac{b_s}{t_s}\right)\pi \end{cases} \tag{4.94}$$

Flux reversal permanent magnet machines often have a large stator slot opening ratio (b_s/t_s) to enhance torque density, and according to Eq. (4.29), $\mathcal{F}_{eh}$ is substantially increased. Consequently, the cogging torque of the flux reversal permanent magnet machine is generally higher than that of the vernier permanent magnet machine with the same pole ratio.

3. **Interior FMMs**

Interior vernier permanent magnet machines, flux switching permanent magnet machines, and doubly-salient permanent magnet machines are all categorized as interior FMMs. Two cogging torque-related issues concerning these machines need to be addressed:

(1) Do interior permanent magnets increase cogging torque compared to FMMs with surface-mounted permanent magnets?
(2) Do interior FMMs exhibit lower cogging torque compared to conventional interior permanent magnet synchronous machines?

To answer these questions, it is essential to understand specific characteristics of interior FMMs:

(1) The excitation MMFs of interior FMMs are characterized by square waves with positive and negative symmetry, similar to surface-mounted FMMs. However, unlike surface-mounted FMMs that are single-side salient, resulting in fewer specific permeance harmonics, interior FMMs are double-side salient. This double-sided saliency leads to more specific permeance harmonics and, consequently, more cogging torque components.
(2) Interior FMMs have a smaller equivalent air gap length which increases the magnitude of high-order specific permeance harmonics, further enhancing the cogging torque.
(3) Interior FMMs feature additional excitation MMFs that influence both the no-load magnetic field and cogging torque.

While the first two characteristics of interior FMMs typically increase cogging torque, the impact of the third characteristic requires further investigation. The additional excitation MMF not only affects harmonic magnitude but also influences harmonic period. Taking the flux switching permanent magnet machine as an example, the air gap flux density can be expressed as:

$$B_{\mathrm{e}}(\theta, t) = \frac{\mu_0}{g} F_{\mathrm{e}}(\theta)\lambda_{\mathrm{s}}(\theta)\lambda_{\mathrm{r}}(\theta, t) \tag{4.95}$$

The square of the air gap flux density can be computed by squaring both sides of Eq. (4.95). After squaring, F_{e}^2, λ_{s}^2, and λ_{r}^2 have numerous harmonics, and their interactions result in cogging torque. By calculating F_{e}^2 can refer to the content of additional MMF of spoke-magnetization permanent magnets in Sect. 2.2.5. Given that the magnetic potential of each ferromagnetic pole alternates Z_{r} times periodically with the rotation of the rotor, and the position of each ferromagnetic pole is also periodic, it can be derived that:

$$\mathcal{F}_{\mathrm{s},n}(\theta_{\mathrm{m}}) = \mathcal{F}_{\mathrm{s}0} + \sum_{h=1}^{\infty} \mathcal{F}_{\mathrm{s}h} \cos\left[hZ_{\mathrm{r}}\left(\theta_{\mathrm{m}} - \theta_{\mathrm{s},n}\right)\right] \tag{4.96}$$

where θ_{m} is the relative position of the stator and rotor, $\theta_{\mathrm{s},n} = 2n\pi/Z_{\mathrm{s}}$, $\mathcal{F}_{\mathrm{s},n}$ is the square of the magnetic potential of the nth ferromagnetic pole, $\mathcal{F}_{\mathrm{s}0}$ and $\mathcal{F}_{\mathrm{s}h}$ are the DC term and the hth harmonic term of the Fourier series, respectively. Their detailed calculation method can be referred to in Sect. 2.2.5. Using Eq. (4.96), the square of the excitation MMF can be expressed as:

$$\mathcal{F}_e = \mathcal{F}_{s0} + \sum_{h,h'}^{\infty} \mathcal{F}_{sh} \frac{\sin\left(\frac{h'Z_s+hZ_r}{Z_s}\right)}{\frac{h'Z_s+hZ_r}{Z_s}} \cos\left[\left(h'Z_s + hZ_r\right)\theta - hZ_r\theta_m\right]$$
$$+ \sum_{h,h'}^{\infty} \mathcal{F}_{sh} \frac{\sin\left(\frac{h'Z_s-hZ_r}{Z_s}\right)}{\frac{h'Z_s-hZ_r}{Z_s}} \cos\left[\left(h'Z_s - hZ_r\right)\theta + hZ_r\theta_m\right] \quad (4.97)$$

Equation (4.97) consists of two components: the square of excitation MMF excluding the additional MMF (first component) and the square of additional MMF (second component). When multiplied by $\lambda_s{}^2$ and $\lambda_r{}^2$, with harmonic pole pairs of $2l_sZ_s$ and l_rZ_r, respectively, only the time-varying term with a pole pair number of zero can produce cogging torque. If only the first term is considered, the period of cogging torque is n_c, which is the least common multiple of $2Z_s$ and Z_r. If the additional MMF is also considered, the pole pair numbers should satisfy the following equation:

$$h'Z_s + \mathrm{sgn}_1 hZ_r + 2\mathrm{sgn}_2 l_sZ_s + \mathrm{sgn}_3 l_rZ_r = 0 \quad (4.98)$$

where sgn_i $(i = 1, 2, 3)$ can take values of 1 or $-$ 1. Subsequently, the period of cogging torque can be calculated as $|\mathrm{sgn}_1 h + \mathrm{sgn}_3 l_r|Z_r$, which can be further simplified as the least common multiple of Z_s and Z_r. Consequently, compared to the situation when the additional MMF is excluded, the period of cogging torque is halved. Thus, the additional MMF introduces low-order cogging torque components, resulting in higher cogging torque for interior FMMs compared to surface-mounted FMMs.

4.4.3 Reluctance Torque Pulsation

Rotors for FMMs can be classified into three types: surface-mounted permanent magnet rotor, interior permanent magnet rotor, and salient rotor. The first type of rotor does not produce any reluctance torque component, while the latter two types both generate reluctance torque. The reluctance torque of the second type of rotor is complex due to the presence of additional MMF. However, it can be considered comparable to the salient rotor for qualitative studies. Hence, in this subsection, only the reluctance torque of the salient rotor is analyzed.

Section 4.2 describes the generation of average torque in FMMs. Most FMMs comprise numerous torque components, each associated with an ideal basic model. Assuming that the stator and rotor frequencies do not align (i.e., the injected current frequency does not match the fundamental frequency), the basic model does not produce average torque but instead generates sinusoidal pulsating torque. Following this concept, the armature MMF harmonic is modulated by the salient rotor, producing a rotating magnetic field in the air gap. This magnetic field can, in turn, interlink with the armature winding. If the frequency of the induced flux linkage in the armature winding matches that of the armature current, average reluctance torque is produced; otherwise, reluctance torque pulsation occurs.

The rotor of FMMs can either be an interior permanent magnet rotor or a salient rotor. Therefore, the reluctance torque generation may correspond to either the single-layer modulation unit model illustrated in Fig. 4.3b or the double-layer modulation unit model shown in Fig. 4.3c. To generate torque, the slot-pole combination of the corresponding basic model must satisfy:

$$l_s Z_s + l_r Z_r + \mathrm{sgn}_a P_{a1} = P_{a2} \tag{4.99}$$

where Z_s and Z_r denote the numbers of stator and rotor teeth, respectively. P_{a1} and P_{a2} represent the pole pair numbers of the two interacting armature MMFs sgn that can take values of 1 or $-$ 1, and l_s and l_r can be any positive or negative integers. When $l_s = 0$, it corresponds to the single-layer modulation unit model; otherwise, it corresponds to the double-layer modulation unit model. l_r cannot be zero; otherwise, the rotor lacks a modulation unit, and no reluctance torque will be generated.

Based on the winding theory, the pole pair number of armature MMF that can be generated by the three-phase armature winding is:

$$P = \begin{cases} 3mn_p + P_a, \ Z_s/n_p \text{ is odd} \\ 6mn_p + P_a, \ Z_s/n_p \text{ is even} \end{cases} \tag{4.100}$$

where m can be any integer (i.e., $m = 0, \pm 1, \pm 2, \ldots$) and n_p is the number of winding periods in one circumference. Equation (4.99) deduces that the frequency of the reluctance torque generated by the corresponding basic model is:

$$\omega = \left(n_r l_r + \mathrm{sgn}_a - 1\right)\omega_a \tag{4.101}$$

where n_r denotes the number of rotor salient poles in one electrical cycle. When $\omega = 0$, the corresponding reluctance torque is the average torque; otherwise, it is the pulsating torque. Next, Eqs. (4.99)–(4.101) are utilized to study the reluctance torque pulsation of the synchronous reluctance machine and the vernier reluctance machine.

In the synchronous reluctance machine, the excitation MMF has P_a pole pairs, and the rotor salient pole number $Z_r = 2P_a$. Substituting into Eq. (4.97), one can get $l_s = 0$, $l_r = 1$, and $\mathrm{sgn}_a = -1$. Based on Eq. (4.99), it can be found that the above combination can only produce average torque when $n_r = 2$. It means that the period when the rotor rotates by two salient poles corresponds to one electrical cycle, which is consistent with the relationship between the frequency and speed of the synchronous reluctance machine. In the vernier reluctance machine, if $Z_s > Z_r$, one can have $Z_s - Z_r = 2P_a$. Then, substituting into Eq. (4.101), it can be determined that $n_r = -2$. It means that the rotor must rotate in the opposite direction, and similarly, the period when the rotor rotates by two salient poles corresponds to one electrical cycle.

The reluctance torque of the basic model corresponding to Eq. (4.99) can be calculated as:

$$T_{\text{rel}} = \begin{cases} \frac{18}{\pi}\frac{\mu_0}{g_{\text{ef}}}\lambda_{\text{s}0}\lambda_{\text{r}|l_{\text{r}}|}N_{\text{sa}}^2 k_{\text{wa}P_{\text{a1}}}k_{\text{wa}P_{\text{a2}}}I_{\text{a1}}^2\frac{|l_{\text{r}}|Z_{\text{r}}}{P_{\text{a1}}P_{\text{a2}}}\cos(n_{\text{r}}l_{\text{r}} + \text{sgn}_{\text{a}} - 1)\omega_{\text{a}}t, \ l_{\text{s}} = 0 \\ \frac{9}{\pi}\frac{\mu_0}{g_{\text{ef}}}\lambda_{\text{s}|l_{\text{s}}|}\lambda_{\text{r}|l_{\text{r}}|}N_{\text{sa}}^2 k_{\text{wa}P_{\text{a1}}}k_{\text{wa}P_{\text{a2}}}I_{\text{a1}}^2\frac{|l_{\text{r}}|Z_{\text{r}}}{P_{\text{a1}}P_{\text{a2}}}\cos(n_{\text{r}}l_{\text{r}} + \text{sgn}_{\text{a}} - 1)\omega_{\text{a}}t, \ l_{\text{s}} \neq 0 \end{cases} \quad (4.102)$$

It can be observed that the reluctance torque is inversely proportional to the pole pair number of the two MMFs (i.e., P_{a1} and P_{a2}) and the equivalent air gap length g_{ef}. Additionally, as the order $|l_{\text{s}}|$ and $|l_{\text{r}}|$ increase, the corresponding specific permeances decrease. Hence, the reluctance torque of each basic model mainly depends on the values of P_{a1}, P_{a2}, and l_{s}, while the increase of l_{r} and the decrease of $\lambda_{\text{r}|l\text{r}|}$ cancel each other, and hence, the relationship with the reluctance torque is not significant. Similar to the cogging torque, a reluctance torque factor can be defined using Eq. (4.102):

$$k_{\text{rel}} = \begin{cases} \frac{2}{|P_{\text{a1}}P_{\text{a2}}|}, \ l_{\text{s}} = 0 \\ \frac{1}{|l_{\text{s}}P_{\text{a1}}P_{\text{a2}}|}, \ l_{\text{s}} \neq 0 \end{cases} \quad (4.103)$$

For machines of similar size and air gap length, the larger the k_{rel} is, the higher the reluctance torque pulsation will be. Therefore, the magnitude of reluctance torque pulsation depends on the product of P_{a1}, P_{a2}, and l_{s}, which is closely related to the slot-pole combination. To illustrate this, we will use the interior vernier permanent magnet machine as an example and analyze reluctance torque pulsation using Eqs. (4.99)–(4.103).

The pole pair number of the equivalent specific permeance for the interior vernier permanent magnet machine is $Z_{\text{r}} = 2P_{\text{r}}$. When the winding pole pair number is chosen as $P_{\text{a}} = Z_{\text{s}} - P_{\text{r}}$, the armature MMF pole pair number is $P = (6m + 1)P_{\text{a}}$ if an integer-slot overlapping winding is adopted. Moreover, $n_{\text{r}} = -2$ in Eq. (4.101). Substituting the above data into Eq. (4.97) yields the following results:

(1) When $\text{sgn}_{\text{a}} = 1$, Eq. (4.99) can be simplified as follows:

$$6(m_1 - m_2)(Z_{\text{s}} - P_{\text{r}}) - l_{\text{s}}Z_{\text{s}} = 2l_{\text{r}}P_{\text{r}} \quad (4.104)$$

According to the pole ratio definition of the vernier permanent magnet machine, we can derive:

$$\begin{aligned} Z_{\text{s}} &= (1 + \text{PR})P_{\text{a}} \\ P_{\text{r}} &= \text{PR}P_{\text{a}} \end{aligned} \quad (4.105)$$

Upon substituting the above into Eq. (4.104):

$$m_1 - m_2 - \frac{\text{PR} + 1}{6}l_{\text{s}} = \frac{l_{\text{r}}}{3}\text{PR} \quad (4.106)$$

When PR is an integer, due to the constraint of the slot-pole combination, (PR + 1)/3 must be a positive integer, which is denoted as z. By substituting it into Eq. (4.104):

$$2(m_1 - m_2) - zl_s = \frac{2l_r}{3}(3z - 1) \tag{4.107}$$

Regardless of the value of z, the values of $|m_1|$, $|m_2|$, and $|l_s|$ will not exceed two, thus leading to a small k_{rel} and a low reluctance torque pulsation.

(2) Using the same approach, Eq. (4.97) can be simplified for $sgn_a = -1$:

$$6(m_1 + m_2)(6z + 1) = 6zl_r + 3zl_s - 2l_r - 2 \tag{4.108}$$

When $z = 1$, implying a corresponding pole ratio of 2, there exists a solution of Eq. (4.108): $m_1 = m_2 = 0$, $l_s = -2$, $l_r = 2$, which will result in a large k_{rel} and thus a high reluctance torque pulsation. However, when $z \geq 2$, i.e., PR ≥ 5, no solution exists where $|m_1|$, $|m_2|$, $|l_s|$ can exceed 2, so the reluctance torque pulsation remains small.

Generally, for vernier permanent magnet machines with pole ratios exceeding five and employing integer-slot overlapping windings, the reluctance torque pulsations are negligible, even with an interior permanent magnet structure. However, for vernier permanent magnet machines with low pole ratios and fractional slot concentrated windings or auxiliary-tooth structures, it is necessary to estimate the reluctance torque pulsations using Eq. (4.99) to determine if pulsation components are generated by both low-order and high-magnitude MMF and specific permeance.

Compared to the interior vernier permanent magnet machine, the flux reversal permanent magnet machine provides a variety of armature winding types and exhibits more MMF harmonics. Nevertheless, the reluctance torque pulsations are generally negligible due to its large equivalent air gap length. For the flux switching permanent magnet machine, when employing fractional slot concentrated winding, the armature MMF harmonics are abundant, and the equivalent air gap length is small. Consequently, the reluctance torque pulsation is closely tied to the selection of slot-pole combinations and cannot be generalized.

4.5 Power Factor

In Sect. 4.2.3, the power factor of the basic model of FMM is analyzed. It is concluded that the effective excitation field of the basic model is much lower than that of a conventional permanent magnet machine, resulting in a significantly lower power factor. Actual FMMs consist of multiple basic models, and the armature winding includes both harmonic leakage inductance and slot leakage inductance. Therefore, the power factor of actual FMMs requires further study. The slot leakage inductance of FMMs is typically lower than that of the conventional permanent magnet machine due to the larger slot opening ratio. Moreover, the harmonic leakage inductance of FMMs differs from that of the conventional permanent magnet machine because of the flux modulation effect. The characteristics of harmonic leakage inductances are discussed in detail below.

In FMMs, there is not only the non-modulated armature magnetic field generated by the armature MMF but also the modulated armature magnetic field produced by the flux modulation effect. Both fields can interlink with the armature winding, making the harmonic leakage inductance components particularly complex. Similar to the generation of reluctance torque pulsation, the generation of harmonic inductance can be viewed as an interaction between two armature MMFs. Thus, the mutual coupling can also be analyzed using Eq. (4.99). The key difference is that this interaction does not necessarily produce torque; that is, the l_r in Eq. (4.99) can be zero. Using winding theory, the inductance magnitude resulting from the mutual coupling of the P_{a1} and P_{a2} pole pair MMF can be calculated as follows:

$$L_{P_{a1}P_{a2}} = \begin{cases} \frac{6}{\pi}\frac{\mu_0}{g}N_{as}^2\frac{k_{waP_{a1}}k_{waP_{a2}}}{P_{a1}P_{a2}}r_g l_{stk}\lambda_{s0}\lambda_{r0}, & l_s = l_r = 0 \\ \frac{3}{\pi}\frac{\mu_0}{g}N_{as}^2\frac{k_{waP_{a1}}k_{waP_{a2}}}{P_{a1}P_{a2}}r_g l_{stk}\lambda_{s|l_s|}\lambda_{r|l_r|}, & \text{either } l_s \text{ or } l_r \text{ is } 0 \\ \frac{3}{2\pi}\frac{\mu_0}{g}N_{as}^2\frac{k_{waP_{a1}}k_{waP_{a2}}}{P_{a1}P_{a2}}r_g l_{stk}\lambda_{s|l_s|}\lambda_{r|l_r|}, & l_s \neq 0, l_r \neq 0 \end{cases} \tag{4.109}$$

In Eq. (4.109), the first line represents the inductance generated by the non-modulated MMF components. In this scenario, P_{a1} must equal P_{a2}, and this harmonic leakage inductance remains a DC value. The second line indicates the harmonic leakage inductance produced by the single-layer modulation unit, where P_{a1} and P_{a2} do not necessarily have to be equal. When only the stator is slotted, this inductance remains a DC value. Conversely, when only the rotor is slotted, the inductance could manifest as either a DC value or a pulsating value. The third line denotes the harmonic leakage inductance generated by the double-layer modulation unit, which can also fluctuate between being a DC value or a pulsating value. Moreover, beyond harmonic leakage inductance, the DC component of the q-axis inductance is a crucial parameter for calculating the power factor. Therefore, subsequent sections will explore inductance components and power factor considerations for various FMM types.

4.5.1 Vernier Permanent Magnet Machine

In a vernier permanent magnet machine with a surface-mounted permanent magnet structure, when the pole ratio (PR) is greater than or equal to 5, the inductance does not include a pulsating term. According to Eq. (4.109), the magnetizing inductance is generated by the interaction between the P_a-pole-pair fundamental MMF and the DC term of the specific permeance. The harmonic leakage inductance is produced by interactions between the fundamental and harmonic MMFs or between two harmonic

MMFs. Given that the pole pair number of harmonic MMF is at least $5P_a$, the fundamental and harmonic MMFs interact through specific permeance harmonics. Moreover, the large equivalent air gap length of the vernier permanent magnet machine ensures that the fundamental specific permeance $\lambda_{s1} \leq 0.25$, and the higher order harmonics of specific permeance are even smaller. Additionally, a coefficient of 0.5 is introduced by the flux modulation effect. Consequently, the harmonic leakage inductance generated by interactions between the fundamental and harmonic MMFs amounts to approximately 3% of the magnetizing inductance. Similarly, the harmonic leakage inductance generated by the interaction between two harmonic MMFs is only $1/5^2 = 4\%$ of the magnetizing inductance. Furthermore, the slot leakage inductance of the vernier permanent magnet machine is also minimal. Overall, in high pole ratio vernier permanent magnet machines, the total leakage inductance accounts for up to 20% of the magnetizing inductance, and thus, the leakage inductances are often neglected in the calculation of the power factor. Based on this assumption and Eqs. (4.21)–(4.23) of the basic model, the ratio of the excitation flux linkage to armature flux linkage in the vernier permanent magnet machine is [21, 22]:

$$\frac{\psi_{e1}}{\psi_{a1}} \approx \frac{\frac{F_{e1}\lambda_{s0}}{P_e} + \frac{F_{e1}\lambda_{s1}}{2P_a}}{\frac{F_{a1}\lambda_{s0}}{P_a}} = \frac{F_{e1}}{F_{a1}}\left(\frac{1}{PR} + \frac{\lambda_{s1}}{2\lambda_{s0}}\right) \tag{4.110}$$

For high pole ratios, the coefficients within the parentheses on the right-hand side of Eq. (4.110) are less than one. The higher the pole ratio, the lower the coefficient. Quantitatively, for a conventional fractional slot concentrated winding permanent magnet machine, ψ_{e1}/ψ_{a1} is about three under natural cooling conditions. Substituting this value into Eq. (4.9) yields a power factor of 0.95, indicating that the conventional permanent magnet machine has a very high power factor and can fully utilize the inverter. For the vernier permanent magnet machine with a pole ratio of five, the value within the parentheses on the right side of Eq. (4.110) is approximately $(1/5 + 0.3/2) = 0.35$, so ψ_{e1}/ψ_{a1} is reduced to 1.05, and the power factor decreases to only 0.7. When the pole ratio increases to 11, the power factor further declines to about 0.5.

The low power factor of the vernier permanent magnet machine can be explained as follows: Firstly, the P_e pole pair excitation unit generates a magnetic field with P_e pole pairs of high magnitude and another magnetic field with P_a pole pairs of lower magnitude. Both configurations produce no-load flux linkages, and after considering the pole ratio and flux density magnitude, the flux linkages generated by the two configurations are nearly equal. In contrast, for the conventional permanent magnet machine with a P_a pole pair excitation unit, its no-load flux linkage is primarily generated by a P_a pole pair magnetic field. According to flux modulation theory, the P_a pole pair magnetic field of the conventional permanent magnet machine is approximately six times that of the vernier permanent magnet machine. Consequently, the no-load flux linkage of the vernier permanent magnet machine is inferior to that of the conventional permanent magnet machine with the same armature winding pole pair (P_a), leading to a significant decrease in power factor.

4.5.2 Flux Reversal Permanent Magnet Machine

When the pole ratio of the flux reversal permanent magnet machine is high, its armature inductance primarily consists of the magnetizing inductance, similar to the vernier permanent magnet machine. The key distinction is the absence of a non-modulated no-load flux linkage component in the flux reversal permanent magnet machine. Therefore, ψ_{e1}/ψ_{a1} can be represented as:

$$\frac{\psi_{e1}}{\psi_{a1}} \approx \frac{\frac{F_{eh}\lambda_{r1}}{2P_a}}{\frac{F_{a1}\lambda_{r0}}{P_a}} = \frac{F_{eh}}{F_{a1}}\frac{\lambda_{r1}}{2\lambda_{r0}} \tag{4.111}$$

where F_{eh} is the excitation MMF with the highest pole ratio.

Compared to the vernier permanent magnet machine, the power factor of the flux reversal permanent magnet machine is further reduced due to the lack of non-modulated working harmonics. Additionally, in Eq. (4.111), it is observed that the power factor of the flux reversal permanent magnet machine is also related to $\lambda_{r1}/\lambda_{r0}$, which is influenced by the pole ratio. However, this relationship between the power factor and the pole ratio is not as significant as that of the vernier permanent magnet machine, which is directly reflected in Eq. (4.110).

4.5.3 Flux Switching Permanent Magnet Machine

Compared to the flux reversal permanent magnet machine, the flux switching permanent magnet machine has a smaller equivalent air gap length, resulting in larger values of $\lambda_{s1}/\lambda_{s0}$ and $\lambda_{r1}/\lambda_{r0}$, which are close to one. Therefore, the power factor decrease due to the specific permeance is mitigated. In contrast to the conventional interior spoke-magnetization permanent magnet synchronous machine, the flux switching permanent magnet machine exhibits a lower power factor primarily due to the coefficient of 1/2 or 1/4 introduced by the flux modulation effect.

4.5.4 Electrically Excited Flux Modulation Machine

Contrary to permanent magnet machines, electrically excited machines have a weaker excitation capacity. Generally, $F_{e1}/F_{a1} \approx 1$, which is only 1/3 of that of permanent magnet excitations. Consequently, conventional electrically excited synchronous machines and synchronous reluctance machines have low power factors and require additional regulation of the current angle to improve the power factor. For electrically excited FMMs, the power factor is further reduced due to the high pole ratio and low specific permeance magnitudes.

In summary, all FMMs suffer from the defect of a low power factor. It is important to note that this defect not only leads to increased inverter capacity, system weight, volume, and cost but also makes FMMs prone to saturation and exhibit low overload torque capability at high current density.

4.6 Summary of This Chapter

In this chapter, based on flux modulation theory, three main electromagnetic characteristics of various types of FMMs are analyzed including torque density, torque pulsation, and power factor.

1. For the most crucial performance index, torque density, analytical expressions for the average torque of various FMMs were derived. Most FMMs exhibited multiple working harmonics, functioning as integrations of multiple basic models, each with a corresponding torque amplification factor (i.e., pole ratio). This resulted in multiple torque components for FMMs. The chapter also examined the torque characteristics of different FMMs separately:
   (a) For the vernier permanent magnet machine with the combination of $P_f = P_e + P_a$, the average torque primarily consisted of two components: non-modulated and modulated. The former approximated the torque of the conventional permanent magnet machine, while the latter could even exceed the non-modulated component at high pole ratios, significantly increasing torque density.
   (b) Transverse-flux permanent magnet machines, theoretically having an infinite pole ratio, exhibit the highest torque density.
   (c) Flux reversal and flux switching permanent magnet machines, with low pole ratios and fractional slot concentrated windings, these machines featured rich torque components. At high pole ratios, the recommended integer-slot overlapping winding predominantly contributed to the torque. Although lacking non-modulated components, their torque density was lower than that of the vernier permanent magnet machine but higher than that of conventional permanent magnet machines. Their simple and robust rotor structure, devoid of windings or permanent magnets, made them suitable for applications requiring high reliability or low rotor cost.
   (d) For electrically excited vernier machines, excitation capability was inversely proportional to the pole pair number of the excitation unit. The torque expression was not related to the pole ratio. For electrically excited vernier reluctance machines, the torque expression was proportional to the pole ratio of $Z_r/(P_eP_a)$. However, due to the use of fractional slot concentrated winding, the pole ratio value was small, resulting in output torques inferior to those of electrically excited synchronous and synchronous reluctance machines.

(e) Brushless doubly-fed reluctance machines and vernier reluctance machines, with overlapping windings and much lower pole pair numbers, these machines exhibited a very high torque amplification factor (i.e., pole ratio), leading to superior torque densities compared to synchronous reluctance machines.
(f) Flux modulation induction and asynchronous induction machines had lower mutual inductances between armature and excitation windings than their self inductances, implying lower torque density than conventional induction machines. The asynchronous induction machine's simple rotor structure allowed for the convenient use of series resistance or capacitor methods for torque regulation.

2. Torque pulsations in FMMs are categorized into three types: excitation torque pulsation, cogging torque, and reluctance torque pulsation. Detailed analyses for vernier permanent magnet, flux reversal permanent magnet, and flux switching permanent magnet machines are provided:

(a) Excitation torque pulsation depends on the ratio of back-EMF harmonics to the fundamental wave. Vernier permanent magnet machines exhibit low excitation torque pulsations due to a lower pole ratio amplification factor of back-EMF harmonics compared to the fundamental wave. Conversely, flux reversal and flux switching permanent magnet machines, characterized by lower fundamental back-EMFs and richer back-EMF harmonics, exhibit higher excitation torque pulsations.
(b) Cogging torque depends on the harmonic contents of the squared excitation MMF, squared specific permeance, and the goodness factor. Surface-mounted vernier permanent magnet machines, with low harmonic contents and a small goodness factor, experience minimal cogging torque. Moreover, flux reversal and flux switching permanent magnet machines, despite having small goodness factors, possess rich harmonic contents, resulting in larger cogging torque.
(c) Reluctance torque pulsation is closely related to the slot-pole combinations of FMMs. For vernier permanent magnet machines, the reluctance torque pulsation is negligible with surface-mounted permanent magnets in the excitation unit and minimal with interior permanent magnets. Flux reversal permanent magnet machines exhibit negligible reluctance torque pulsation due to a large equivalent air gap length. However, flux switching permanent magnet machines may demonstrate significant reluctance torque pulsation with specific slot-pole combinations.

For surface-mounted vernier permanent magnet machines, the reluctance torque is essentially non-existent, and both excitation torque pulsation and cogging torque are minimal, resulting in low overall torque pulsation. In contrast, flux reversal and flux switching permanent magnet machines experience relatively high torque pulsations, which are sensitive to slot-pole combinations. This sensitivity necessitates careful consideration during the design phase to optimize performance and minimize undesirable torque pulsations.

3. The vernier permanent magnet machine exhibits a low power factor, influenced by factors such as the pole ratio, the 1/2 coefficient introduced by the flux modulation process, and the low fundamental specific permeance. However, compared to other FMMs, vernier permanent magnet machines have a relatively higher power factor due to the presence of both modulated and non-modulated components of no-load flux linkage. Under natural cooling conditions, the power factor of the vernier permanent magnet machine typically ranges from 0.5 to 0.7, while the power factors of flux reversal and flux switching permanent magnet machines are even lower. Electrically excited FMMs, constrained by their excitation capability, experience a further reduction in power factor.
4. In summary, the vernier permanent magnet machine, transverse-flux permanent magnet machine, vernier reluctance machine, and brushless doubly-fed reluctance machine demonstrate clear advantages in average torque, making them suitable for applications with stringent torque density requirements. Flux reversal and flux switching permanent magnet machines, electrically excited vernier reluctance machines, and asynchronous induction machines do not exhibit distinct advantages in torque density and torque pulsation compared to conventional machines. However, their simpler rotor structures, devoid of permanent magnets and windings, render them promising for applications requiring high reliability or low rotor cost, such as urban railway transit, electromagnetic launch, and renewable power generation.

References

1. Zhang P, Sizov G, Demerdash N (2011) Comparison of torque ripple minimization control techniques in surface-mounted permanent magnet synchronous machines. In: IEEE International electric machines & drives conference (IEMDC), Niagara Falls. IEEE, pp 188–193
2. Raziee SM, Misir O, Ponick B (2017) IEEE Trans Magn 53(10):1
3. Li DW, Qu RH, Li J et al (2016) IEEE Trans Ind Appl 52(1):125
4. Ren X, Li DW, Qu RH et al (2020) IEEE Trans Ind Electron 67(8):6248
5. Wu LL, Qu RH, Li DW et al (2015) IEEE Trans Ind Appl 51(5):3707
6. Li DW (2015) Research on flux modulation permanent magnet machine (in Chinese). Huazhong University of Science and Technology, Wuhan
7. Zou TJ (2018) Research on theory and topologies of high-performance flux modulation permanent magnet machines (in Chinese). Huazhong University of Science and Technology, Wuhan
8. Li DW, Gao YT, Qu RH et al (2018) IEEE Trans Ind Appl 54(1):172
9. Gao YT, Qu RH, Li DW et al (2017) IEEE Trans Ind Appl 53(3):2110
10. Gao YT (2017) Theoretical analysis and topology investigation of flux reversal permanent magnet machines (in Chinese). Huazhong University of Science and Technology, Wuhan
11. Li DW, Qu RH, Li J et al (2016) IEEE Trans Energy Convers 31(1):106
12. Lin MX, Li DW, Ren X et al (2021) IEEE Trans Ind Electron 68(5):3707
13. Jia SF, Qu RH, Li J (2015) IEEE Trans Magn 51(11):1

14. Xie KF (2020) Permanent magnet machines theory and topology study based on electromagnetic superimposition (in Chinese). Huazhong University of Science and Technology Press, Wuhan
15. Li HM, Li QF (1982) Electromagnetic reduction machine (in Chinese). Machinery Industry Press, Beijing
16. Zhu L, Jiang SZ, Zhu ZQ et al (2009) IEEE Trans Magn 45(4):2023
17. Fang L, Li DW, Shi CJ et al (2022) IEEE J Emerg Sel Top Power Electron 10(2):1766
18. Shi CJ, Qu RH, Li DW et al (2021) IEEE Trans Ind Electron 68(6):4748
19. Shi CJ, Qu RH, Gao YT et al (2018) IEEE Trans Magn 54(11):1
20. Li R, Shi CJ, Qu RH et al (2021) IEEE Trans Magn 57(2):1
21. Zhao Y, Li DW, Ren X et al (2022) IEEE J Emerg Sel Top Power Electron 10(3):2934
22. Yu ZX, Kong WB, Li DW et al (2019) IEEE Trans Ind Appl 55(5):4643

Chapter 5
High Performance Flux Modulation Machine Topologies

The FMM has attracted considerable interest as a focal point in research due to its torque amplification properties. One line of inquiry within the FMM field focuses on increasing its output torque. To address this issue, scholars have proposed numerous innovative structural topologies for the FMM. This chapter explores the technical concepts and methodologies aimed at enhancing the performance of the FMM, based on flux modulation theory. Additionally, the FMM topologies associated with each performance enhancement strategy will be analyzed. Although it is not feasible to extensively cover all the new FMM topologies in this chapter, their core concepts will be clarified.

5.1 Torque Boosting Concept for Flux Modulation Machines

Chapter 3 provided an overview of the fundamental topologies and operational principles of various types of FMMs. A clear pattern emerges, indicating the presence of multiple working magnetic field harmonics within the air gap of most FMMs. For example, in the vernier permanent magnet machine, there exists a non-modulated magnetic field with P_e pole pairs and a modulated magnetic field of $P_f - P_e$, both producing tooth harmonics at the same electrical frequencies. These harmonics can induce a fundamental back-EMF in the armature winding, which generates the average electromagnetic torque. Consequently, the vernier permanent magnet machine can be viewed as a combination of two basic models: a standard permanent magnet machine and a single-side/single-layer FMM. Furthermore, the armature winding for both models is characterized as an ideal winding, capable of linking solely with a single working magnetic field. Consequently, the output torque of the vernier permanent magnet machine can be determined by summing the torque contributions of the standard permanent magnet machine and the single-side/single-layer

R. Qu et al., *Flux Modulation Machine*, https://doi.org/10.1007/978-981-95-4746-3_5

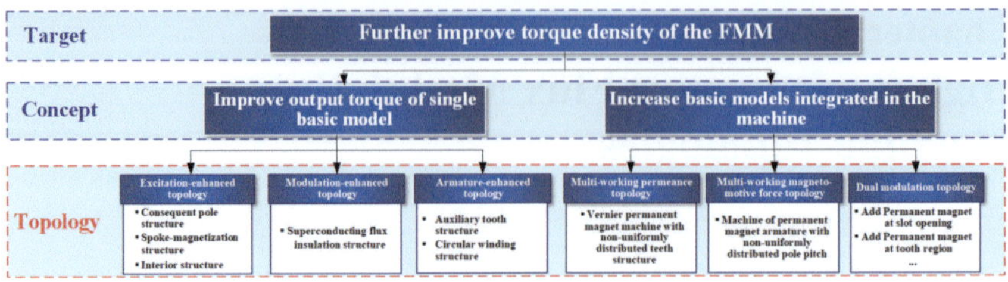

Fig. 5.1 Concepts and methods for improving the torque density of FMM

FMM, respectively. Similar characteristics apply to the flux switching permanent magnet machine and the flux reversal machine.

Based on the analysis above, two technical strategies are suggested to further enhance the torque density of the FMM. First, increasing the output torque of one of the basic models is proposed to improve the overall torque density of the machine. Second, an alternative approach involves generating additional working magnetic fields in the air gap—effectively integrating more working magnetic fields, which can also contribute to improving the machine's overall torque density [1–3].

Both concepts will be explored in detail in this subsection. Concerning the first concept, each basic model consists of three units—excitation, modulation, and armature units, as previously mentioned. To increase the machine's torque density, the performance of at least one of these units must be enhanced. This first performance improvement concept is further divided into three specific methods: enhancing excitation capability, improving modulation effects, and strengthening the armature's current-carrying capacity. For the second concept, it is crucial to construct new working MMF or new working specific permeance to expand the number of basic models. The machine is designated as the multi-working MMF FMM when only a new working MMF is introduced, the multi-working permeance FMM when only a new working specific permeance is introduced, and the hybrid FMM when both new working MMF and new working specific permeance are incorporated into the FMM. In summary, the strategies for improving the torque density of the FMM are outlined in Fig. 5.1.

5.2 Excitation-Enhanced Flux Modulation Machines

With ongoing enhancements in permanent magnet performance, permanent magnet machines have seen extensive adoption in recent years. Consequently, several innovative topologies for permanent magnet machines have been proposed. In the excitation unit, beyond the surface-mounted topology, internal structures such as the consequent-pole topology and spoke-magnetization topology have gradually been developed. These novel arrangements of permanent magnets aim to enhance excitation capability. The FMM, particularly the vernier permanent magnet machine, features a similar structure to the regular permanent magnet machine. Consequently, numerous scholars have sought to apply these new excitation topologies to the FMM, achieving significant results.

5.2.1 Consequent-Pole Flux Modulation Machine

Equations (2.36) and (2.51) calculate the no-load air gap flux density for the surface-mounted permanent magnet machine and the consequent-pole permanent magnet machine, respectively. When the pole-arc coefficient of the machine equals 1, the waveforms of no-load air gap flux density for both topologies are square waves with positive and negative symmetry. The ratio of their corresponding amplitudes is given by:

$$\frac{B_{\mathrm{em\_CP}}}{B_{\mathrm{em\_SPM}}} = \frac{1 + \mu_{\mathrm{r}}\frac{g}{h_{\mathrm{m}}}}{1 + \mu_{\mathrm{r}}\frac{2g}{h_{\mathrm{m}}}} \tag{5.1}$$

where $B_{\mathrm{em\_SPM}}$ represents the amplitude of the surface-mounted permanent magnet machine's no-load air gap flux density, while $B_{\mathrm{em\_CP}}$ denotes that of the consequent-pole permanent magnet machine. It is evident that the right side of Eq. (5.1) yields a result between 0.5 and 1. Consequently, for the consequent-pole structure, half of the permanent magnet material is conserved, leading to a reduction in the level of the no-load magnetic field. However, this level can still be maintained at more than half of that observed in the surface-mounted permanent magnet machine. For instance, assuming a relative permeability $\mu_{\mathrm{r}} = 1$, and the thickness of the permanent magnet is four times the air gap length (i.e., $g/h_{\mathrm{m}} = 0.25$), the result of Eq. (5.1) is 5/6. Under these conditions, even though the consequent-pole permanent magnet machine is larger and heavier, it uses less permanent magnet material and is less costly, while delivering the same output torque. Based on this analysis, it is clear that the consequent-pole permanent magnet machine does not fit into the category of excitation-enhanced topology. Additionally, as described in Sect. 2.2.5, the consequent-pole permanent magnet FMM generates additional rotor MMF, which can diminish the air gap's working magnetic field. This raises the question of whether the performance of such a topology is inferior to that of the surface-mounted permanent magnet machine.

It is important to note that the preceding analysis does not compare the two topologies under equivalent conditions. As depicted in Fig. 5.2, the analysis of the surface-mounted permanent magnet FMM in Sect. 2.2.3 is conducted considering the equivalent air gap formed by the permanent magnet and air gap, designating plane 1 as the equivalent plane for the rotor's magnetic potential. Conversely, the analysis of the consequent-pole permanent magnet FMM's additional rotor MMF in Sect. 2.2.5 utilizes plane 2 as the magnetic potential of the rotor. Due to these differences in selected air gaps, substantial variations exist not only in their MMFs but also in their specific air gap permeances. From the standpoint of MMF, the surface-mounted

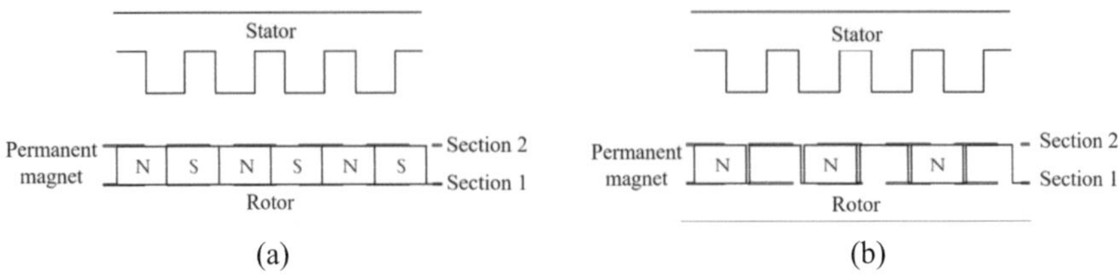

Fig. 5.2 Plane of analysis for a vernier permanent magnet machine. **a** Surface-mounted permanent magnet vernier machine. **b** Consequent-pole permanent magnet vernier machine

topology, which does not generate additional rotor MMF, is ideal for performance improvement. However, in terms of specific permeance, the equivalent air gap of the consequent-pole topology is advantageous for flux modulation.

Overall, the performance of the two topologies cannot be directly compared unless the MMF or specific permeance of both topologies is standardized. However, selecting a plane with the same magnetic potential for both topologies poses a challenge, and in practice, it is only feasible to choose a plane with the same specific permeance. Assuming that plane 1 in Fig. 5.2 is selected as the plane of analysis for both topologies, the MMF of the permanent magnet becomes a square wave with positive and negative symmetry for both topologies. The permanent magnet and the entire air gap are then integrated as the electromagnetic air gap. In this scenario, the specific permeance function from the MMF/specific permeance model is the specific permeance of the electromagnetic air gap. However, for the consequent-pole topology, part of the magnetic pole is replaced with an equipotential iron core, and the air gap magnetic field on the iron core cannot be directly determined using flux modulation theory. Therefore, only plane 2 can be selected as the plane of analysis for both topologies.

The flux modulation effect of the consequent-pole topology is discussed in Sect. 2.2.5, where one of the most significant conclusions is highlighted: with the rotation of the permanent magnet, the slot-tooth position it faces continually changes, resulting in the variation of the magnetic circuit's reluctance. This variation introduces additional MMF at the surface of the permanent magnet, consequently weakening the air gap magnetic field. For the permanent magnet in the surface-mounted topology, the reluctance of the air gap it faces also continuously alternates. Therefore, additional MMF must be introduced at the surface of the permanent magnet (i.e., plane 2). In this scenario, if only the additional MMF introduced by the N-poles is considered, there appears to be no difference between the surface-mounted and consequent-pole topologies. Further analysis is required to ascertain whether the additional MMF introduced by S-poles enhances or weakens such an effect.

Figure 5.3 depicts the additional MMF on the surface of permanent magnets caused by stator slotting in a surface-mounted permanent magnet machine. When an N-pole permanent magnet is fully aligned with a stator tooth, the magnetic potential on its surface is minimally affected by the slotting, resulting in the minimum additional MMF (with a phase position of 180°). Conversely, the adjacent S-pole

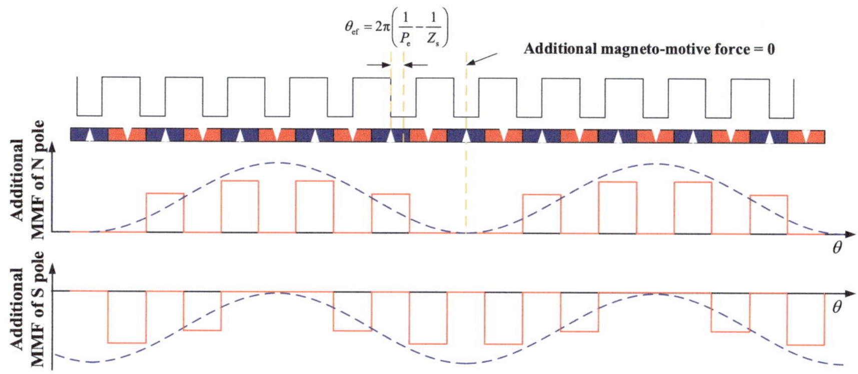

Fig. 5.3 Additional MMF on the surface of the permanent magnet in a surface-mounted vernier permanent magnet machine

permanent magnet faces a stator slot, which results in the maximum additional MMF. Moreover, as the direction of magnetization is opposite between the S-pole and N-pole permanent magnets, the phase position of their additional MMF aligns accordingly.

Based on the analysis provided, let us assume that the outer surface of the permanent magnets (plane 2) is selected as the analysis plane. In this configuration, permanent magnets of different polarities enhance the additional MMF for the surface-mounted permanent magnet FMM. In contrast, for the consequent-pole permanent magnet FMM, the S-pole is interconnected with an iron core of equipotentiality, which inhibits the generation of additional MMF. Consequently, the consequent-pole permanent magnet FMM exhibits lower additional MMF. Referring to Eq. (2.66), despite the relatively small portion of the regular excitation F_{Pe}, F_{Pa} experiences a significant reduction, leading to an increase in its working flux density. A finite element analysis is used to compare the no-load flux densities of the surface-mounted permanent magnet vernier machine (SPMVM) and the consequent-pole permanent magnet vernier machine (CPPMVM), as illustrated in Fig. 5.4. In this simulation model, the working flux density with a low pole number in the consequent-pole topology is 1.2 times that of the surface-mounted topology. However, for the working flux density with a high pole number, the decrease in F_{Pa} does not compensate for the decrease in F_{Pe}. Consequently, B_{Pa} of the consequent-pole topology is only 0.7 times that of the surface-mounted topology.

Figure 5.5a compares the no-load back-EMF of the consequent-pole vernier permanent magnet machine and the surface-mounted vernier permanent magnet machine. It is apparent that the working magnetic field with a low pole number significantly amplifies the pole ratio, consequently enhancing the performance of the consequent-pole machine. As a result, the back-EMF value is 1.1 times that of the surface-mounted vernier permanent magnet machine. Additionally, the back-EMF can be further increased by raising the pole-arc coefficient of the permanent

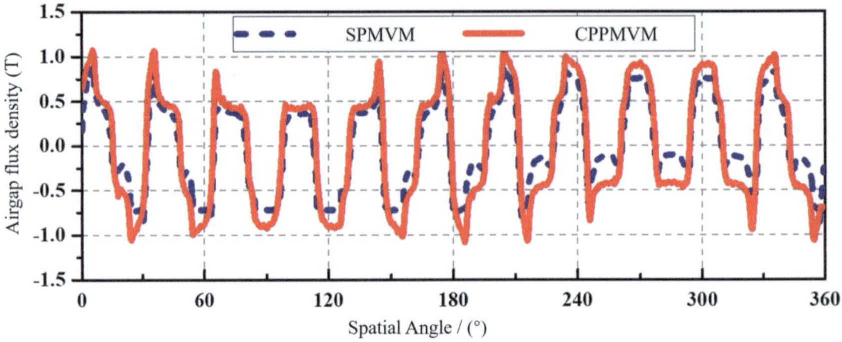

(a)

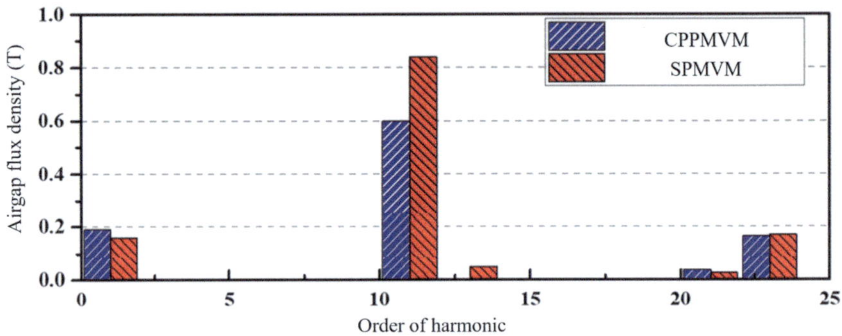

(b)

Fig. 5.4 Comparison between the consequent-pole topology's air gap magnetic field and surface-mounted topology's air gap magnetic field. **a** Waveform. **b** Spectrum

magnet. For instance, when the pole-arc coefficient is set at 1.32, the no-load back-EMF of the consequent-pole vernier permanent magnet machine can be increased by 30%. Figure 5.5a also examines the output torque of both machines, showing a similar pattern of variation with the back-EMF. In summary, the consequent-pole can be considered an excitation-enhanced topology in the vernier permanent magnet machine. However, this topology improves the machine's performance by reducing the non-working MMF rather than by increasing the working magnetic motive force [4, 5].

Similarly, according to Sect. 3.4, the permanent magnet flux reversal machine typically utilizes the surface-mounted topology, as depicted in Fig. 5.6a. By employing the surface-mounted topology, the consequent-pole permanent magnet flux reversal machine can be constructed by replacing all the permanent magnets of a specific polarity with an iron core, as shown in Fig. 5.6b.

Figures 5.7 and 5.8 compare the no-load back-EMF and the output torque of both topologies. It is observed that the consequent-pole permanent magnet flux

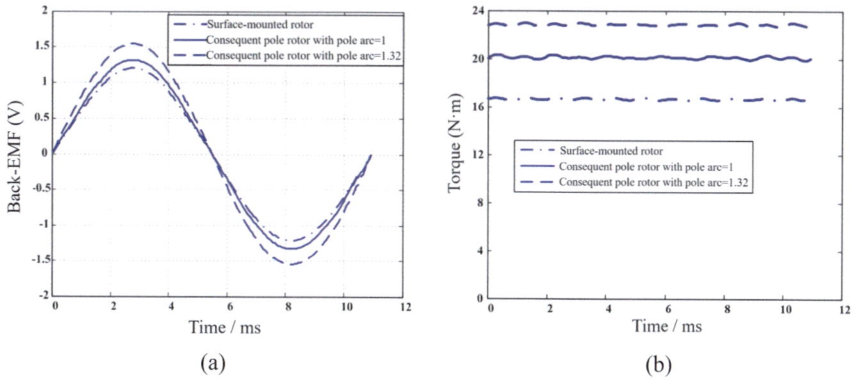

Fig. 5.5 Comparison between the consequent-pole vernier permanent magnet machine's performance and surface-mounted vernier permanent magnet machine's performance. **a** No-load back-EMF. **b** Electromagnetic torque

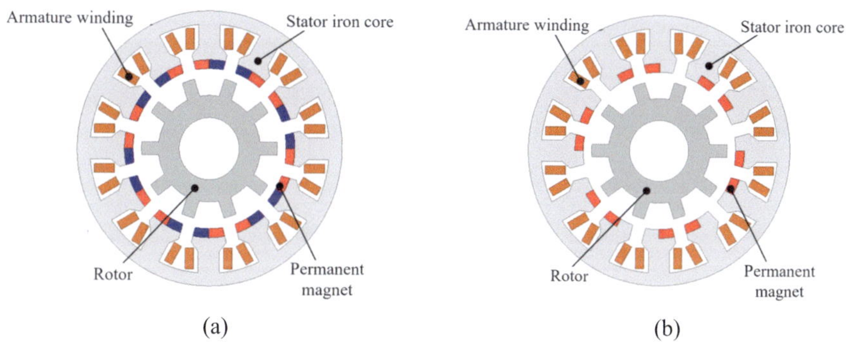

Fig. 5.6 Topologies of permanent magnet flux reversal machine. **a** Surface-mounted permanent magnet flux reversal machine. **b** Consequent-pole permanent magnet flux reversal machine

reversal machine can enhance the back-EMF and torque by 30%, nearly matching the performance improvements seen in the vernier permanent magnet machine [6].

5.2.2 *Spoke-Magnetization Flux Modulation Machine*

In the realm of permanent magnet machines, the spoke-magnetization configuration is notable for its strong excitation capability. This arrangement enhances the no-load magnetic field through a flux-focusing effect, thereby increasing torque density. Consequently, this configuration is also adopted in FMMs. When the permanent magnet is positioned on the rotor side, it constitutes a spoke-magnetization vernier permanent magnet machine, as depicted in Fig. 5.9a. Conversely, positioning the

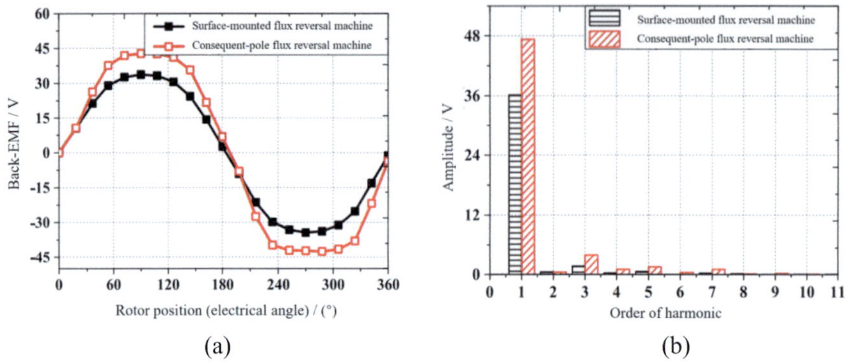

Fig. 5.7 Comparison between the surface-mounted flux reversal machine's no-load back-EMF and consequent-pole flux reversal machine's no-load back-EMF. **a** No-load back-EMF. **b** Electromagnetic torque

Fig. 5.8 Comparison between the surface-mounted flux reversal machine's output torque and consequent-pole flux reversal machine's output torque

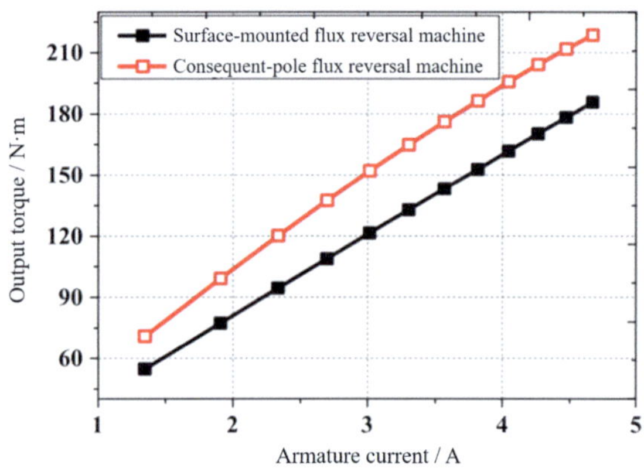

permanent magnet on the stator side results in a permanent magnet flux switching machine, as shown in Fig. 5.9b. Nonetheless, the spoke-magnetization FMM does not meet expected torque levels. Despite some advantages over the surface-mounted structure in vernier permanent magnet machines when the pole ratio is relatively low, the torque density of the spoke-magnetization vernier permanent magnet machine becomes less competitive compared to that of the standard permanent magnet machine as the pole ratio increases. Additionally, the torque advantage of the permanent magnet flux switching machine is minimal compared to the surface-mounted stator permanent magnet machine, i.e., the permanent magnet flux reversal machine. The suboptimal performance of the spoke-magnetization FMM can be explained through the lens of overall magnetic circuit structure and flux modulation theory.

Analyzing the overall magnetic circuit structure, the FMM features multiple working harmonics. Although the working magnetic field with a low pole number

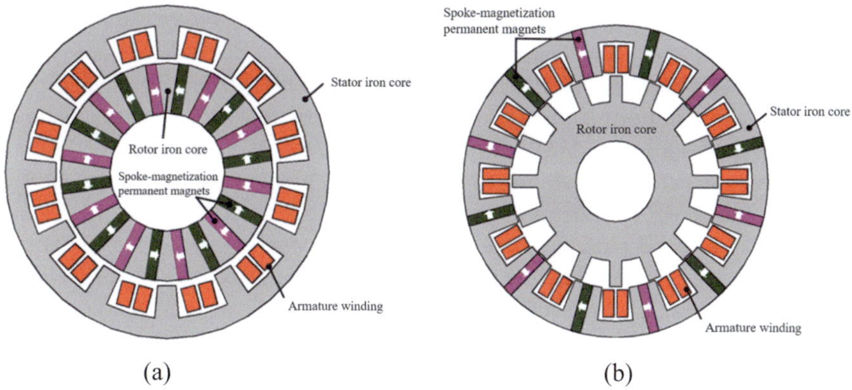

Fig. 5.9 Spoke-magnetization FMMs. **a** Spoke-magnetization permanent magnet vernier machine. **b** Permanent magnet flux switching machine

among them has a low amplitude, their pole ratio amplification coefficient is relatively high, contributing significantly to the no-load back-EMF and electromagnetic torque. However, in the spoke-magnetization structure, the excitation unit's iron core intersects with the permanent magnet. Consequently, the pole pitch of the working magnetic field with a low pole number is comparatively high, necessitating that the magnetic field traverse multiple permanent magnets to complete the circuit, as shown in Fig. 5.10. Given that the permeability of the permanent magnet is similar to that of air, the reluctance of the magnetic circuit for the magnetic field with a low pole number is relatively high. This leads to a significant reduction in magnetic field intensity, a phenomenon known as the magneto-barrier effect. With an increase in the pole ratio, the magnetic field with a low pole number must pass through more

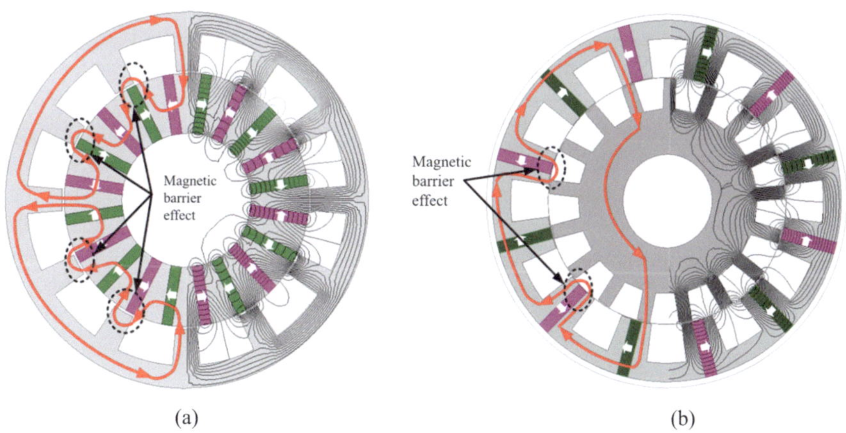

Fig. 5.10 Magnetic barrier effect in FMMs. **a** Spoke-magnetization permanent magnet vernier machine. **b** Permanent magnet flux switching machine

permanent magnets to complete the circuit, intensifying this effect. Consequently, the torque of the spoke-magnetization FMM experiences a marked decrease [7].

In Sect. 2.2.5, the rotor MMF of the spoke-magnetization permanent magnet structure under a non-uniform air gap is analyzed. Similar to the consequent-pole structure, there is an additional MMF of $P_e - P_f$ pole pairs, which weakens the FMM's working magnetic field. According to Eq. (2.75), the higher the pole ratio, the larger the additional MMF, and the more pronounced the effect of weakening [8]. Therefore, the key to enhancing the performance of the spoke-magnetization FMM lies in mitigating the additional MMF, and two technical methods have been developed to achieve this. First, all the ferromagnetic poles with odd numbers are connected together, and the same process is repeated for ferromagnetic poles with even numbers. In this configuration, the ferromagnetic poles are compelled to maintain the same magnetic potential, resulting in what is known as the connected spoke-magnetization FMM. By eliminating the variation of specific permeance described in Eq. (5.2), the additional MMF can be effectively mitigated, leading to the development of a dual stator spoke array permanent magnet vernier machine (DSSAPMVM).

5.2.3 *Connected Spoke-Magnetization Flux Modulation Machine*

Figure 5.11a depicts the topology of the spoke array permanent magnet vernier machine with alternate flux bridges (SAPMVMB). In contrast to the regular spoke-magnetization vernier permanent magnet machine, this topology incorporates a flux bridge and iron yoke beneath every other ferromagnetic pole, hence the designation "alternate" in its name. The inclusion of these additional flux bridges and iron yokes facilitates the construction of a complete magnetic circuit for the working magnetic field with a low pole number, as shown in Fig. 5.11b. This approach significantly reduces the magnetic barrier effect within the machine [9, 10]. Equivalent magnetic circuit of the SAPMVMB is shown in Fig. 5.12.

Compared to Fig. 2.28, two distinct differences are observed in the SAPMVMB. First, the ferromagnetic poles with even numbers are connected to the flux bridges. This connection ensures that these poles share the same magnetic potential, leading to a reduction in F_{Pa}. Under these conditions, the working magnetic field, particularly the field with a low pole number, is significantly enhanced. Additionally, the iron yokes positioned beneath the ferromagnetic poles contribute to flux leakage, represented by R_r. The nodal equation for this magnetic circuit can be derived as follows:

$$2\frac{F_{\mathrm{r},2n} - F_{\mathrm{r}} - F_{\mathrm{mag}}}{R_{\mathrm{mag}}} + \frac{F_{\mathrm{r},2n}}{R_{\mathrm{g},2n}} + \frac{F_{\mathrm{r},2n} - F_{\mathrm{r}}}{R_{\mathrm{r}}} = 0 \tag{5.2}$$

Similar to the methodology described in Sect. 2.2.5, it is assumed that $F_{\mathrm{r,2n}}$ only contains the DC and fundamental terms, which can be expressed as follows:

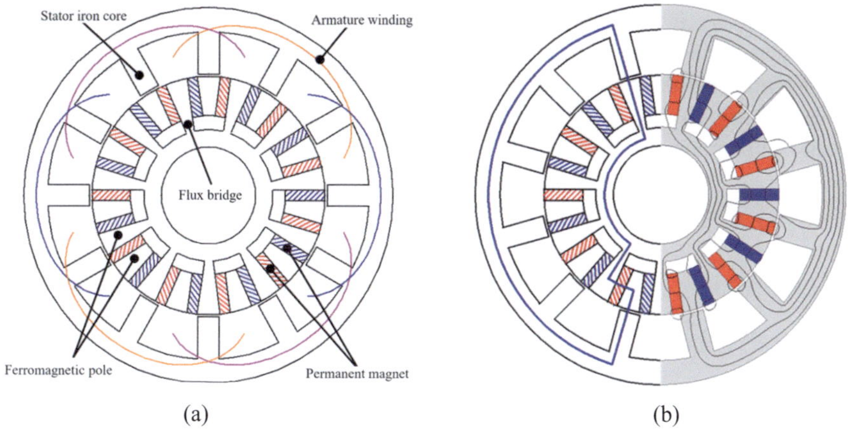

Fig. 5.11 Spoke array permanent magnet vernier machine with alternate flux bridges. **a** Machine topology. **b** Flux line distribution

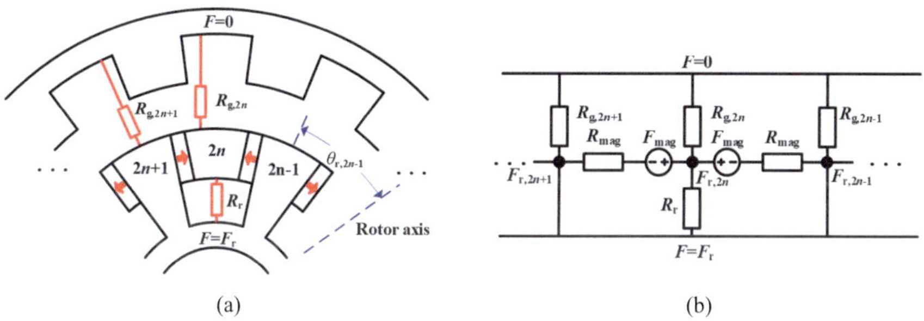

Fig. 5.12 Equivalent magnetic circuit of the SAPMVMB. **a** Formation of magnetic circuit. **b** Structure of magnetic circuit

$$F_{r,2n} = F_{r0} + F_{r1} \cos Z_s\left(\theta_{r,2n} + \theta_m\right) \tag{5.3}$$

Moreover, in accordance with the principle of magnetic flux continuity:

$$\sum_{n=1}^{P_e}\left(\frac{F_{r,2n}}{R_{g,2n}} + \frac{F_r}{R_{g,2n-1}}\right) = 0 \tag{5.4}$$

where the expression for $R_{g,2n}$ can be referred to in Eq. (2.67). By substituting Eq. (5.3) into Eq. (5.4):

$$F_r = -\left(F_{r0} + \frac{\lambda_{lump\_1}}{\lambda_{lump\_0}} F_{r1}\right) \approx -F_{r0} \tag{5.5}$$

In Eq. (5.5), it is assumed that F_{r1} is considerably smaller than F_{r0}, allowing F_r to approximately equal F_{r0}. This simplification introduces minimal error while significantly facilitating the mathematical calculations.

By substituting Eq. (5.5) into Eq. (5.2) and further simplifying, the following result can be obtained:

$$\begin{aligned}&\left[\left(4+2\frac{R_{\text{mag}}}{R_{\text{r}}}\right)F_{\text{r0}}-2F_{\text{mag}}+\frac{\mu_0}{g}r_{\text{g}}l_{\text{stk}}R_{\text{mag}}\left(F_{\text{r0}}\lambda_{\text{lump\_0}}+\frac{F_{\text{r1}}\lambda_{\text{lump\_1}}}{2}\right)\right]\\&+\left\{\left(2+\frac{R_{\text{mag}}}{R_{\text{r}}}\right)F_{\text{r1}}+\frac{\mu_0}{g}r_{\text{g}}l_{\text{stk}}R_{\text{mag}}\left(F_{\text{r1}}\lambda_{\text{lump\_0}}+F_{\text{r0}}\lambda_{\text{lump\_1}}\right)\right\}\cos Z_{\text{s}}\left(\theta_{\text{r},2n}+\theta_{\text{m}}\right)\\&+\frac{\mu_0}{g}r_{\text{g}}l_{\text{stk}}R_{\text{mag}}\frac{F_{\text{r1}}\lambda_{\text{lump\_1}}}{2}\cos 2Z_{\text{s}}\left(\theta_{\text{r},2n}+\theta_{\text{m}}\right)=0\end{aligned} \tag{5.6}$$

Upon neglecting the term with doubled frequency and adopting the solving approach similar to Eq. (2.73):

$$\begin{aligned}F_{\text{r0}}&\approx\frac{2F_{\text{mag}}}{4+2\frac{R_{\text{mag}}}{R_{\text{r}}}+\frac{\mu_0}{g}r_{\text{g}}l_{\text{stk}}R_{\text{mag}}F_{\text{r0}}\lambda_{\text{lump\_0}}}\\F_{\text{r1}}&\approx-\frac{2F_{\text{mag}}}{4+2\frac{R_{\text{mag}}}{R_{\text{r}}}+\frac{\mu_0}{g}r_{\text{g}}l_{\text{stk}}R_{\text{mag}}\lambda_{\text{lump\_0}}}\times\frac{\frac{\mu_0}{g}r_{\text{g}}l_{\text{stk}}R_{\text{mag}}\lambda_{\text{lump\_1}}}{2+\frac{R_{\text{mag}}}{R_{\text{r}}}+\frac{\mu_0}{g}r_{\text{g}}l_{\text{stk}}R_{\text{mag}}\lambda_{\text{lump\_0}}}\end{aligned} \tag{5.7}$$

Comparing Eq. (5.7) with Eq. (2.75), it becomes evident that the presence of leakage reluctance R_r results in a reduction of F_{r0}, with the degree of reduction depending on the ratio R_{mag}/R_r. F_{r0} experiences a pronounced decline when R_r is relatively small (indicating greater flux leakage). Additionally, the expression for F_{r1} omits the term "$\cos(Z_s/P_e\pi)$" in its denominator. This implies that, even at a relatively high pole ratio, F_{r1} will not undergo a steep increase. Essentially, there is no direct correlation between the additional MMF and the pole ratio in this topology.

Based on the magnetic potential of each ferromagnetic block, the rotor's MMF can be derived:

$$\begin{aligned}F_{\text{e}}=&\frac{4F_{\text{r0}}}{\pi}\sum_{h=1,3\ldots}^{\infty}\cos hP_{\text{e}}(\theta-\theta_{\text{m}})\\&+2F_{\text{r1}}\sum_{h=1,3\ldots}^{\infty}\frac{\sin\frac{(hP_{\text{e}}+Z_{\text{s}})\pi}{2P_{\text{e}}}}{\frac{(hP_{\text{e}}+Z_{\text{s}})\pi}{2P_{\text{e}}}}\cos[(hP_{\text{e}}+Z_{\text{s}})\theta-hP_{\text{e}}\theta_{\text{m}}]\\&+2F_{\text{r1}}\sum_{h=1,3\ldots}^{\infty}\frac{\sin\frac{(hP_{\text{e}}-Z_{\text{s}})\pi}{2P_{\text{e}}}}{\frac{(hP_{\text{e}}-Z_{\text{s}})\pi}{2P_{\text{e}}}}\cos[(hP_{\text{e}}-Z_{\text{s}})\theta-hP_{\text{e}}\theta_{\text{m}}]\end{aligned} \tag{5.8}$$

Comparing Eq. (5.8) with Eq. (2.77), the ripple term F_{r1} is present only in half of the ferromagnetic poles' magnetic potential. Moreover, F_{r1} in this topology is significantly smaller compared to that in the SAPMVMB. As a result, this structure substantially diminishes the overall MMF. Therefore, employing the alternate flux

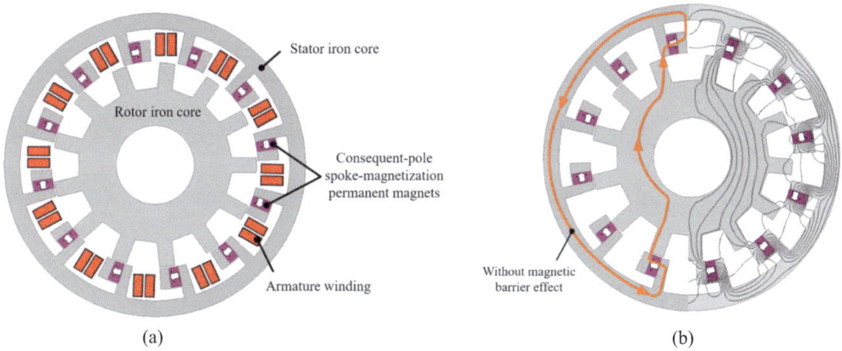

Fig. 5.13 Permanent magnet flux switching machine with alternate flux bridges. **a** Machine topology. **b** Flux line distribution

bridge can significantly enhance the torque performance of the spoke array permanent magnet vernier machine, particularly at high pole ratios. However, it is important to note that using the alternate flux bridge does not consistently lead to performance improvement, especially when the machine's pole ratio is relatively low. In such cases, the additional MMF is relatively low, making the weakening effect of the flux bridge negligible. Furthermore, the introduction of leakage reluctance by the flux bridge reduces F_{r0}, thereby contributing to a reduction in torque.

Given that the permanent magnet flux switching machine is part of the spoke-magnetization structure, it faces similar challenges. Figure 5.13 depicts a permanent magnet flux switching machine with alternate flux bridges (PMFSMB), sharing a similar design concept with SAPMVMB [11]. However, the PMFSMB uses homopolar permanent magnets to accommodate the winding. Additionally, the stator yoke of this machine acts as the flux path for the magnetic field with a low pole number. The PMFSMB demonstrates commendable torque output performance when its pole ratio is high.

While the PMFSMB significantly reduces the additional MMF by connecting the homopolar ferromagnetic poles, a portion of each ferromagnetic pole is still exposed to the air, indicating the persistence of some additional MMF. To further mitigate this effect, the spoke array claw pole permanent magnet vernier machine (SACPPMVM) is proposed, as illustrated in Fig. 5.14 [12]. The stator structure of this machine resembles that of the regular vernier permanent magnet machine, with only the rotor displayed in the figure for clarity. Each ferromagnetic pole's half side is connected to a flux bridge extending from one side of the rotor claw disc, while the opposite side connects to a flux bridge extending from the other side of the rotor claw disc.

Figure 5.15 illustrates the equivalent magnetic circuit of the SACPPMVM. The diagram indicates that since the ferromagnetic poles with the same polarity are interconnected, only two levels of magnetic potential, F_{r+} and F_{r-}, exist in the ferromagnetic poles. Adhering to the principle of flux continuity, the following relationship can be established:

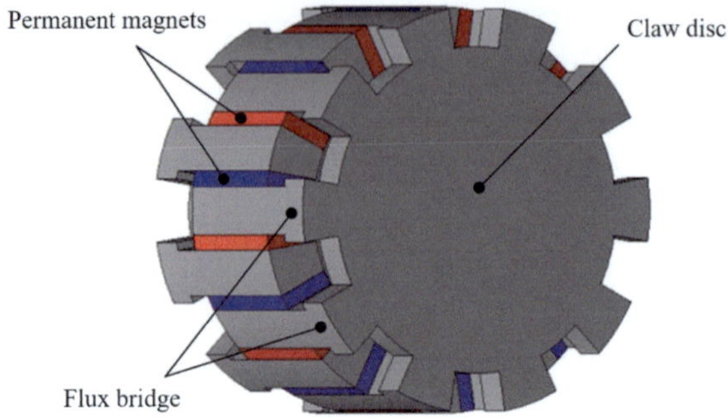

Fig. 5.14 Rotor structure of SACPPMVM

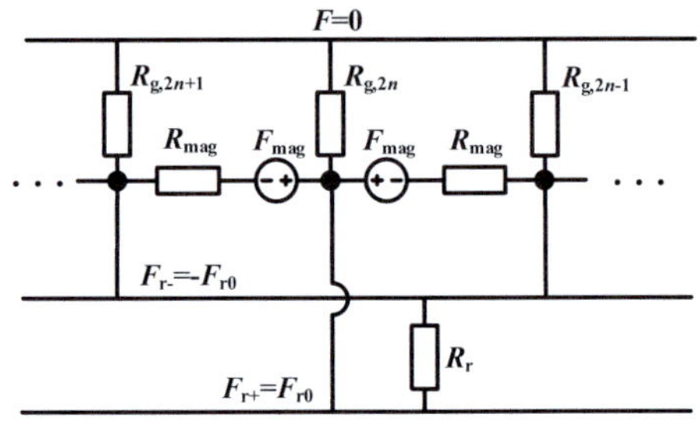

Fig. 5.15 Equivalent magnetic circuit of SACPPMVM under no-load condition

$$\sum_{n=1}^{P_e}\left(\frac{F_{r+}}{R_{g,2n}}+\frac{F_{r-}}{R_{g,2n-1}}\right)=0 \tag{5.9}$$

It can be derived that $F_{r+} = -F_{r-}$. Setting $F_{r+} = F_{r0}$ and applying the principle of flux continuity for all the ferromagnetic poles with positive polarity allows for further calculations:

$$\sum_{n=1}^{P_e}\left(\frac{2F_{r0}-F_{mag}}{R_{mag}}+\frac{F_{r0}}{R_{r,2n}}\right)+\frac{2F_{r0}}{R_r}=0 \tag{5.10}$$

It can be solved that

$$F_{r0}=\frac{2F_{mag}}{4+2\frac{R_{mag}}{P_eR_r}+\frac{\mu_0}{g}r_gl_{stk}R_{mag}F_{r0}\lambda_{lump\_0}} \tag{5.11}$$

The MMF of its rotor does not contain an additional term, which can be expressed as:

$$F_e=\frac{4F_{r0}}{\pi}\sum_{h=1,3\ldots}^{\infty}\cos hP_e(\theta-\theta_m) \tag{5.12}$$

This characteristic enables the SACPPMVM to achieve the highest torque density among FMMs with spoke-magnetization topology. However, two significant drawbacks are associated with this type of machine. Firstly, its magnetic circuit possesses a 3-D structure, which prohibits the use of laminated silicon steel sheets; only soft magnetic composite (SMC) or solid steel are viable materials. SMC presents challenges in terms of processing, and achieving precise dimensions is not guaranteed. The use of solid silicon can lead to significant eddy current losses. Additionally, the leakage flux between the rotor claw disc and ferromagnetic poles is relatively substantial, potentially exceeding that in the flux bridge structure. This results in a decrease in F_{r0} and compromises the machine's torque density. Therefore, it is impractical to utilize the SACPPMVM when the pole ratio is low.

To directly demonstrate the enhanced performance of the aforementioned spoke-magnetization FMMs, particularly the utility of the flux bridge, four vernier permanent magnet machines—SPMVM, SAPMVM, SAPMVMB, and SACPPMVM—are selected for a quantitative comparison. These machines share identical slot-pole combinations, outer diameters, stack lengths, air gap dimensions, and stator sizes. The no-load air gap flux densities of these four machines are analyzed in Fig. 5.16a, b. Due to the magnetic barrier effect, the working magnetic flux density of low pole number, B_{Pa}, in the SAPMVM does not show improvement compared to that of the SPMVM. In contrast, for the SAPMVMB and SACPPMVM, despite the presence of the magnetic barrier effect, the reduction in the additional MMF F_{Pa}, allows B_{Pa} to gradually increase. However, the working magnetic flux density of high pole number, B_{Pe}, follows an opposite trend. With the application of the spoke-magnetization topology, B_{Pe} in the SAPMVM is significantly augmented due to the flux-focusing effect. Meanwhile, the adoption of the flux bridge structure in the SAPMVMB and SACPPMVM results in a gradual decrease in B_{Pe} due to increased leakage flux. The no-load back-electromotive force (back-EMF) of these four vernier permanent magnet machines is compared in Fig. 5.16c. The performance of the SPMVM serves as the benchmark. Due to the magnetic barrier effect, the back-EMF in the SAPMVM exhibits almost no increase. With the introduction of the alternate flux bridge, the back-EMF in the SAPMVMB increases by 24%. Moreover, with the complete elimination of the additional magnetic potential ripple, the back-EMF in the SACPPMVM is enhanced by 45%, convincingly demonstrating the advantage of this technical approach.

5.2.4 Double-Stator Spoke-Magnetization Vernier Permanent Magnet Machine

The configuration of the DSSAPMVM is illustrated in Fig. 5.17a. This machine is designed with a double-stator structure, incorporating a spoke-magnetization rotor positioned between two stators. Both the inner and outer stators are designed with an open-slot structure to enhance the flux modulation effect. These stators have equal

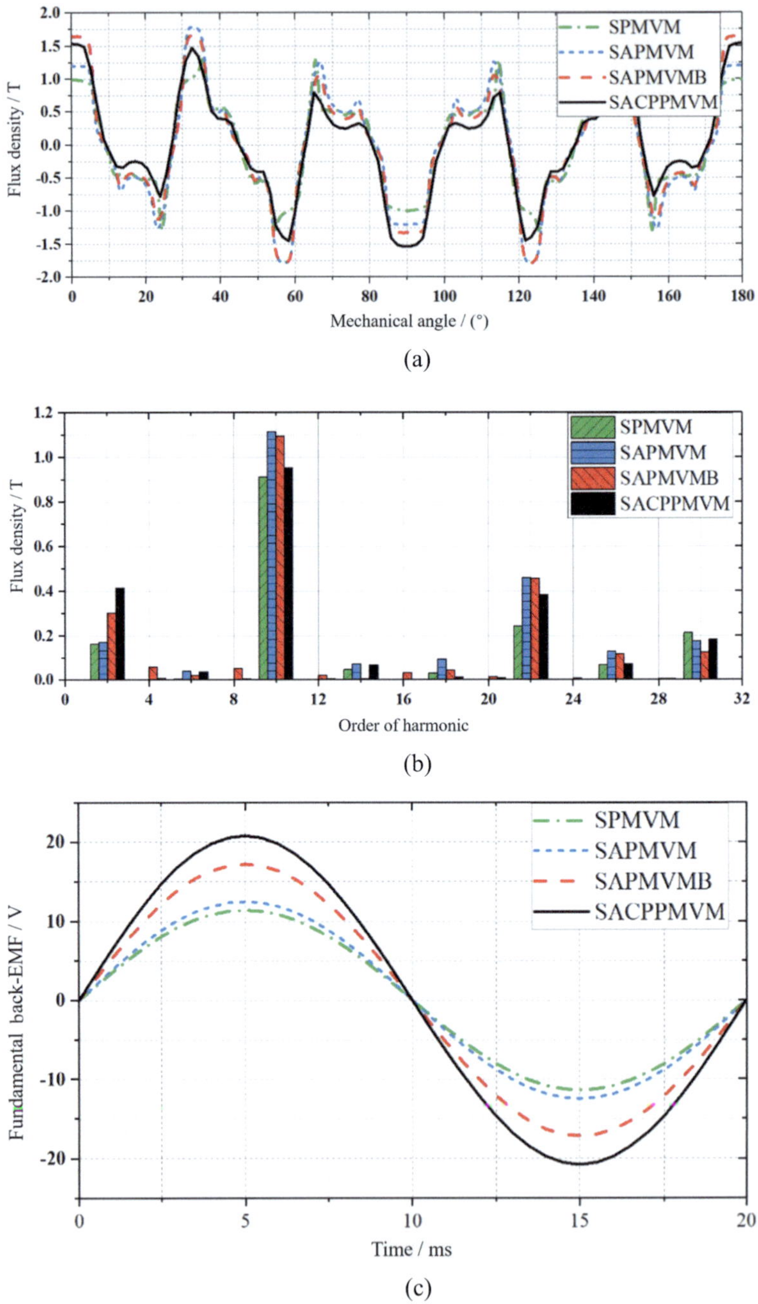

Fig. 5.16 Comparison of four vernier permanent magnet machines' performance: **a** waveform of no-load air gap flux density, **b** spectrum of no-load air gap flux density, **c** no-load back-EMF

tooth counts and are offset by a half-tooth-pitch relative to each other, positioning the teeth of the outer stator opposite the slots of the inner stator. Two sets of armature windings, each with the same pole pair number and adhering to the flux modulation relationship, are installed in the inner and outer stators. The phase difference between the inner and outer armature windings is specified as follows:

$$\theta_{\mathrm{io}} = P_{\mathrm{e}}\frac{\pi}{Z_{\mathrm{s}}} = \frac{P_{\mathrm{e}}}{P_{\mathrm{e}} + P_{\mathrm{a}}}\pi \tag{5.13}$$

The phase difference is approximately 180°. In this arrangement, the two windings should be connected in series, but with reverse polarity, as depicted in Fig. 5.17b [13, 14].

The equivalent magnetic circuit of the DSSAPMVM is illustrated in Fig. 5.18a. This machine features an inner air gap and an outer air gap, leading to a reluctance between each rotor's ferromagnetic pole and the inner air gap, as well as between each pole and the outer air gap. Due to the staggered arrangement of inner and outer rotor teeth, these two reluctances alternate in opposite phases. According to Eq. (2.67), these reluctances can be described as:

$$R_{\mathrm{og},n} = \frac{1}{\frac{\mu_0}{g} r_{\mathrm{g}} l_{\mathrm{stk}}} \frac{1}{\lambda_{\mathrm{lump\_0}} - \lambda_{\mathrm{lump\_1}} \cos Z_{\mathrm{s}}\left(\theta_{\mathrm{r},n} + \theta_{\mathrm{m}}\right)} \tag{5.14}$$

$$R_{\mathrm{ig},n} = \frac{1}{\frac{\mu_0}{g} r_{\mathrm{g}} l_{\mathrm{stk}}} \frac{1}{\lambda_{\mathrm{lump\_0}} + \lambda_{\mathrm{lump\_1}} \cos Z_{\mathrm{s}}\left(\theta_{\mathrm{r},n} + \theta_{\mathrm{m}}\right)} \tag{5.15}$$

As shown in Fig. 5.18a, these two reluctances are in parallel with each other. Consequently, they can be represented as a single equivalent reluctance:

$$R_{\mathrm{g},n} = \frac{1}{\frac{1}{R_{\mathrm{ig},n}} + \frac{1}{R_{\mathrm{og},n}}} = \frac{1}{\frac{\mu_0}{g} r_{\mathrm{g}} l_{\mathrm{stk}}} \frac{1}{2\lambda_{\mathrm{lump\_0}}} = \frac{R_{\mathrm{g}}}{2} \tag{5.16}$$

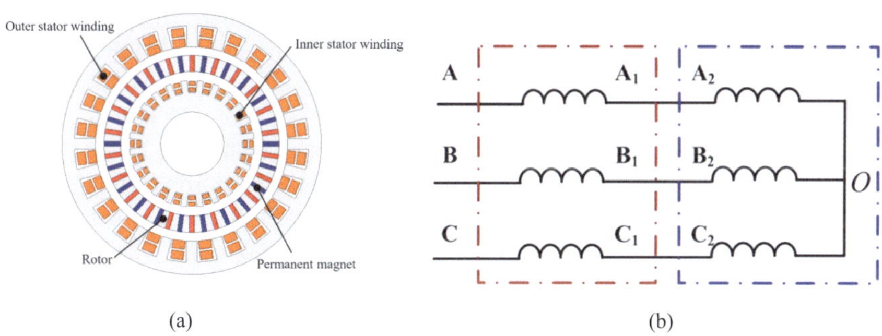

Fig. 5.17 Topology of DSSAPMVM: **a** electromagnetic structure, **b** connection of winding

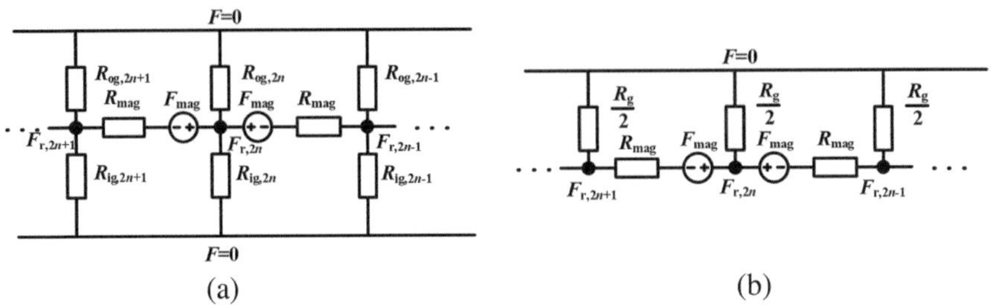

Fig. 5.18 Equivalent magnetic circuit of DSSAPMVM. **a** Actual magnetic circuit structure. **b** Simplified magnetic circuit structure

The equivalent reluctance is a constant, $R_g/2$, which is half the value of the reluctance when there is no slotting and only a single-sided air gap is present.

According to Eq. (5.16), the simplified model of DSSAPMVM's magnetic circuit is displayed in Fig. 5.18b. It is evident that since each ferromagnetic pole corresponds to a constant reluctance $R_g/2$, there is no additional MMF. The rotor's MMF for this machine also aligns with Eq. (5.12) but with a different value for F_{r0}:

$$F_{r0} = \frac{2F_{mag}}{4 + 2\frac{\mu_0}{g} r_g l_{stk} R_{mag} F_{r0} \lambda_{lump\_0}} = \frac{F_{mag}}{2 + \frac{\mu_0}{g} r_g l_{stk} R_{mag} F_{r0} \lambda_{lump\_0}} \tag{5.17}$$

Compared to the regular spoke-magnetization topology, the air gap specific permeance, located in the denominator of the rotor magnetic potential's expression, is doubled. This increase indicates the introduction of additional leakage flux, equivalent to the value of the air gap reluctance, which further impacts the machine's performance. Consequently, DSSAPMVM is not practical when the pole ratio is low, such as in configurations like the fractional-slot concentrated-winding topology.

For a DSSAPMVM with an outer diameter of 620 mm, a stack length of 50 mm, and a pole ratio of 11, the relationship between its output torque and current angle is shown in Fig. 5.19, when the machine's total electrical loading is 220 A/cm (outer stator winding = 120 A/cm, inner stator winding = 100 A/cm).

The machine exhibits a significant portion of reluctance torque. The current angle is approximately 17° when the output torque reaches its maximum value of around 1030 N m. In this scenario, the machine achieves a torque density of 68 N m/L, which is more than twice that of the conventional surface-mounted vernier permanent magnet machine. These simulation results demonstrate that the DSSAPMVM possesses a superior torque density. Furthermore, thanks to its robust field excitation capability and the ability to adjust the current angle, DSSAPMVM can achieve a power factor of 0.85 when the output torque is at its maximum. For a standard vernier permanent magnet machine with a pole ratio of 11, such a value is typically limited to 0.6. Moreover, if the current angle is increased further, albeit with a slight sacrifice in output torque, the power factor can reach an impressive value of 0.91 while still maintaining a torque density level of 66 N m/L. In summary, DSSAPMVM

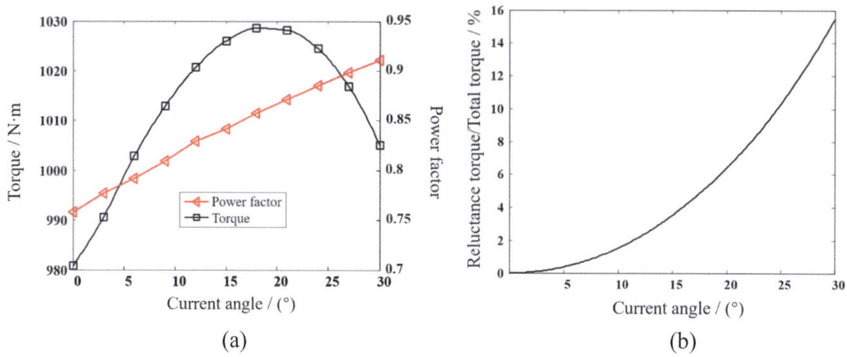

Fig. 5.19 Relationship between DSSAPMVM's electromagnetic performance and current angle. **a** Relationship between current angle and output torque/power factor. **b** Relationship between current angle and proportion of reluctance torque

capitalizes on the FMM's advantage of high torque density while overcoming the traditional drawback of a relatively low power factor, demonstrating its exceptional potential for engineering applications.

5.2.5 *Other Excitation-Enhanced Flux Modulation Machines*

In Sect. 5.2.5, it is noted that the implementation of an alternate flux bridge structure beneath the spoke-magnetization permanent magnet resolves the magnetic barrier issue in the spoke-magnetization vernier permanent magnet machine. However, the narrow root of the flux bridge is prone to saturation, limiting the overload capability of machines with this topology. This limitation is addressed by skewing the bottom of the permanent magnets outward from the flux bridge, thereby increasing its width. The adjacent permanent magnets in this topology have bottoms that nearly touch, effectively eliminating the leakage flux path [15]. Consequently, the iron core replaces the physical air magnetic barrier, resulting in a V-shaped vernier permanent magnet machine, termed the consequent-pole V-shaped permanent magnet vernier machine (CPVPMVM), as illustrated in Figs. 5.20 and 5.21.

Building upon the CPVPMVM, the iron core of the consequent pole can be replaced by a V-shaped permanent magnet with an opposite magnetization direction, thus forming a complete V-shaped permanent magnet vernier machine (VPMVM), as depicted in Fig. 5.22. Similar configurations, as shown in Figs. 5.21 and 5.22, have been proposed for conventional permanent magnet machines. However, for vernier permanent magnet machines, it is essential to position the iron yoke beneath the permanent magnet to mitigate the magnetic barrier effect. With this arrangement and the enhanced flux-focusing effect, machines with these two V-shaped topologies

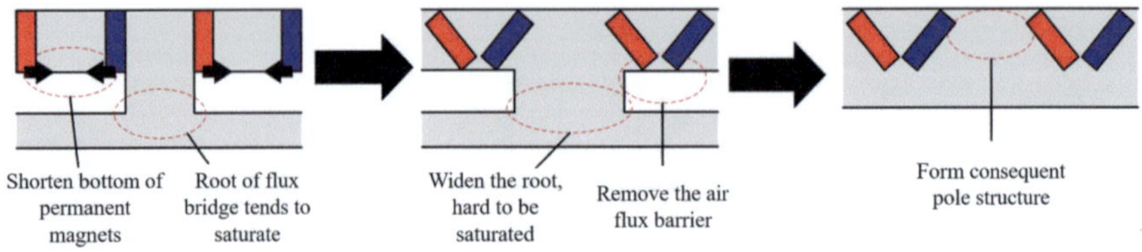

Fig. 5.20 Evolution process from SAPMVMB to CPVPMVM

Fig. 5.21 Topology of CPVPMVM

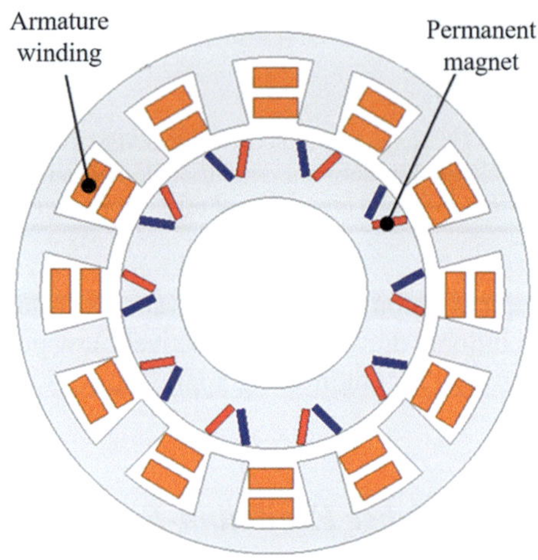

achieve higher torque densities compared to spoke-magnetization machines with alternate flux bridges.

Fig. 5.22 Topology of VPMVM

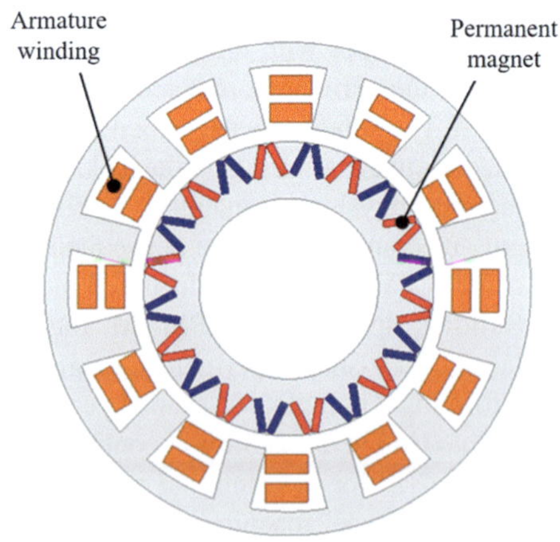

5.3 Modulation-Enhanced Flux Modulation Machines

The modulation-enhanced FMM attains high performance by employing technical strategies to improve the working specific permeance of the modulation unit. In the single-side slotting FMM, the working specific permeance primarily consists of the fundamental wave portion, which is determined by the difference between the specific permeance at the tooth region and the slot region, as illustrated in Fig. 2.12. In FMMs that utilize an open-slot structure for both stator and rotor, such as the permanent magnet flux switching machine and the vernier permanent magnet machine, the air gap specific permeance is equivalent to the product of the two single-side specific permeances involving stator and rotor slotting. Consequently, the overall machine's performance is directly proportional to the fundamental wave portion of the single-side slotting's specific permeance. The relationship between the specific permeance's fundamental wave term and the air gap's dimensional parameters is analyzed in Figs. 2.14–2.16. It is observed that enhancing the specific permeance's fundamental wave term can be achieved by reducing the air gap thickness, decreasing the slot number, and optimizing the tooth width. However, these methods are generally impractical from an engineering perspective. The air gap thickness is constrained by factors such as machining accuracy, operating speed, and air gap radius, which are difficult to modify. Moreover, reducing the slot number typically results in a lower pole ratio, adversely affecting the output torque. Additionally, the tooth width is usually optimized during the initial design phase, providing limited further enhancements to the modulation effect. As analyzed, improving modulation capability offers limited practical potential for enhancing FMM torque density.

However, the introduction of superconducting materials presents an additional possibility. The Meissner effect in superconducting materials leads to the expulsion of magnetic fields when in the superconducting state. Positioning superconducting material at the stator notch reduces the specific permeance to zero, thereby enhancing the difference in specific permeance between the tooth region and slot region. This improvement ultimately augments the modulation effect [16].

As an example, Fig. 5.23 illustrates a topology of the modulation-enhanced vernier permanent magnet machine. The sole distinction from the standard SPMVM is the incorporation of a thin film of superconducting material at the stator notch. In Fig. 5.24, the finite element simulation result for this topology's specific permeance demonstrates that the use of superconducting material at the stator notch reduces the specific permeance in this region to zero, while maintaining the specific permeance in the tooth region nearly unchanged. Consequently, the fundamental term of the machine's air gap specific permeance significantly increases from 0.3 to 0.5, with only a slight reduction in the DC term. A comparison of the back-EMF of the two topologies in Fig. 5.25 shows that the fundamental term of back-EMF increases by 41%, attributable to the enhanced modulation effect. In Sect. 4.5, it was noted that the armature inductance of the FMM is relatively large, contributing to a relatively low power factor. The armature inductance arises from the interaction between the armature MMF of P_a pole pairs and the DC term of the air gap specific permeance.

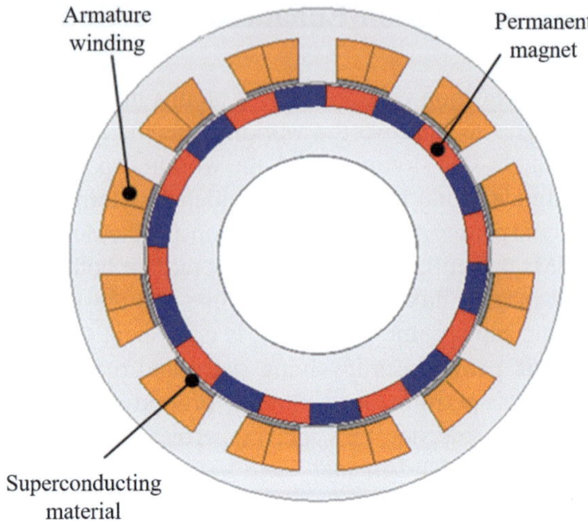

Fig. 5.23 Topology of modulation-enhanced vernier permanent magnet machine

For the modulation-enhanced vernier permanent magnet machine mentioned above, the decrease in the specific permeance's DC term results in a reduction of the armature inductance from 65.8 to 52 μH. Finally, a comparison of the output torque and power factor of the two topologies in Fig. 5.26 reveals that, with the same armature current, the torque of the modulation-enhanced vernier permanent magnet machine increases by 41%. Moreover, due to the increased back-EMF and decreased armature inductance, the power factor of the modulation-enhanced vernier permanent magnet machine declines more gradually with the increase of the armature current. As a result, the power factor of the modulation-enhanced vernier permanent magnet machine can be maintained at 0.9, while that of the SPMVM drops to around 0.65.

Currently, superconducting material remains relatively expensive and requires stringent temperature conditions, which limits the practical application of the modulation-enhanced FMM to more of an idealized model. Nevertheless, this ideal model illustrates that the torque density of the FMM can be improved by enhancing the modulation effect. Moreover, this approach effectively addresses a notable concern in FMMs—the relatively low power factor. By providing insights into the maximum torque density and power factor achievable under extreme conditions, this model offers valuable guidance for ongoing FMM research.

5.4 Armature-Enhanced Flux Modulation Machines

To increase the pole ratio and enhance output torque, the rotor's pole number in FMMs is typically much larger than that of the armature. However, in certain FMM configurations, such as vernier permanent magnet machines and vernier reluctance machines, where stator teeth function as the modulation unit and the tooth number is

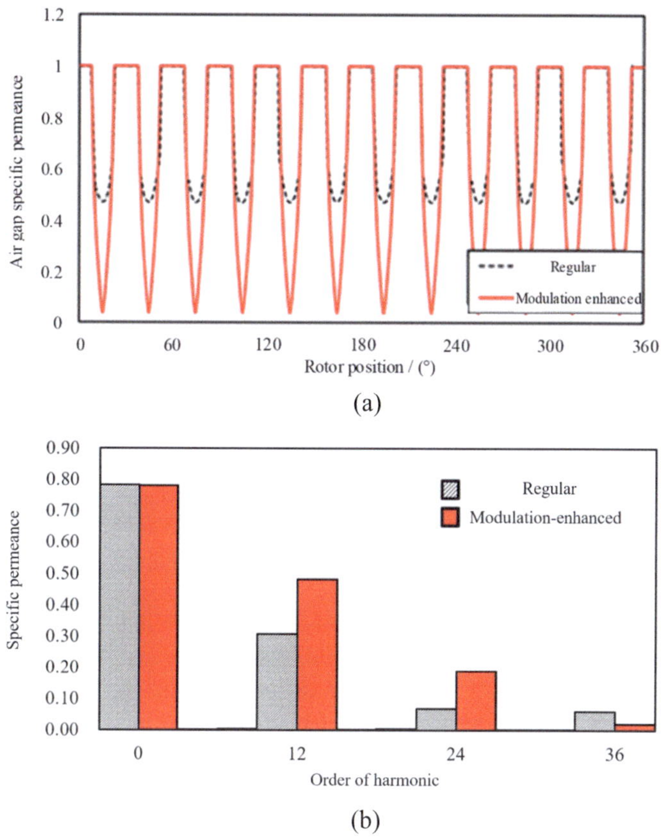

Fig. 5.24 Comparison of modulation-enhanced vernier permanent magnet machine's and regular vernier permanent magnet machine's specific permeance: **a** waveform, **b** spectrum

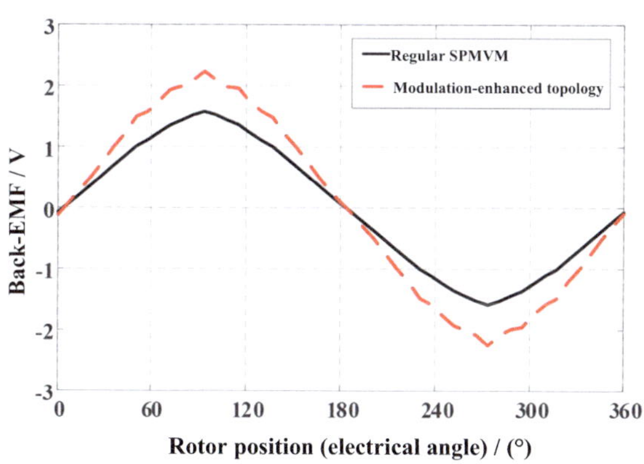

Fig. 5.25 Comparison of modulation-enhanced vernier permanent magnet machine's and regular vernier permanent magnet machine's no-load back-EMF

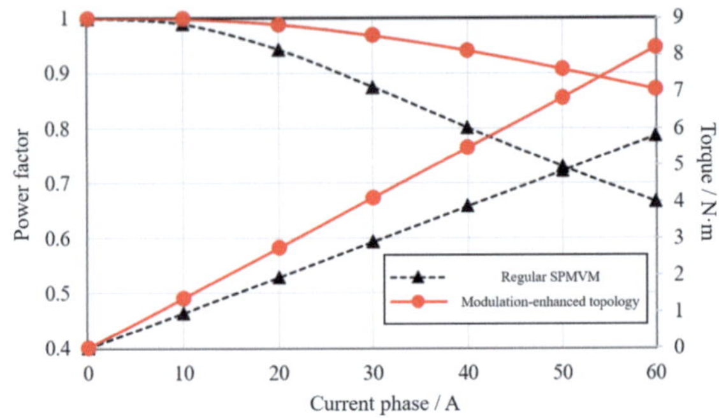

Fig. 5.26 Comparison of modulation-enhanced vernier permanent magnet machine's and regular vernier permanent magnet machine's performance

relatively large, the armature adopts a stack winding that spans multiple slots. This arrangement results in a relatively long end-turn, increasing the overall volume of the machine. Such an increase compromises the high torque density advantage of FMMs and escalates losses, thereby reducing machine efficiency.

To address this challenge, some researchers have proposed FMMs with an auxiliary tooth structure. This design decouples the modulation unit from the slot number, significantly reducing the number of slots. In such configurations, FMMs with a high pole ratio can utilize fractional-slot concentrated winding to shorten the end-turn length. For additional details about this topology, please refer to Sect. 3.3.1. However, the auxiliary tooth structure introduces additional teeth and an iron yoke, reducing the slot area and air gap radius. Consequently, while this topology might diminish the output torque somewhat, it optimizes the performance of the armature winding. Some scholars have further proposed enhancing the design of vernier permanent magnet machines with circular windings to improve armature winding performance.

The topology of a vernier permanent magnet machine with circular winding is illustrated in Fig. 5.27a. Unlike the stack winding structure, the circular winding is wrapped around the stator yoke, with its terminus closely adhering to the end plane of the stator yoke. This configuration substantially reduces the extension length, markedly enhancing the compactness of the machine. However, there are also disadvantages associated with the circular winding structure. Firstly, slotting on the inner and outer surfaces of the yoke is necessary to accommodate the conductor, which consequently diminishes the slot area and air gap radius—similar drawbacks to those encountered with the auxiliary tooth structure. Moreover, the conductor on the outer side of the yoke does not contribute to torque production due to the magnetic shielding effect of the iron core, rendering it functionally equivalent to the end of a stack winding. Although it does not encroach upon the space of the end region, it incurs significant copper losses. This indicates that the performance of circular winding might be inferior to that of stack winding, particularly when the aspect ratio is substantial.

To address the issue of the outer conductor's lack of contribution, an additional rotor can be installed on the outer side of the stator, forming a double-rotor vernier permanent magnet machine with circular winding, as depicted in Fig. 5.27b.

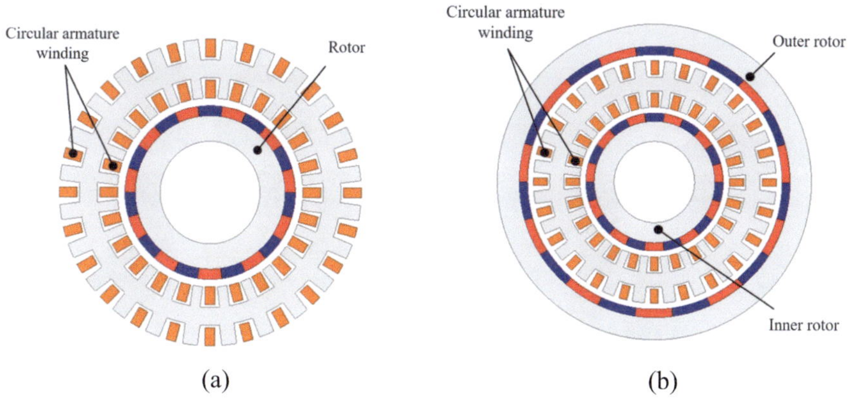

Fig. 5.27 Topologies of vernier permanent magnet machines with circular winding: **a** single-rotor structure, **b** double-rotor structure

These two rotors possess the same pole pair number and rotate at identical speeds, producing torque coaxially. However, the inner air gap is further reduced, resulting in mismatched torque output capabilities between the inner and outer rotors. In this context, the topology can be modified to create an axial-flux vernier permanent magnet machine, as shown in Fig. 5.28 [17]. In this design, the conductors on the left and right sides of the stator interact with the rotors positioned on each side, respectively. The end winding of this topology is exceedingly short, contributing to the machine's enhanced compactness. Moreover, the symmetrical structure of this double-rotor configuration helps avoid issues related to unbalanced output torque.

The performance of the axial-flux vernier permanent magnet machine utilizing circular winding is compared to that of a conventional axial-flux permanent magnet machine through a three-dimensional finite element simulation. To fully capitalize on the advantages of circular winding, the slot number for the stator of the axial-flux vernier permanent magnet machine is set at 24, the rotor's pole pair number is 22, and the armature winding's pole pair number is 2, resulting in a pole ratio of 11. In contrast, the conventional axial-flux permanent magnet machine features 24 stator slots, 10 rotor pole pairs, and employs a fractional-slot concentrated-winding topology for its armature. Consequently, both machines share the same axial length and outer diameter.

The no-load back-EMF of both the axial-flux vernier permanent magnet machine and the regular axial-flux permanent magnet machine is compared in Fig. 5.29. Under the same line load, the average torque of the axial-flux vernier permanent magnet machine is found to be over 1.8 times that of the regular axial-flux permanent magnet machine. Additionally as shown in Fig. 5.30, the power factor of the axial-flux vernier permanent magnet machine decreases sharply with an increase in current, aligning with the electromagnetic characteristics of vernier machines, which exhibit a pronounced armature reaction. Under the rated line load, the power factor of the axial-flux vernier permanent magnet machine can drop as low as 0.64, while the

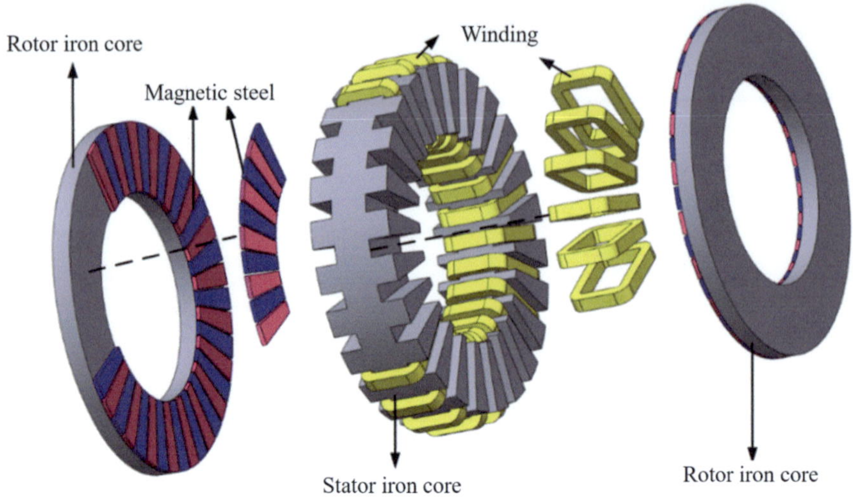

Fig. 5.28 Topology the double-stator axial flux vernier permanent magnet machine with circular winding

Fig. 5.29 Comparison of the no-load back-EMF

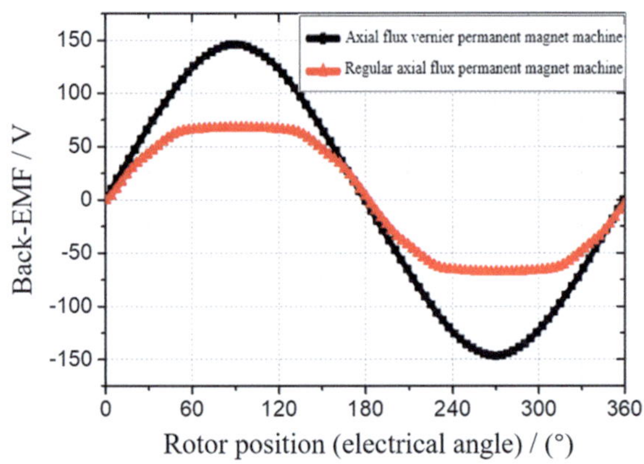

power factor of the regular axial-flux permanent magnet machine can be maintained at 0.97. These findings highlight the inherent challenges associated with the FMM technology (Fig. 5.29).

5.5 Integrated Flux Modulation Theory

In conventional machines, the magnetic field with a specific pole pair number is primarily utilized, and harmonic magnetic fields are minimized as much as possible. For FMMs, multiple working magnetic fields are utilized to enhance the

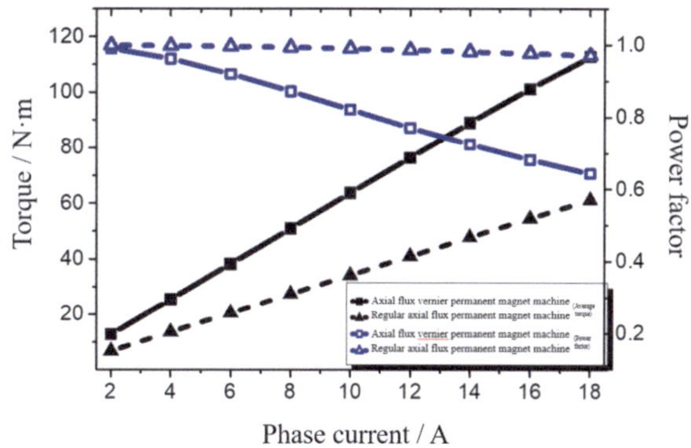

Fig. 5.30 The variation of the average torque with the change of the phase current

machine's output torque and power. Therefore, in addition to increasing the amplitude of a specific working magnetic field, the output torque can also be amplified by augmenting the number of working magnetic fields. Machines that employ this strategy are referred to as integrated FMMs.

5.5.1 *Mathematical Characterization of Integrated Flux Modulation Machine*

The topologies of FMMs are diverse and are challenging to classify directly. However, all adhere to the principles of flux modulation theory. To facilitate a theoretical examination of the integrated FMM, the machine's topology is abstracted into a mathematical characterization based on the MMF-specific permeance model. Mathematical equations are utilized to ensure clarity and comprehensiveness in the study and classification of the integrated FMM's topologies. Several assumptions are made to simplify the analysis:

(1) Only the permanent magnet machine topology is considered in this and subsequent analyses, excluding hybrid excitation and electrically excited machines.
(2) The additional MMF caused by slotting is disregarded during the mathematical characterization. Although slotting can attenuate the working magnetic field, it does not alter the fundamental operation of the machine, thus it is acceptable to omit it for qualitative analysis.
(3) The machine is assumed to have a single electrical port and a single mechanical port, indicating it operates at a specific rotation speed and electrical frequency.
(4) All working magnetic fields are considered as tooth harmonics to enhance the effect of integrated modulation.

With these assumptions in place, the expressions for the stator's and rotor's MMF are formulated. The MMF of the permanent magnet is categorized into two types: homopolar and multipolar. All FMMs mentioned previously employ multipolar MMF, which requires no further elaboration here. For the homopolar MMF, it

(a) (b)

Fig. 5.31 Different forms of the FMM: **a** homopolar magneto-motive force, **b** multi-polar magneto-motive force

is aligned in a uniform direction, as illustrated in Fig. 5.31a. According to the principle of flux continuity, the magnetic circuit of a homopolar machine must traverse a three-dimensional structure to complete a closed loop.

The MMF sources mentioned can theoretically coexist, and the multipolar MMF may comprise segments with varying pole pair numbers. Therefore, the MMF of the stator can be uniformly represented as follows:

$$F_{\mathrm{es}}(\theta) = F_{\mathrm{es0}} + \sum_{h=1}^{\infty} F_{\mathrm{es}h} \cos P_{\mathrm{es}h}\theta \tag{5.18}$$

where the subscript "s" represents the MMF produced by the stator. In previous sections, F_{s} and F_{r} represent the magnetic potential of the stator and rotor, respectively. However, they do not represent the MMF, but represent F_{e} directly, while the F_{es} in Eq. (5.18) represents the excitation MMF of the permanent magnet on the stator. The difference between these two parameters should be noted. $F_{\mathrm{es}h}$ represents the amplitude of the hth working magnetic field, and $P_{\mathrm{es}h}$ represents the corresponding pole pair number. It is important to note that F_{eh} in the previous section denotes the amplitude of the MMF for the hth harmonic, which has a pole pair number h times that of the fundamental wave. However, such a MMF might not work. Moreover, to emphasize the working principle of the machine, $F_{\mathrm{es}h}$ in Eq. (5.18) only represents a working MMF, and it is not necessary for $P_{\mathrm{es}h}$ to be h times P_{es1}. Additionally, the value of $F_{\mathrm{es}h}$ can be negative, corresponding to the MMF's spatial phase of 180°.

Equation (5.18) is applicable to all the stator-excitation FMMs. For example, for flux-switching permanent magnet machine, one only needs to set $F_{\mathrm{es0}} = 0$ and $P_{\mathrm{es}h} = hZ_{\mathrm{s}}/2$. Similarly, the function of the stator's working specific permeance can be represented as:

$$\lambda_{\mathrm{s}}(\theta) = \lambda_{\mathrm{s0}} + \sum_{l=1}^{\infty} \lambda_{\mathrm{s}l} \cos P_{\mathrm{fs}l}\theta \tag{5.19}$$

In contrast to the stator, the rotating MMF and harmonics of specific permeance on the rotor side can introduce alternating frequencies. Since this section does not address multiple electrical ports, the machine operates under a singular working frequency. Therefore, the working MMF of the rotor is distinct, and so is the pole pair number for the rotor's specific permeance. In cases where only the working MMF is present in the rotor, its pole pair number can be represented by P_{e}, similar to the configuration in a vernier permanent magnet machine. Additionally, the rotor typically features a salient-pole structure with Z_r pole pairs when only the working specific permeance is present in the rotor. However, when both the working MMF and the working specific permeance are present in the rotor, the two earlier representations are no longer accurate. In such instances, the pole pair number of the rotor is denoted by P_r:

$$\begin{aligned} F_{er}(\theta, t) &= F_{er0} + F_{er1}\cos P_r(\theta - \Omega_r t) \\ \lambda_r(\theta, t) &= \lambda_{r0} + \lambda_{r1}\cos P_r(\theta - \Omega_r t) \end{aligned} \tag{5.20}$$

The formulation of the MMF in Eq. (5.20) differs from that in previously analyzed machines, introducing a new DC term. This scenario is based on the assumption of homopolar excitation. As previously analyzed, another multipolar working harmonic cannot coexist in the MMF simultaneously. Furthermore, there are scenarios where only one side is slotted, such as a vernier permanent magnet machine featuring a smooth rotor and a slotted stator. In such configurations, the rotor's specific permeance is inherently 1.

With the aforementioned assumptions, the no-load air gap magnetic flux density of the FMM can be uniformly represented as follows:

$$\begin{aligned} B_e(\theta, t) = \frac{\mu_0}{g}\Bigg\{ & F_{es0}\lambda_{s0}\lambda_{r0} + F_{er0}\lambda_{s0}\lambda_{r0} + F_{es0}\lambda_{r0}\sum_{l=1}^{\infty}\lambda_{sl}\cos P_{fsl}\theta \\ & + F_{er0}\lambda_{r0}\sum_{l=1}^{\infty}\lambda_{sl}\cos P_{fsl}\theta + \lambda_{s0}\lambda_{r0}\sum_{h=1}^{\infty}F_{esh}\cos P_{esh}\theta \\ & + \frac{\lambda_{r0}}{2}\sum_{h,l}F_{esh}\lambda_{sl}\cos(P_{esh} \pm P_{fsl})\theta + F_{es0}\lambda_{s0}\lambda_{r1}\cos P_r(\theta - \Omega_r t) \\ & + \frac{F_{es0}}{2}\sum_{l=1}^{\infty}\lambda_{sl}\lambda_{r1}\cos\left[(P_r \pm P_{fsl})\theta - P_r\Omega_r t\right] + F_{er0}\lambda_{s0}\lambda_{r1}\cos P_r(\theta - \Omega_r t) \\ & + \frac{F_{er0}}{2}\sum_{l=1}^{\infty}\lambda_{sl}\lambda_{r1}\cos\left[(P_r \pm P_{fsl})\theta - P_r\Omega_r t\right] + \frac{\lambda_{s0}}{2}\sum_{h=1}^{\infty}F_{esh}\lambda_{r1}\cos\left[(P_r \pm P_{esh})\theta - P_r\Omega_r t\right] \\ & + \frac{1}{4}\sum_{h,l}F_{esh}\lambda_{sl}\lambda_{r1}\cos\left[(P_r \pm P_{esh} \pm P_{fsl})\theta - P_r\Omega_r t\right] + \lambda_{s0}\lambda_{r0}F_{er1}\cos P_r(\theta - \Omega_r t) \\ & + \frac{\lambda_{s0}}{2}F_{er1}\lambda_{r1}\cos 2P_r(\theta - \Omega_r t) + \frac{\lambda_{s0}}{2}F_{er1}\lambda_{r1} + \frac{\lambda_{r0}}{2}\sum_{l=1}^{\infty}F_{er1}\lambda_{sl}\cos\left[(P_r \pm P_{fsl})\theta - P_r\Omega_r t\right] \\ & + \frac{1}{4}\sum_{l=1}^{\infty}F_{er1}\lambda_{sl}\lambda_{r1}\cos\left[(2P_r \pm P_{fsl})\theta - 2P_r\Omega_r t\right] + \frac{1}{2}\sum_{l=1}^{\infty}F_{er1}\lambda_{sl}\lambda_{r1}\cos P_{fsl}\theta \Bigg\} \end{aligned} \tag{5.21}$$

It can be observed that the potential air gap magnetic fields within the FMM can be categorized into 18 distinct groups. However, not all of these groups contribute as working magnetic fields. Initially, the magnetic fields from groups one to six, fifteen, and eighteen remain stationary, indicating their non-contribution to electromagnetic energy conversion, and thus should be excluded from consideration. Additionally,

Table 5.1 Air gap working magnetic field harmonic of the FMM

Number	Pole pair number	Amplitude	Angular frequency
1	P_r	$\frac{\mu_0}{g}F_{es0}\lambda_{s0}\lambda_{r1}$	$P_r\Omega_r$
2	$\lvert P_r \pm P_{fsl}\rvert$	$\frac{\mu_0}{2g}F_{es0}\lambda_{sl}\lambda_{r1}$	$P_r\Omega_r$
3	P_r	$\frac{\mu_0}{g}F_{er0}\lambda_{s0}\lambda_{r1}$	$P_r\Omega_r$
4	$\lvert P_r \pm P_{fsl}\rvert$	$\frac{\mu_0}{2g}F_{er0}\lambda_{sl}\lambda_{r1}$	$P_r\Omega_r$
5	$\lvert P_r \pm P_{esh}\rvert$	$\frac{\mu_0}{2g}F_{esh}\lambda_{s0}\lambda_{r1}$	$P_r\Omega_r$
6	$\lvert P_r \pm P_{esh} \pm P_{fsl}\rvert$	$\frac{\mu_0}{4g}F_{esh}\lambda_{sl}\lambda_{r1}$	$P_r\Omega_r$
7	P_r	$\frac{\mu_0}{g}F_{er1}\lambda_{s0}\lambda_{r0}$	$P_r\Omega_r$
8	$\lvert P_r \pm P_{fsl}\rvert$	$\frac{\mu_0}{2g}F_{er1}\lambda_{sl}\lambda_{r0}$	$P_r\Omega_r$

groups fourteen and seventeen, which introduce frequencies resulting from simultaneous contributions of the rotor's MMF and the rotor's specific permeance, produce a magnetic field frequency that is twice that of the fundamental frequency. Consequently, these groups are also unsuitable for electromagnetic conversion and should likewise be excluded. After discarding the aforementioned ten groups, the remaining eight groups of magnetic fields are identified as capable of functioning as working magnetic fields, as summarized in Table 5.1.

In any given FMM configuration, the air gap working magnetic field consists either of a single group or a combination of multiple groups from the aforementioned eight eligible groups. Given that the essence of the integrated machine design is to augment the working magnetic field, it is prudent to first analyze the working magnetic fields within conventional FMM configurations to lay the foundation for further enhancements.

5.5.2 *Magnetic Field Analysis of Conventional Flux Modulation Machine*

The term "conventional FMM" refers to a machine topology characterized by a limited number of working magnetic fields, a straightforward structure, and clearly defined operational principles. As its working magnetic field includes at least one group mentioned in Table 5.1, this subsection will employ the eight groups listed in Table 5.1 as the foundational framework to investigate the most fundamental FMM topology.

The MMF of the first group's working magnetic field is generated by homopolar permanent magnets situated on the stator side. For the specific permeance of this group, the stator contributes to the DC component, while the rotor contributes to the fundamental wave component. When these characteristics are applied to a specific machine topology, it features a permanent magnet magnetized axially and positioned at the midpoint of the stator yoke. Therefore, a complete magnetic circuit is formed with the stator and rotor iron cores. This topology includes a rotor with a salient-pole structure, divided into two parts, each staggered by half a tooth pitch. The purpose of this design is to ensure that the magnetic field remains in the same phase on both sides, modulated by the rotor teeth, as illustrated in Fig. 5.32a. The stator in this topology features a semi-closed slot to enhance the DC component of the specific permeance function. The overall machine topology is depicted in Fig. 5.32b. Since the permanent magnet in this machine is located on the stator side and produces the homopolar field excitation MMF, this design is termed a stator-excitation homopolar machine.

The working magnetic field of the second group is analogous to that of the first group, with its MMF also supplied by a homopolar permanent magnet on the stator side. However, the distinction lies in the stator's working specific permeance, which is an alternating component. In this section, only the topology with a single stator and a single rotor is analyzed, based on prior assumptions. Due to the absence of

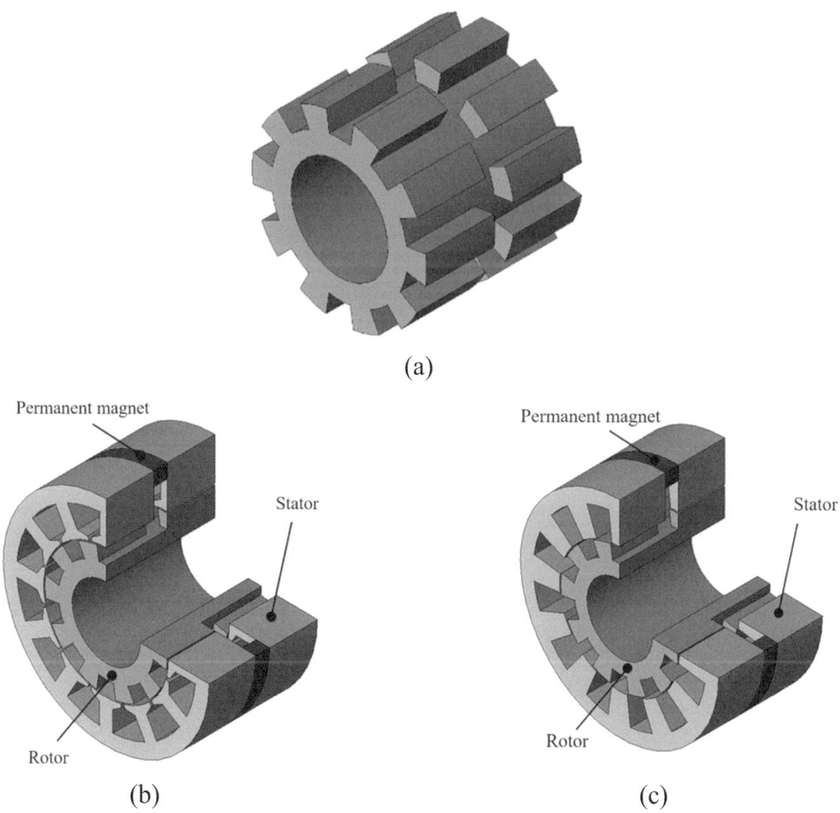

Fig. 5.32 Topologies of the stator excitation homopolar machine: **a** machine topology, **b** semi-closed slot topology, **c** open slot topology

an independent modulation ring, this alternating component can only be introduced through stator slotting. Therefore, there is a limitation to the working pole pair number of this machine:

$$P_{\mathrm{fs}l} = l n_{\mathrm{f}} Z_{\mathrm{s}}, \quad l = 1, 2 \ldots \tag{5.22}$$

When $n_{\mathrm{f}} = 1$, the machine exhibits a straight-tooth structure. For $n_{\mathrm{f}} > 1$, it adopts an auxiliary-tooth structure. According to Eq. (5.22), the pole pair numbers of the working magnetic fields for the first and second groups are harmonics of each other, sharing the same electrical frequency. The presence of the DC term in specific permeance necessitates the simplest FMM topology for the second group's working magnetic field, achieved by integrating the working magnetic fields of the first and second groups. This integrated topology is shown in Fig. 5.32c, differing from Fig. 5.32b only by its open-slot stator configuration. Consequently, besides the DC term of specific permeance, harmonics are introduced and actively contribute to the working magnetic field.

The working magnetic fields of the third and fourth groups arise from a mechanism almost identical to the first two groups. The key difference is that the homopolar field excitation originates from the rotor, leading to the term rotor excitation homopolar machine, with the permanent magnet positioned on the rotor side. However, since the homopolar magnetic field's form is independent of its rotation, it does not affect the modulation principle. The corresponding topology is illustrated in Fig. 5.33.

The working magnetic field of the fifth group appears as the air gap rotating magnetic field, sourced from the multi-polar MMF modulated by the fundamental wave of the rotor's specific permeance. This results in a flux reversal machine with uniformly distributed permanent magnets. Similar to the second group, the sixth group's working magnetic field cannot exist independently in a topology with a single rotor and a single stator. However, by combining it with the fifth group's

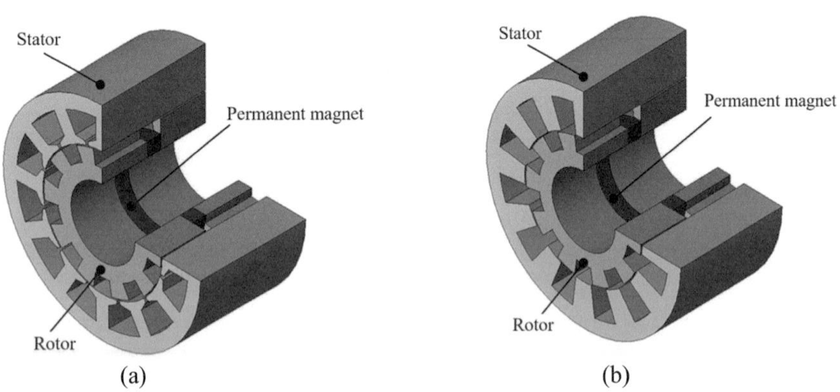

Fig. 5.33 Topologies of the rotor excitation homopolar machine. **a** Semi-closed slot topology. **b** Open slot topology

working magnetic field, a flux reversal machine with uniformly distributed permanent magnets and an open-slot stator can be constructed. The conventional flux-switching permanent magnet machine, flux reversal permanent magnet machine, and permanent magnet double-salient machine fall into this category. Despite their relatively simple structures, these machine types exhibit multiple working MMF harmonics, adding to the magnetic field's complexity. Therefore, they do not represent the simplest topology for the working magnetic field harmonics.

The seventh group's working magnetic field originates from the multi-polar MMF on the rotor side. This specific permeance represents the unmodulated DC term of both the stator and rotor specific permeances, aligning with the conventional permanent magnet synchronous machine. In contrast, the eighth group's working magnetic field is an alternating component that cannot exist independently. It only manifests when combined with the seventh group's magnetic field, forming the topology of the vernier permanent magnet machine.

In summary, the simplest topology of the FMM for each group's working magnetic field has been identified, and the fundamental method of integrated flux modulation will be analyzed further below.

5.5.3 Basic Method of Integrated Flux Modulation

As mentioned at the beginning of this chapter, integration is achieved by modifying the machine's topology to increase the number of working magnetic fields. According to the classification in Table 5.1, there are two possibilities for the additional working magnetic field for each type of machine discussed. First, if the additional working magnetic field is of the same type that the machine inherently possesses but with a different pole pair number, it is classified as inner-group integration. Second, if the additional working magnetic field is of a different type than what the machine inherently possesses, it is classified as intergroup integration [18].

First, we explore inner-group integration. Topologies corresponding to the first, third, and seventh groups in Table 5.1 cannot adopt this form of integration. Their field excitation MMF and air gap specific permeance are entirely fixed and cannot be expanded further. For example, the additional working magnetic field from inner-group integration in the first group should conform to its mathematical characterization. However, the field excitation for this group is the homopolar MMF on the stator side, which cannot be modified. Although the working magnetic field can be enhanced by connecting permanent magnets in series, this belongs to an excitation-enhanced topology, which is unrelated to integration. Additionally, the stator's specific permeance is fixed as a DC term, making it impossible to introduce an additional DC term. While theoretically possible to add the rotor's specific permeance harmonic, this method would introduce multiple electrical frequencies. This contradicts the initial assumption and requires the stator to have multiple sets of windings, leading to a reduction in torque.

For the remaining groups, machines corresponding to these groups can increase the number of working magnetic fields through inner-group integration. When the working magnetic field is generated by the multi-polar stationary specific permeance, as in the second, fourth, sixth, and eighth groups, the integrated FMM can be designed by increasing the specific permeance harmonic. These machines are called multi-working permeance FMMs. Alternatively, when the working magnetic field is generated by multi-polar stationary MMF, as in the fifth and sixth groups, the integrated FMM can be designed by increasing the MMF harmonic. These machines are termed multi-working MMF FMMs.

Compared to inner-group integration, intergroup integration is more complex and involves various combination approaches. First, we will introduce two types of new machines constructed through inner-group integration. Subsequently, we will explore the integrated topology formed by intergroup integration.

5.5.4 Multi-Working Permeance Flux Modulation Machine

Using the vernier permanent magnet machine as an example, the construction process of the multi-working permeance topology is shown in Fig. 5.34 [19].

To increase the stator's working specific permeance, its modulation unit must shift from a uniformly distributed structure to a non-uniformly distributed one. Additionally, to maintain winding symmetry, the main tooth traditionally used to wind the coil cannot serve as the machine's modulation unit. Consequently, the multi-working permeance can only be generated by the non-uniformly distributed auxiliary tooth extending from beneath the uniformly distributed main tooth. The homopolar, flux switching, and flux reversal machines with multi-working permeance are depicted in Fig. 5.35. Among these, the homopolar machine employs the second and fourth groups' magnetic fields, while the flux switching and flux reversal machines utilize the sixth group's magnetic field. Similar to the vernier permanent magnet machine, these three machines establish the multi-working permeance harmonic topology by designing the non-uniformly distributed auxiliary tooth. Furthermore, each main tooth of the flux switching machine is divided into two parts by the permanent magnet, resembling the auxiliary tooth structure. Hence, additional working specific permeance can be introduced by enlarging the slot opening to form a C-shaped iron core.

5.5.5 Multi-Working Magneto-Motive Force Flux Modulation Machine

The multi-working MMF FMM faces two primary challenges: selecting the pole pair number of the working MMF and constructing the corresponding topology.

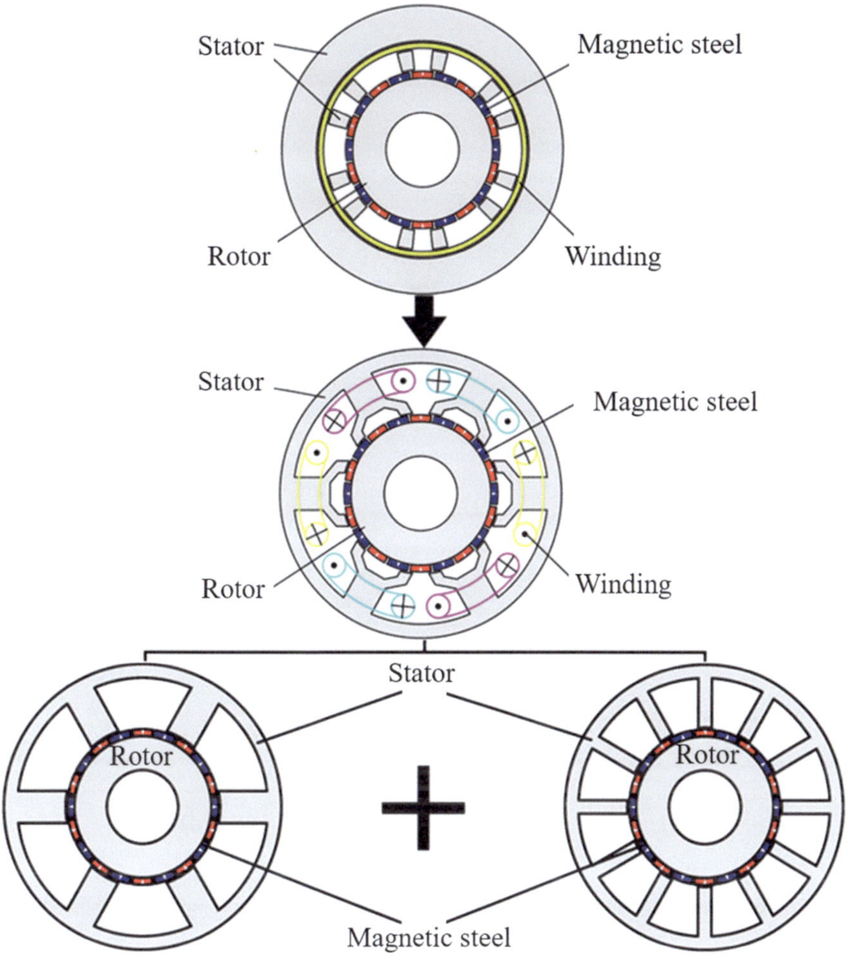

Fig. 5.34 Construction process of the multi-working permeance vernier permanent magnet machine

To maximize the utilization of working magnetic field harmonics, these magnetic fields are typically required to be tooth harmonics of each other. Therefore, it can be deduced that the pole pair number of the corresponding source of the MMF needs to satisfy the following equation:

$$|P_{\mathrm{es}h} \pm P_{\mathrm{es1}}| = n_h Z_{\mathrm{s}} \quad n_h = 1, 2, 3 \ldots \tag{5.23}$$

where n_h denotes a natural number, and $P_{\mathrm{es}h}$ represents the pole pair number of the working MMF. P_{es1} refers to the pole pair number of the working MMF when $h = 1$. According to Eq. (5.23), it is not mandatory for the pole pair number of each working MMF to be proportional to the tooth count Z_{s}. The sole condition is

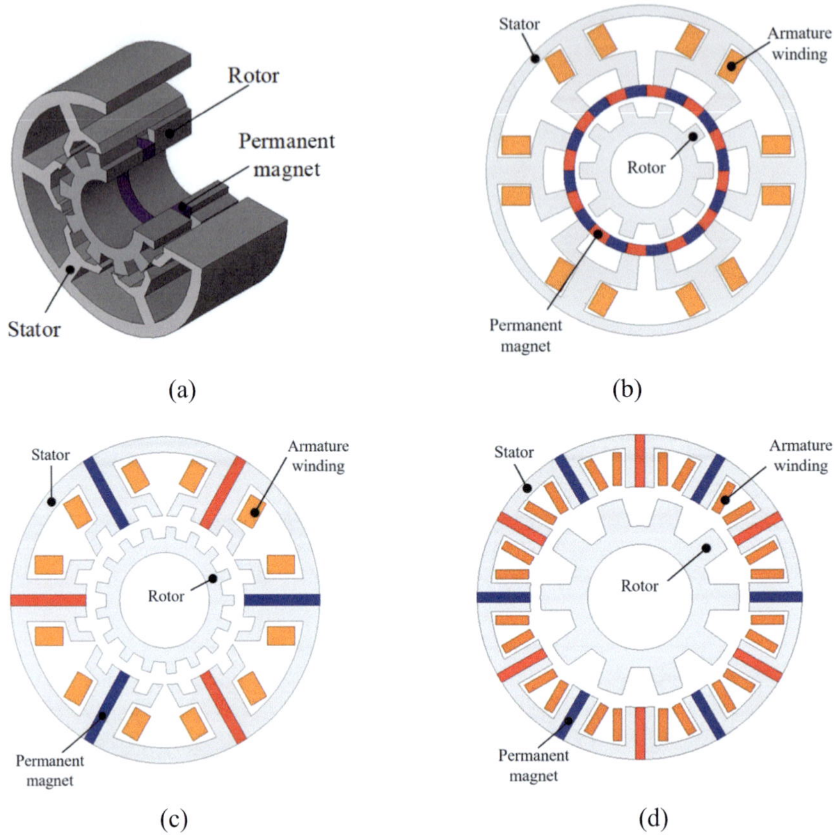

Fig. 5.35 Topologies of the multi-working permeance FMM: **a** homopolar machine with non-uniformly distributed auxiliary teeth, **b** flux reversal machine with non-uniformly distributed auxiliary teeth, **c** flux switching machine with non-uniformly distributed auxiliary teeth [20], **d** flux switching machine with C-shaped core [21]

that the sum or the difference of these two parameters should be a multiple of Z_s. This requirement may lead to a highly asymmetrical distribution of the machine's permanent magnets. In addition to working harmonics, the MMF produced by the permanent magnets is likely to generate numerous non-working harmonics, thereby inducing parasitic effects, such as asymmetry in the back-EMF of the three phases. For instance, the permanent magnet double-salient machine encounters a similar challenge. Conversely, this problem does not manifest in the flux switching permanent magnet machine, as the fundamental wave of its MMF has a pole pair number of $Z_s/2$, thus ensuring structural symmetry. Moreover, if the pole pair number of the MMF's fundamental wave is selected as Z_s, the machine's overall structure remains relatively symmetrical. Consequently, there are two primary selection strategies for the working pole pair number of the multi-working MMF (FMM):

$$P_{esh} = n_{eh}Z_s \quad \text{or} \quad P_{esh} = \frac{(2n_{eh} + 1)Z_s}{2} \tag{5.24}$$

The structure of the multi-working MMF FMM manifests in two variations. One variation involves the non-uniform distribution of permanent magnets. Figure 5.36 illustrates two types of multi-working MMF flux reversal machine topologies. The permanent magnets in these topologies occupy identical spatial positions to those in the flux reversal machine with uniformly distributed magnets; however, their configuration differs. This method of configuration is referred to as "MMF integration." In Fig. 5.36a, the permanent magnets beneath adjacent teeth exhibit the same configuration but differ in polarity, known as an "antisymmetric magnetic pole structure." The pole pair numbers of the MMF generated by such a structure are multiples of $Z_s/2$, marked as $hZ_s/2$. In Fig. 5.36b, the permanent magnets beneath adjacent teeth maintain identical shapes and polarities, described as a "symmetric magnetic pole structure." The pole pair numbers of the MMF generated by this structure are multiples of hZ_s.

The second approach involves the addition of new permanent magnets to the foundational structure. Both the original and the newly introduced permanent magnets function as independent field excitation structures, collectively enhancing the machine's operational magnetic field. This method of integration is referred to as "excitation unit integration." Owing to the requirement for additional space to accommodate the new permanent magnets, machines employing excitation unit integration typically feature a stator with an open-slot design. In this design, a gap is present between the main and auxiliary teeth, which allows for the embedding of the permanent magnet. For machine topologies that align with the magnetic fields of the fifth and sixth groups listed in Table 5.1, both the flux switching permanent magnet machine and the flux reversal permanent magnet machine employ an open-slot stator structure. However, in the flux reversal permanent magnet machine, the

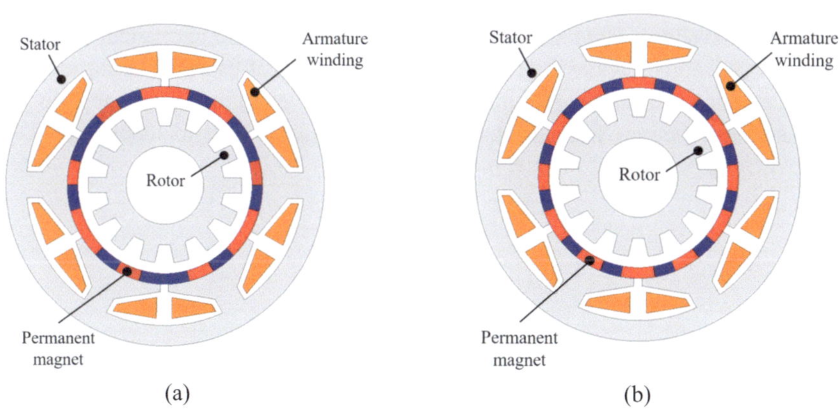

Fig. 5.36 Topologies of multi-working MMF flux reversal machine: **a** antisymmetric magnetic pole structure, **b** symmetric magnetic pole structure

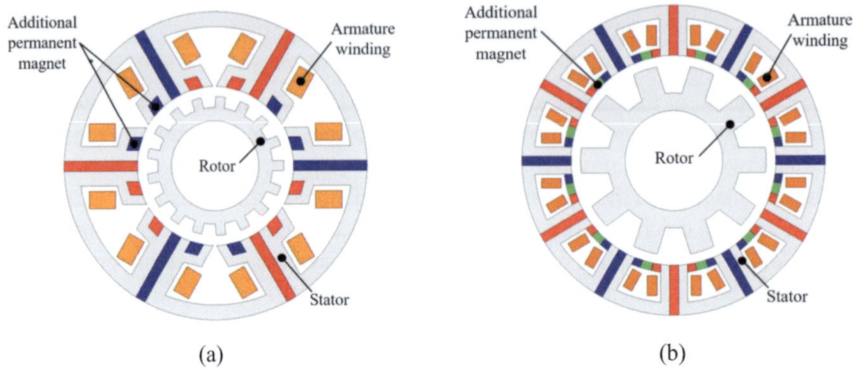

Fig. 5.37 Topologies of the excitation unit integrated flux switching machine: **a** placement of permanent magnet between auxiliary teeth [22], **b** placement of permanent magnet at slot notch

permanent magnet is positioned on the inner surface of the stator, which results in a relatively large equivalent air gap and consequently limits the performance enhancements achievable through this integration topology. Therefore, only the integration topology pertaining to the flux switching permanent magnet machine is discussed in this section, as illustrated in Fig. 5.37. As depicted in Fig. 5.37a, the topology has been evolved from that shown in Fig. 5.35c by incorporating radially magnetized permanent magnets between the auxiliary teeth. Given that the pole pair number for the excitation MMF of the flux switching machine is $Z_s/2$, the pole pair number of the integrated machine adheres to Eq. (5.24). In this scenario, the magnetization direction of two radial permanent magnets on a single U-shaped iron core is identical, whereas the magnetization direction of two radial permanent magnets on adjacent iron cores is opposite. This arrangement of the permanent magnets, maintaining the lowest pole pair number also at $Z_s/2$, introduces a significant number of odd-order harmonics that contribute to enhancing the output torque. Figure 5.37b shows the topology evolved from that in Fig. 5.35d by placing a Halbach permanent magnet at the stator slot notch, where the magnetization direction of two permanent magnets at adjacent slot notches is opposite, aligning with the combination of the pole pair number. Here, the new permanent magnet's pole pair number matches that of the flux switching permanent magnet machine, which is $Z_s/2$.

5.5.6 Intergroup Magnetic Fields Based Integrated Flux Modulation Machine

This section discusses the integration of working magnetic fields from various groups and the corresponding new machine topologies.

1. Non-integrable magnetic fields across groups one to four.

It has been established that the working magnetic field of the first to the fourth group does not support the formation of an integrated machine. This is because the fundamental machine topology corresponding to the second group's working magnetic field inherently includes the working magnetic field of the first group. Therefore, any attempt to integrate the magnetic fields of these two groups would be redundant. This redundancy also extends to the integration of the third and fourth groups' magnetic fields. Furthermore, any integration attempt among these four groups, such as between the first and third groups' magnetic fields, invariably leads to a topology characterized by homopolar permanent magnets on both the stator and rotor sides. This results in a serial magnetic circuit configuration, which fails to enhance the integrated topology. Thus, there is no practical benefit in pursuing integration among the magnetic fields of these four groups.

2. Homopolar-multipolar stationary excitation integrated machine.

The working magnetic fields of the first and fifth groups originate from the interaction between the DC term of the stator's specific permeance and the fundamental wave term of the rotor's specific permeance. The key difference lies in the type of MMF utilized: the first group employs a homopolar MMF on the stator side, whereas the fifth group uses a multipolar MMF on the stator side. In considering an integrated machine topology that combines the magnetic fields of these two groups, it is imperative that the multipolar MMF satisfies Eq. (5.25). This ensures that the working magnetic fields generated by these two magnetic motive forces are harmonically related as tooth harmonics of one another, optimizing the machine's functional efficacy and structural integrity.

$$P_{esh} = n_{eh} Z_s, \quad h = 1, 2, 3 \ldots \tag{5.25}$$

The equation stipulates that the pole pair number of the integrated machine must be an integral multiple of the stator tooth number. This requirement dictates the specific characteristics of the stator: the stator's specific permeance is represented by a DC term, the stator slot opening is relatively small, and the multipolar permanent magnets are confined to placement on the inner surface of the stator. Figure 5.38a depicts each stator tooth surface equipped with a pole pair of permanent magnets. To ensure the phase of the flux linkage generated by the two portions of permanent magnets corresponds with that of the back-EMF, several criteria must be met during the placement of the permanent magnets:

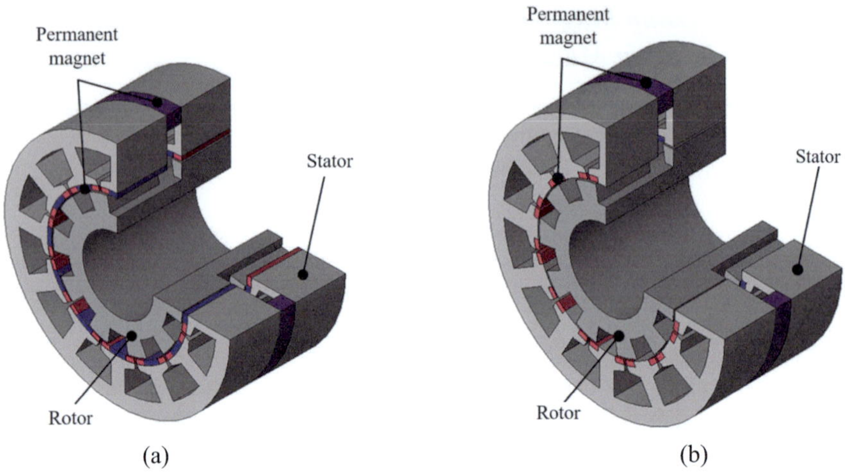

Fig. 5.38 Topologies of the homopolar-multipolar stationary excitation integrated machine. **a** Stator with surface-mounted permanent magnet. **b** Stator with consequent-pole permanent magnet

(a) Permanent magnets of a specific polarity are positioned in the middle of the teeth to maximize flux linkage when the stator tooth is precisely aligned with the rotor tooth.
(b) Permanent magnets of an alternate polarity are situated on both sides of the teeth.
(c) Given that the rotor teeth in the homopolar machine are staggered relative to each other, the axially aligned permanent magnets on the stator's surface must exhibit opposite polarities.

From the described configuration, it is evident that the excitation MMFs of this machine type are stationary and manifest in both homopolar and multipolar forms. Consequently, this topology is designated as the "homopolar-multipolar stationary excitation integrated machine."

The analysis extends to the integrated machine topology resulting from the interaction of the working magnetic fields of the first and sixth groups. The sixth group's working magnetic field is characterized by the stator's multipolar specific permeance and multipolar MMF, which necessitates an open slot structure in the stator to accommodate these features. Additionally, the inclusion of the second group's working magnetic field is crucial because the air gap specific permeance must also include a DC term, thus integrating the harmonic of the fifth group's working magnetic field. This complex integration encompasses four distinct types of working magnetic fields within the machine. To ensure harmonious operation and effective integration aligning all working magnetic fields as tooth harmonics of each other, both the MMF and the specific permeance of the stator are required to be integral multiples of the tooth number:

$$P_{esh} = n_{eh}Z_s \quad h = 1, 2, 3 \ldots$$
$$P_{fsl} = n_{fl}Z_s \quad l = 1, 2, 3 \ldots \tag{5.26}$$

The detailed machine topology can be derived from the structure illustrated in Fig. 5.38a by substituting the permanent magnet located directly beneath each stator tooth with an iron core. In this revised configuration, the stator adopts a consequent-pole structure, wherein the ferromagnetic pole at the tooth tip produces alternating specific permeance. Let P_{es1} and P_{fs1} denote the lowest pole pair numbers of the MMF and specific permeance, respectively. According to Eq. (5.26), the simplest viable combination of pole pair numbers is:

$$P_{es1} = P_{fs1} = Z_s \tag{5.27}$$

By integrating the working magnetic fields of the third and fourth groups with those of the fifth and sixth groups, a homopolar-multipolar stationary excitation integrated machine can be constructed. In this design, the homopolar permanent magnet is positioned on the rotor side. It is important to note that stationary or rotary characteristics do not apply to homopolar excitation; thus, the machine's topology remains similar to that depicted in Fig. 5.38.

1. Homopolar-multipolar rotatory excitation integrated machine

The focus now shifts to the integrated machine formed by combining the working magnetic fields of the first and seventh groups. The working magnetic field of the first group configures a homopolar machine with the rotor serving as the modulation unit, whereas the seventh group's working magnetic field constitutes a conventional permanent magnet machine with the rotor as the excitation unit. Previous analyses indicate that both groups' working pole pair numbers are equivalent, which necessitates a consequent-pole structure for the rotor. In a homopolar machine, the rotor is typically divided into two parts axially, staggered relative to each other by half-tooth pitches, ensuring that the multipolar excitation structure maintains consistent polarity. The magnetization directions of the permanent magnets in these two parts are opposite, as shown in Fig. 5.39a. Unlike the stationary excitation machine presented in Fig. 5.38, the multipolar excitation component of this machine is rotatory, thereby earning the designation "homopolar-multipolar rotatory excitation integrated machine."

Similar to the homopolar-multipolar stationary excitation integrated machine with an open-slot structure, integrating the working magnetic fields of the first group with those of the eighth group results in a machine whose topology mirrors that of Fig. 5.39a but includes an open-slot stator. This topology falls under the category of the homopolar-multipolar rotatory excitation integrated machine, as illustrated in Fig. 5.39b. Since the integration of the working magnetic fields of the third and fourth groups with those of the seventh and eighth groups leads to an identical machine topology, this scenario is not reiterated here .

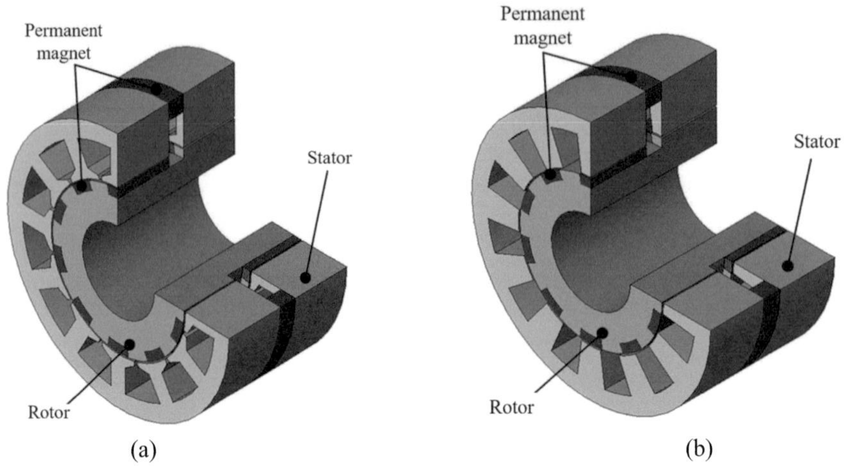

Fig. 5.39 Topologies of the homopolar-multipolar rotatory excitation integrated machine: **a** stator with semi-closed slot structure, **b** stator with open slot structure

2. Bilateral excitation integrated flux modulation machine

The integration of the working magnetic fields of the fifth to the eighth groups has not yet been analyzed extensively. Notably, the integration of the fifth and seventh groups' working magnetic fields results in a stator structure with surface-mounted permanent magnets, as depicted in Fig. 5.40a. However, this configuration does not exploit the modulation effect of the stator, which leads to suboptimal performance. Therefore, the integration of the working magnetic fields of the fifth and eighth groups will be the primary focus in the subsequent discussion. The fifth group's working magnetic field is associated with the flux reversal machine that features uniformly distributed permanent magnets. The eighth group's working magnetic field, in contrast, relates to the FMM with a stationary modulation unit. According to Table 5.1, the pole pair numbers of these two working magnetic fields are denoted by $|P_{esh} \pm P_r|$ and $|P_{fsl} \pm P_r|$, respectively. To ensure that these magnetic fields are tooth harmonics of each other, the following conditions, detailed in Eq. (5.28), must be met:

$$\begin{aligned}|P_{esh} \pm P_{es1}| &= n_h Z_s \\ |P_{fsl} \pm P_{fs1}| &= n_l Z_s \\ |P_{esh} \pm P_{fsl}| &= n_{hl} Z_s\end{aligned} \tag{5.28}$$

Given that only a single main harmonic is present in both the stator's working MMF and specific permeance, satisfying the third function in Eq. (5.28) is imperative. From the perspective of topology construction, the simplest strategy is to design a structure akin to the consequent-pole configuration, with $P_{es1} = P_{fs1} = Z_s$ [23]. Figure 5.40b, c show two examples of such topologies.

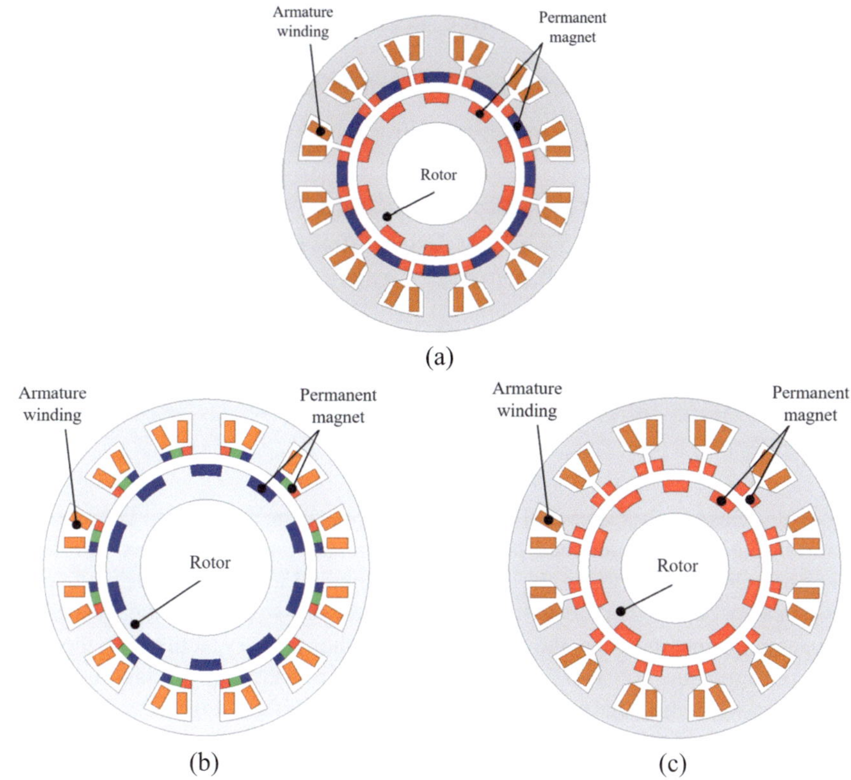

Fig. 5.40 Topologies of the bilateral excitation integrated FMM. **a** Stator with surface-mounted permanent magnet. **b** Placing Halbach permanent magnet at slot notch. **c** Stator with consequent-pole permanent magnet

In this section, we analyze the MMF-specific permeance model to investigate the potential configurations of the FMM's working magnetic field. By applying the enumeration method, we provide insights into the integration method and the constraint relationship between the pole pair number of the working MMF and the pole pair number of the specific permeance. It is important to understand that the presented topologies are merely illustrative examples of each integration approach, and numerous other topologies remain unlisted. Nonetheless, the mathematical integration forms can be classified within the comprehensive categorization outlined in this section. We refrain from conducting a detailed analysis of the performance of each integrated machine here. Sections 5.6–5.8 will delve into the working principles and performance characteristics of three selected typical machine topologies.

5.6 Multi-Working Permeance Vernier Permanent Magnet Machine

5.6.1 Basic Topology

The configurations of the multi-working permeance vernier permanent magnet machine are depicted in Fig. 5.41a. This design differs from the conventional vernier permanent magnet machine, which features auxiliary teeth, in that it exhibits unequal spacing between each auxiliary tooth. In the standard vernier permanent magnet machine, the spacing between each auxiliary tooth is uniform, and the pitch angle is calculated as follows:

$$\theta_{f0} = \frac{2\pi}{P_f} = \frac{2\pi}{n_f Z_s} \tag{5.29}$$

where n_f represents the number of auxiliary teeth below the main tooth. The topology shown in Fig. 5.41a, which incorporates a six-slot stator, twelve auxiliary teeth, and ten rotor pole pairs, serves as an illustrative example. Furthermore, the pitch angle between adjacent auxiliary teeth θ_{f0} is 30°. For the multi-working permeance vernier permanent magnet machine, the spacing between the auxiliary teeth below the same main tooth remains constant but is denoted by θ_f instead of θ_{f0}. Additionally, the spacing between the adjacent auxiliary teeth beneath different main teeth (i.e., the slot notch of the armature winding) is modified. When $\theta_f > \theta_{f0}$, such a spacing is smaller than θ_{f0}. Conversely, when $\theta_f < \theta_{f0}$, this spacing is larger than θ_{f0}. Therefore, the spacings between each auxiliary tooth vary. The tooth pitch ratio k_f is a critical parameter in this type of machine, defined as:

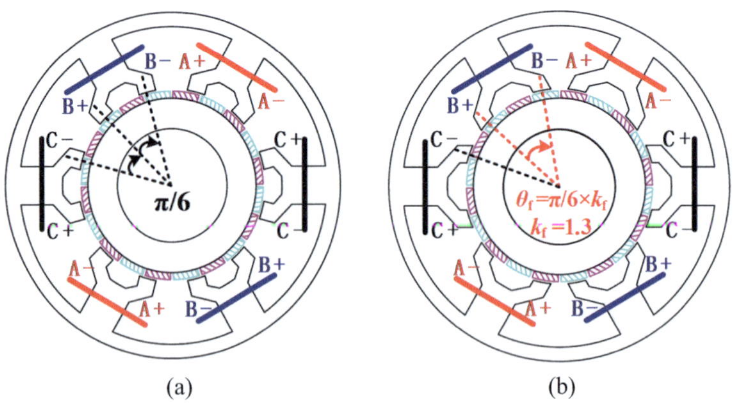

Fig. 5.41 Comparison between two vernier permanent magnet machines with different distribution of auxiliary teeth: **a** regular permanent magnet vernier machine with auxiliary teeth, **b** multi-working permeance permanent magnet vernier machine

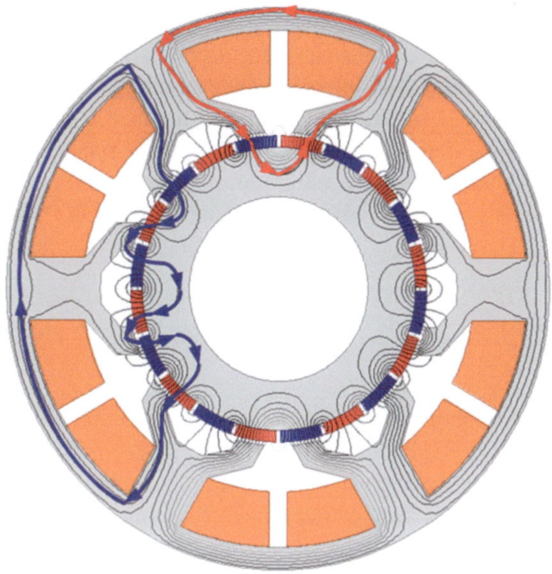

Fig. 5.42 Flux distribution of a multi-working permeance vernier permanent magnet machine under the no-load condition

$$k_{\mathrm{f}} = \frac{\theta_{\mathrm{f}}}{\theta_{\mathrm{f0}}} \quad (5.30)$$

As depicted in Fig. 5.41b and stated in Eq. (5.30), a k_{f} of 1 corresponds to the vernier permanent magnet machine featuring conventional pitch angles for auxiliary teeth. A k_{f} value greater than 1 indicates a smaller slot notch, which affects the placement of the winding. Figure 5.42 illustrates the flux distribution of the machine under a no-load condition.

5.6.2 Flux Density Harmonic Characterization

Compared to the traditional vernier permanent magnet machine with auxiliary teeth, the multi-working permeance vernier permanent magnet machine presented in this section exhibits significant differences. A qualitative analysis will be conducted using the MMF-specific permeance model. When $k_{\mathrm{f}} = 1$, the auxiliary teeth are uniformly distributed, and the specific permeance function can be expressed as follows:

$$\lambda_{\mathrm{s\_reg}} = \lambda_0 + \sum_{l=1}^{\infty} \mathrm{sgn}_l \lambda_{lP_{\mathrm{f}}} \cos(lP_{\mathrm{f}}\theta) = \sum_{l=1}^{\infty} \mathrm{sgn}_l \lambda_{ln_{\mathrm{f}}Z_{\mathrm{s}}} \cos(ln_{\mathrm{f}}\theta) \quad (5.31)$$

where adding "reg" to the subscript of λ_{s} represents that it corresponds to the specific permeance of the conventional uniformly distributed auxiliary teeth topology. "sgn_l"

is the sign function, indicating that the phase of the specific permeance harmonic can be 180°. For ease of subsequent comparisons between this conventional topology and the multi-working permeance topology, the subscript of the harmonic amplitude in Eq. (5.31) indicates its pole pair number. It can be observed from the expression that the minimum pole pair number of the specific permeance is P_f.

When $k_f \neq 1$, the minimum pole pair number (or period number of the spatial specific permeance distribution) decreases from P_f to the stator's main tooth number Z_s. The corresponding expression for the specific permeance is outlined in Eq. (5.32):

$$\lambda_{s\_pro}=\lambda_0+\sum_{l=1}^{\infty}\mathrm{sgn}_l\lambda_{lZ_s}\cos(lZ_s\theta) \tag{5.32}$$

Comparing Eqs. (5.31) and (5.32), it becomes evident that the specific permeance in Eq. (5.32) encompasses not only all the harmonics found in Eq. (5.31) but also additional harmonics whose pole pair numbers are integral multiples of Z_s rather than P_f. This demonstrates that the specific permeance component is more abundant and verifies the working principle introduced in Sect. 5.5.4.

Further analysis of the air gap magnetic field harmonic for such a machine is conducted below. According to the analysis in Sect. 5.2, only the fundamental term of the permanent magnet rotor's MMF participates in the electromechanical energy conversion, which can be expressed as:

$$F_{er}=F_{er1}\cos(P_e\theta-P_e\Omega_e t) \tag{5.33}$$

where F_{er1} represents the fundamental amplitude of the rotor's MMF.

Using Eqs. (5.32) and (5.33), the air gap working magnetic flux density can be stated as Eq. (5.34) by utilizing the principle of the flux modulation:

$$\begin{aligned}B_e(\theta,t)=\frac{\mu_0}{g}\Bigg\{&F_{er1}\lambda_0\cos(P_e\theta-P_e\Omega_e t)+\frac{1}{2}\sum_{l=1}^{\infty}F_{er1}\lambda_{lP_f}\cos[(P_e\pm lP_f)\theta-(P_e\Omega_e)t]\\&+\frac{1}{2}\sum_{\substack{l'=1\\ l'\neq lP_f}}^{\infty}F_{er1}\lambda_{l'Z_s}\cos[(P_e\pm l'Z_s)\theta-(P_e\Omega_e)t]\Bigg\}\end{aligned} \tag{5.34}$$

In Eq. (5.34), the initial term represents the non-modulated magnetic flux density, while the second term corresponds to the magnetic flux density harmonic found in the conventional vernier machine with auxiliary teeth. Upon modifying the stator to incorporate a nonuniform distribution of auxiliary teeth, the third term emerges as the magnetic flux density produced by the modulation of newly identified specific permeance harmonics, previously absent. These magnetic flux density harmonics vary in pole pair numbers and rotational speeds, yet their corresponding spatial alternating frequencies remain identical. If these harmonics can interact with the

Table 5.2 Analysis of multi-working permeance permanent magnet vernier machine's flux density harmonics

Pole pair number	Speed	Frequency
P_e	Ω_e	$\Omega_e \times P_e$
$P_e \pm lP_f$ $l = 1, 2, 3, \ldots$	$\frac{P_e \Omega_e}{P_e \pm lP_f}$	$\Omega_e \times P_e$
$P_e \pm l'Z_s$ $l' = 1, 2, 3, \ldots, l' \neq ln_f$	$\frac{P_e \Omega_e}{P_e \pm l'Z_s}$	$\Omega_e \times P_e$

stator's windings, they facilitate the generation of flux linkage and back-EMF at the same frequency within the windings, as detailed in reference [24]. The characteristics of all magnetic flux density harmonics mentioned in Eq. (5.34) are compiled in Table 5.2.

Table 5.3 outlines the magnetic flux density harmonic content for the multi-working permeance vernier permanent magnet machine, featuring slot number/auxiliary tooth number/rotor's pole pair number combinations of 6/12/10, 6/18/16, and 6/24/22. In the table, entries with a colored base indicate the emergence of new working magnetic fields. As FMMs, the pole pair numbers of the stator windings in these three topologies continue to satisfy the relation $P_a = |P_f \pm P_e|$. Consequently, the armature windings in these three FMMs maintain a six-slot-four-pole configuration.

Moreover, the fundamental back-EMF of this machine type can be expressed as follows:

$$E_a = 2\Omega_e r_g l_{stk} N_{sa} \sin(\omega a t) \times \sum_{l=0}^{\infty} B_{(P_e \pm lZ_s)} \frac{P_e}{P_e \pm lZ_s} k_{wa(P_e \pm lZ_s)} \tag{5.35}$$

where the Ω_e represents rotor's rotation speed, r_g represents the air gap radius, l_{stk} represents the axial length of the stack, N_{sa} represents the serial turn number per phase, ω_a represents the electrical frequency of the fundamental wave, and B and k_{wa} represent the amplitude of the magnetic flux density harmonic and the corresponding

Table 5.3 Analysis of pole pair number of the main harmonics

Rotor's magneto-motive force			Pole pair number of specific permeance	Flux density		
6-24-22	6-18-16	6-12-10	P=6l	6-12-10	6-18-16	6-24-22
			0	10	16	22
			6	4	10	16
		10	12	2	4	10
	16		18	8	2	4
22			24	14	8	2
			30	20	14	8
		–	36	–	20	14
	–		42		–	20

winding factor, respectively. Notably, the subscripts "$P_e \pm lZ_s$" refer to the pole pair numbers of corresponding harmonics.

It is important to note that all vernier permanent magnet machines with auxiliary teeth must comply with Eq. (5.36) to adhere to the flux modulation theory. Given that the pole pair numbers of all working magnetic field harmonics satisfy $P = P_e \pm lZ_s$, the new magnetic flux density harmonics in the multi-working permeance vernier permanent magnet machine are tooth harmonics of the original working harmonics. Consequently, the amplitudes of their winding factors are equivalent, consistent with the conventional FMM analyzed in the fourth chapter. To summarize, Eq. (5.35) can be reformulated as Eq. (5.37), where all the magnetic flux density harmonics correspond to the same amplitude of the winding factor k_{waPa}, which is the winding factor of the fundamental wave.

$$P_f = n_f Z_s = P_e \pm P_a \tag{5.36}$$

$$E_{a1} = 2\Omega_e r_g l_{stk} N_{sa} \sin(\omega a t) \times \sum_{l=0}^{\infty} \left| B_{(P_e \pm lZ_s)} \frac{P_e}{P_e \pm lZ_s} \right| \mathrm{sgn}_l k_{waP_a} \tag{5.37}$$

As indicated in Eq. (5.37), each magnetic flux density harmonic's contribution to the fundamental back-EMF is proportional to its amplitude and the corresponding pole ratio. However, this analysis is merely qualitative, based on the principles of flux modulation. To further elucidate the operational characteristics of this machine type, a quantitative analysis of each magnetic flux density harmonic's contribution to the back-EMF and electromagnetic torque will be conducted in the subsequent subsection.

5.6.3 *Quantitative Calculation of Each Harmonic Contribution*

In the study of the multi-working permeance vernier permanent magnet machine, a pivotal concern is quantifying the contribution of the back-EMF and torque from the new specific permeance created by the non-uniform distribution of auxiliary teeth. Additionally, it is essential to determine whether this contribution can offset the torque reduction caused by the decrease in the original specific permeance. While finite element simulation models can provide the ultimate outcomes for back-EMF and torque, they often fail to disclose the internal variation patterns. This challenge arises because the working magnetic fields generated by each specific permeance are tooth harmonics to each other, producing the fundamental back-EMF in the winding simultaneously, complicating their separation. Although Eq. (5.37) can be employed to analytically calculate the back-EMF contributed by each harmonic, subsequent analyses indicate that these results may not be precise. The underlying reasons are detailed below.

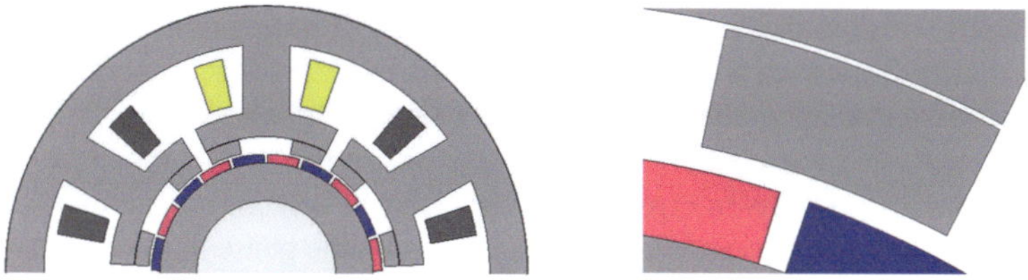

(a) (b)

Fig. 5.43 Schematic diagram of using the virtual air gap method to analyze the multi-harmonic vernier machine: **a** machine model, **b** enlarged view of double air gap structure

The multi-working permeance vernier permanent magnet machine features an auxiliary tooth structure that positions the winding far from the air gap. Observations from Fig. 5.42 reveal that certain magnetic fields, particularly those with higher pole pair numbers, form circuits directly with the auxiliary tooth and iron yoke, rather than interacting with the winding [25]. Consequently, although these magnetic field harmonics are present in Eq. (5.37), they do not effectively contribute to the back-EMF. Furthermore, magnetic fields with relatively high pole pair numbers experience a diminished linkage level due to flux leakage, an effect not accounted for in Eq. (5.37).

To address the challenge of accurately analyzing the contributions of back-EMF and their linkage with the winding in the multi-working permeance vernier permanent magnet machine, the virtual air gap method will be introduced later in this chapter. This method involves introducing a small air gap between the auxiliary teeth and the armature winding, as depicted in Fig. 5.43. The implementation of this virtual air gap is designed to be minimal so as not to detrimentally affect the machine's overall electromagnetic performance. Importantly, because this air gap is positioned close to the winding, it provides a more accurate representation of the magnetic flux density distribution and thus more reliably reflects the harmonic content that effectively links with the winding.

In the virtual air gap model, the separation of the winding and auxiliary teeth allows for the modification of the virtual winding. This adjustment ensures that the working harmonics originally present in the machine are no longer tooth harmonics, enabling the virtual winding to exclusively interact with the magnetic fields of specific pole pair numbers. This configuration facilitates the separation of the back-EMF's content in the finite element model, allowing for a more precise analysis of how each component contributes to the machine's performance.

Figure 5.44 presents the no-load magnetic flux density distribution and corresponding spectral analysis results for the machine configured with a 6/12/10 combination under tooth pitch ratios (k_f) of 1.0 and 1.3. Analyzing the harmonic order, the pole pair numbers of the magnetic flux density harmonics under both constant and variable tooth pitch conditions align with the summaries in Table 5.3. When $k_f = 1.0$, the specific permeance harmonic with 12 pole pairs generates magnetic flux densities of 2 and 22 pole pairs. Meanwhile, the even-order specific permeance harmonic of

24 pole pairs results in a magnetic flux density of 14 pole pairs. The fundamental wave's magneto-motive force interacting with the DC term of the specific permeance notably produces a magnetic flux density of 10 pole pairs.

In the analysis presented, the negative sign above the bar chart for the 14-pole-pair magnetic flux density in Fig. 5.44 indicates that, when this magnetic flux density harmonic is incorporated into the analytical formula, the phase of the resultant back-EMF is opposite to that of the other main magnetic flux density harmonics. This indicates that the back-EMF generated by the 14 pole pairs acts to diminish the overall fundamental back-EMF, thereby influencing the machine's performance. With a k_f of 1.3, the specific permeance harmonics exhibit pole pair numbers of 6, 12, 18, etc. Figure 5.44 depicts that the multi-working permeance vernier permanent magnet machine possesses a greater number of magnetic flux density harmonics that can interact with the stator winding, and the pole pair numbers of these new magnetic flux density harmonics align with the theoretical predictions outlined in Table 5.3.

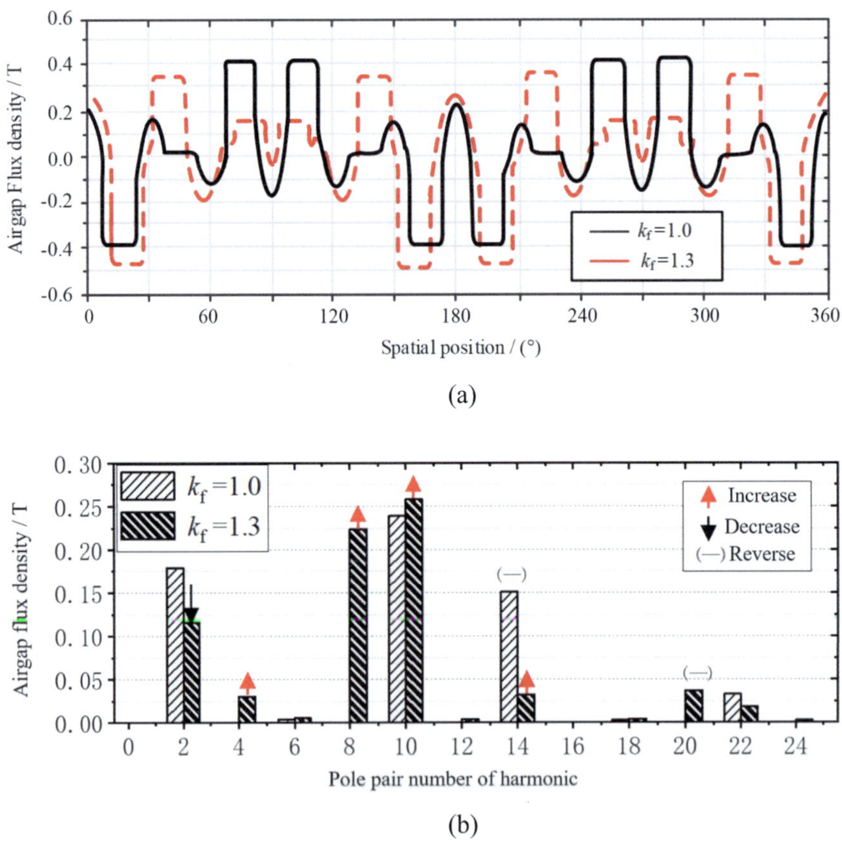

Fig. 5.44 Comparison of outer air gap's magnetic flux density harmonics with different k_f. **a** Comparison of flux density distribution, **b** comparison of flux density harmonic content

To further analyze the component of back-EMF contributed by each harmonic, based on the air gap magnetic field, two methods can be employed:

1. Semi-analytical method: This method uses Eq. (5.37) for calculations. To enhance the accuracy of the results, the magnetic flux density from the virtual air gap is used in the equation, rather than that from the actual air gap. This adjustment ensures that the calculated back-EMF more accurately reflects the effects of the harmonics that truly link with the winding.
2. Finite element simulation: This method involves reconstructing the winding to facilitate the separation of contributions from specific harmonics. The models for this are depicted in Fig. 5.45. The core idea here is to construct a virtual winding that exclusively links with a designated harmonic. This specialized configuration allows for the isolation and measurement of the fundamental back-EMF associated with each harmonic. It is important to note that since the turn number and winding factor of the virtual winding differ from those of the actual winding, a conversion is necessary to accurately interpret the final results.

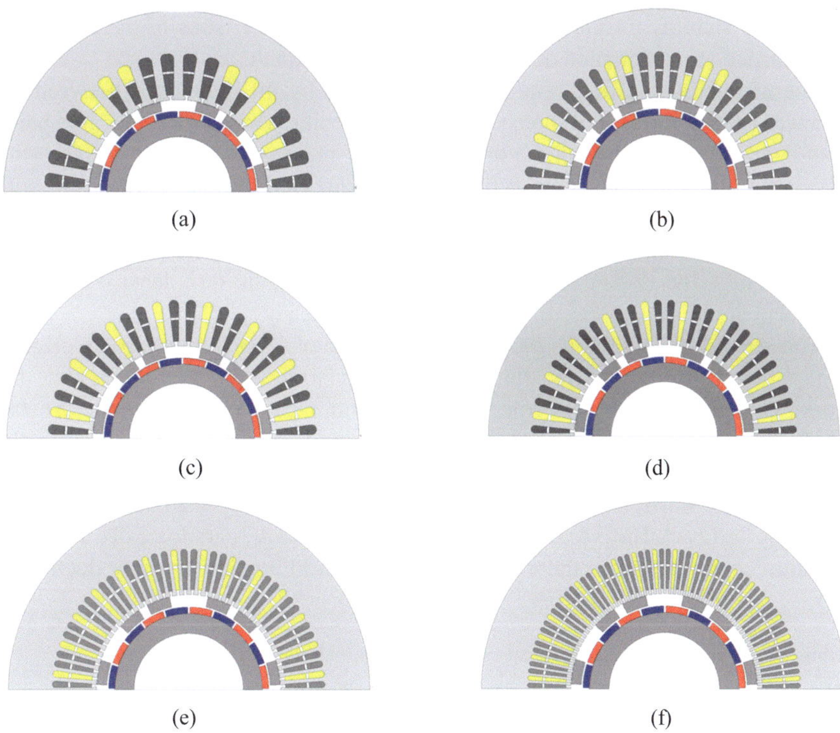

Fig. 5.45 Finite element models used to separate each working harmonic: **a** 36 slots/2 pole pairs, winding pitch = 7, **b** 48 slots/4 pole pairs, winding pitch = 5, **c** 48 slots/8 pole pairs, winding pitch = 3, **d** 60 slots/10 pole pairs, winding pitch = 3, **e** 84 slots/14 pole pairs, winding pitch = 3, **f** 120 slots/20 pole pairs, winding pitch = 3

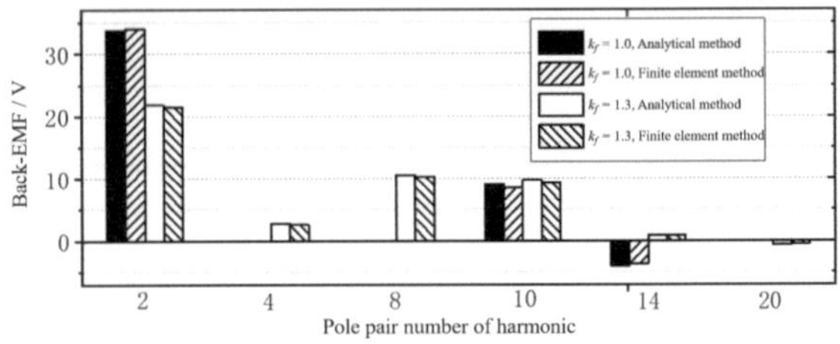

Fig. 5.46 Comparison of fundamental back-EMF with different k_f

The comparative analysis between the results of the analytical method and the finite element simulation method, as depicted in Fig. 5.46 and summarized in Table 5.4, demonstrates a good alignment between the two methodologies, validating their reliability. Based on the results shown in Fig. 5.46, a detailed analysis of how the back-EMF contributed by each spatial harmonic changes with the modification of the spacing between each modulation tooth is undertaken. Initially, the magnetic field of 10 pole pairs represents the non-modulated working magnetic field. The contribution of this field to the back-EMF remains relatively constant due to only minor variations in the DC term of the specific permeance after altering the tooth pitch. This stability ensures that the fundamental characteristics of the machine's performance are maintained despite changes in tooth pitch.

The magnetic field of 2 pole pairs is generated by the interaction between the magneto-motive force of 10 pole pairs and the specific permeance of 12 pole pairs. This interaction is critical in the original vernier permanent magnet machine with auxiliary teeth, where it serves as the main working magnetic field. However, with an increase in k_f to 1.3, resulting in a non-uniform distribution of auxiliary teeth, the amplitude of the specific permeance associated with 12 pole pairs is reduced. Consequently, the back-EMF contributed by the magnetic flux density of 2 pole pairs also diminishes, impacting the machine's efficiency.

The second-order specific permeance harmonic of 24 pole pairs, in its original configuration, reacts with the fundamental magneto-motive force to generate the working magnetic fields of 14 pole pairs. This harmonic typically contributes a back-EMF with a phase opposite to that of the main harmonics, thereby weakening the overall back-EMF. However, with the increased k_f, the phase of this specific permeance harmonic shifts, causing the phases of the corresponding magnetic flux density harmonic and back-EMF to become opposite. This change, paradoxically, results in a positive contribution to the overall back-EMF. Additionally, the introduction of

Table 5.4 Comparison of fundamental back-EMF magnitudes

Parameter	Analytical method	Finite element method
$k_f = 1.0$	38.65 V	38.26 V
$k_f = 1.3$	45.02 V	44.38 V
Back-EMF improvement (%)	16.5%	16.0%

specific permeances of 6 and 18 pole pairs into the air gap, reacting with the fundamental magneto-motive force, results in the generation of magnetic flux densities of 4 and 8 pole pairs, respectively. These changes introduce new components to the back-EMF, potentially enhancing the machine's performance by contributing additional harmonics that may assist in the overall energy conversion process within the machine. To conclude, adjusting the tooth pitch of auxiliary teeth in vernier permanent magnet machines introduces new working magnetic flux density harmonics that can mitigate losses in back-EMF due to reduced amplitudes of the original working harmonics. Furthermore, such adjustments can enhance the overall back-EMF, as demonstrated in the previous example where the back-EMF increased by 16%.

5.6.4 Slot-Pole Combination and Performance Analysis

For any vernier machine with auxiliary teeth, changes in the tooth pitch result in new magnetic flux density harmonics that are tooth harmonics of all original working harmonics. Consequently, all vernier permanent magnet machines with auxiliary teeth can be adapted to a multi-working permeance structure. This new structure must adhere to the flux modulation relationship and maintain the symmetry condition for the three-phase winding, essential for balanced performance:

$$\frac{Z_s}{GCD(Z_s, P_a)} = 3n, \quad n = 1, 2, 3 \ldots \tag{5.38}$$

Additionally, this type of machine requires a sufficiently high winding factor and pole ratio to maintain a robust torque density level under specific line loads. Hence, when determining feasible slot-pole combinations for multi-working permeance vernier permanent magnet machines, certain limitations must be considered.

$$k_{waP_a} \geq 0.866, \quad \frac{P_e}{P_a} \geq 3.5 \tag{5.39}$$

By combining Eqs. (5.38) and (5.39), the feasible slot-pole combinations for machines configured with either 6 or 12 slots are listed in Table 5.5.

Table 5.5 Stator slot number/auxiliary tooth number/stator and rotor's pole pair number combination table

$Z_s=6$					
P_f	12	18		24	
P_m	10	16	14	22	20
P_a	2	2	4	2	4
P_{add}	4	4	2	4	2
k_{waP_a}	0.87	0.87	0.87	0.87	0.87
P_r/P_a	5	8	3.5	11	5

$Z_s=12$										
P_f	24		36				48			
P_m	20	19	32	31	29	28	44	43	41	40
P_a	4	5	4	5	7	8	4	5	7	8
P_{add}	8	7	8	7	5	4	8	7	5	4
k_{waP_a}	0.87	0.93	0.86	0.93	0.93	0.86	0.86	0.93	0.93	0.86
P_r/P_a	5	3.8	8	6.2	4.1	3.5	11	8.6	5.9	5

Within this table, P_a represents the stator's pole pair number as defined in the conventional vernier machine context. P_{add} represents the smallest pole pair number of the new working magnetic flux density harmonic post-tooth pitch modification. k_{waPa} represents the common winding factor for all working magnetic flux density tooth harmonics. The machine topologies with slot number/auxiliary tooth number/rotor's pole pair number/stator's pole pair number combinations of 6/12/10/2, 6/18/16/2, 6/24/22/2, and 12/24/19/5 have been selected for detailed analysis to explore the relationship between their k_f, no-load back-EMF, and output torque. The no-load magnetic field distributions of these topologies are illustrated in Fig. 5.47, and the main design parameters are summarized in Table 5.6.

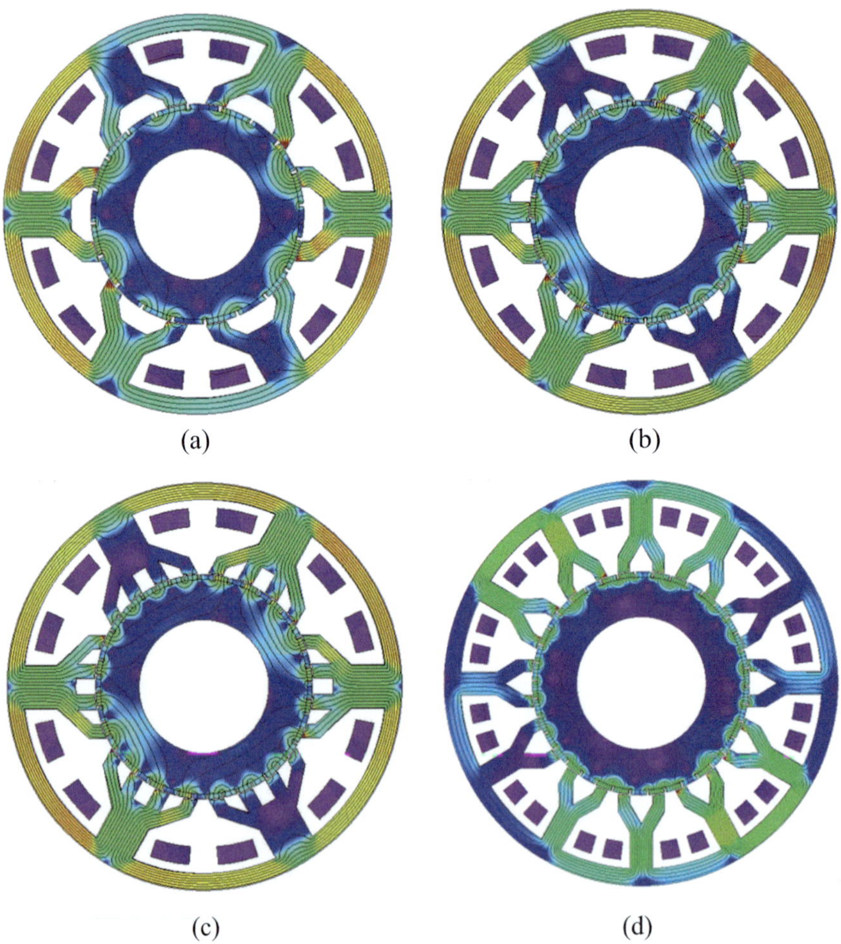

Fig. 5.47 No-load magnetic field distributions of multi-working permeance vernier permanent magnet machines. **a** $Z_s = 6$, $P_f = 12$, $P_e = 10$, $P_a = 2$. **b** $Z_s = 6$, $P_f = 18$, $P_e = 16$, $P_a = 2$. **c** $Z_s = 6$, $P_f = 24$, $P_e = 22$, $P_a = 2$. **d** $Z_s = 12$, $P_f = 24$, $P_e = 19$, $P_a = 5$

Table 5.6 Main design parameters

Parameter	Value	Parameter	Value
Stator outer diameter	124 mm	Stack length	85 mm
Slot depth	14.8 mm	Air gap length	0.5 mm
PM thickness	2.5 mm	Pole arc	0.9
Rotation speed	300 rpm	Winding pitch	1
Electrical loading	375 A/cm	Current density	4.86 A/mm^2
Winding structure	Double layer	Series turns per phase	214
PM remanence	1.235 T	Relative permeability	1.05

Figure 5.48 illustrates the variation in the machine's fundamental back-EMF in response to changes in k_f, the tooth pitch ratio, with the thickness of the permanent magnet set at 2.5 mm and the auxiliary tooth width to average tooth pitch ratio at 0.4. Here, $k_f = 1$ represents the baseline configuration with a conventional equivalent tooth pitch topology. As k_f increases, the back-EMF for each model typically shows an initial increase followed by a decrease, peaking at k_f values of 1.3, 1.18, 1.12, and 1.36 for the respective topologies before declining. When k_f is optimized, under the same magnetic load and with nearly unchanged active material usage, the back-EMF increases by at least 20% compared to the scenario when $k_f = 1$.

Figure 5.49 presents the electromagnetic torque waveforms for each model under a rated line load of 375 A/cm and $I_d = 0$. Similar to the back-EMF findings, the average torque improves by at least 20% when k_f is optimized. Notably, the 6/18/16/2 model, with a pole ratio of 8, achieves the highest average torque following optimization. Conversely, the 12/24/19/5 model, which has a relatively low pole ratio of 3.8, despite showing the largest percentage increase in torque, still produces a lower torque due to a weaker flux modulation effect. Additionally, although the 6/24/22/2 model has the highest pole ratio of 11, its torque level remains below that of the topology with a pole ratio of 8, likely because the large pole pair number of

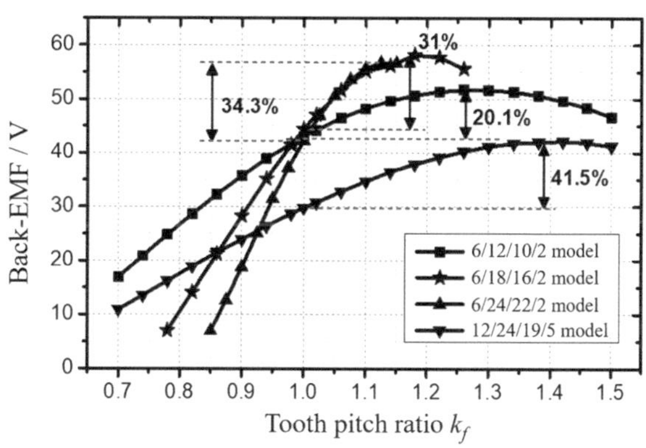

Fig. 5.48 Variation of the fundamental back-EMF with changes in k_f

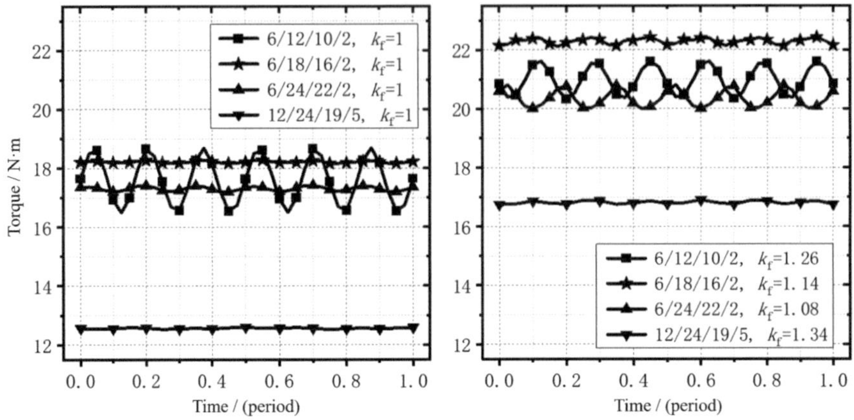

Fig. 5.49 Electromagnetic torque waveforms before and after k_f optimization

the permanent magnets, given the same rotor's outer diameter and air gap diameter, does not optimally contribute to torque production.

5.7 Multi-Working Magneto-Motive Force Flux Reversal Machine

The discussion also mentions two methods for forming a multi-working magnetic motive force FMM. The first method involves modifying the conventional flux-reversal machine to integrate magneto-motive forces of different pole pair numbers. The second method is achieved by integrating different excitation units, typically using a construction based on the conventional flux switching machine, where permanent magnets are embedded at the location of the slot opening. The first method is particularly highlighted as it can enhance the output torque without increasing the usage level of permanent magnets and is thus selected for detailed study in this section.

5.7.1 Construction Method of Permanent Magnet Arrays

The formation of a multi-working magnetic motive force flux-reversal machine critically depends on the arrangement of a permanent magnet array capable of generating multiple magneto-motive force harmonics. While it is possible to create such an array by varying the width of each permanent magnet, this method requires precise optimization of each magnet's width, resulting in a complex design process. Additionally, this approach may not effectively maximize the use of harmonics, prompting

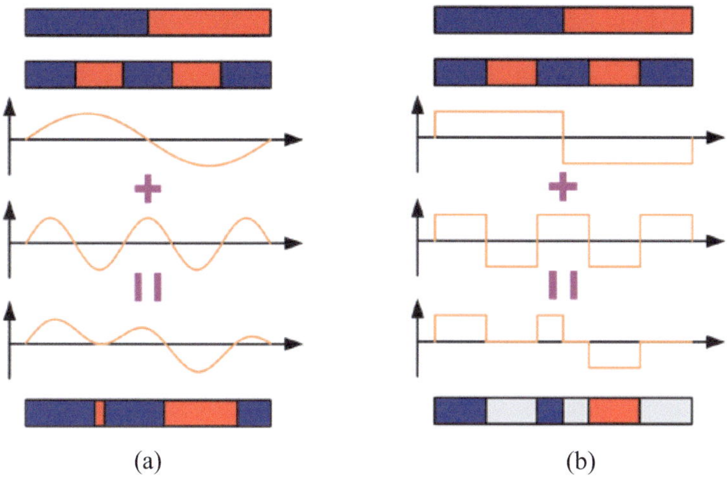

Fig. 5.50 Construction of the multi-magnetic motive force permanent magnet array. **a** Bi-level structure. **b** Tri-level structure

the need for an alternative construction method that can better exploit these elements. Consequently, this section introduces a method for constructing superposed magnetic motive force waveforms. An example of this construction is illustrated in Fig. 5.50.

There are two primary forms for constructing the multi-magnetic motive force permanent magnet array:

1. Bi-level structure: This form involves eliminating the high-order harmonics generated by the permanent magnets by extracting the magneto-motive force content associated with two distinct pole pair numbers. By superposing these two contents, a waveform with a positive and negative asymmetry is created. In regions where the magneto-motive force exceeds zero, permanent magnets of N-polarity are used, while regions where the magneto-motive force is less than zero employ magnets of S-polarity. This approach results in a permanent magnet array with unequal widths, producing a waveform that can have either a fixed positive value (N-polarity) or a fixed negative value (S-polarity). This waveform is akin to the voltage signal produced by a bi-level power electronic controller, hence the designation "bi-level" structure.
2. Tri-level structure: In contrast to the bi-level structure, the tri-level structure involves superposing the square wave magneto-motive force generated by a uniformly distributed permanent magnet array. In the resultant waveform, regions exhibiting a positive value utilize N-polarity permanent magnets, and regions with a negative value use S-polarity magnets. Notably, there are also regions with zero magneto-motive force aside from the aforementioned positive and negative regions. This setup results in what is referred to as the "tri-level" structure.

5.7.2 Air Gap Magnetic Field and Back-Electromotive Force Analysis

In the analysis of the multi-working magnetic motive force flux-reversal machine, it is imperative to select appropriate pole pair numbers and phases for the magneto-motive force. Assume the permanent magnet produces two distinct magneto-motive forces, denoted by pole pair numbers P_{es1} and P_{es2}, with amplitudes F_{es1} and F_{es2}, respectively. This constraint holds even if the actual machine features a greater number of magneto-motive force pole pairs. Under this assumption, the magneto-motive force harmonic can be expressed as:

$$F_{es}(\theta) = F_{es1}\cos(P_{es1}\theta + \theta_{es1}) + F_{es2}\cos(P_{es2}\theta + \theta_{es2}) \tag{5.40}$$

where θ_{es1} and θ_{es2} represent the spatial locations of the magneto-motive forces.

Given that most flux-reversal machines utilize fractional slot concentrated windings, the spatial axis is typically aligned with the center line of the tooth to facilitate subsequent processes. If the rotor tooth aligns with this axis at the initial moment, the specific permeance of the rotor can be represented as follows:

$$\lambda_r(\theta, t) = \lambda_{r0} + \lambda_{r1}\cos Z_r(\theta - \Omega_r t) \tag{5.41}$$

where λ_{r0} and λ_{r1} denote the DC term and the fundamental component of the rotor's specific permeance, respectively. The DC term does not contribute to the generation of the rotational magnetic field; thus, only the fundamental component is operative. Consequently, the operational magnetic field can be described by:

$$\begin{aligned} B_e(\theta, t) = \frac{\mu_0}{g}\lambda_{r1}\Bigg\{ & \frac{F_{es1}}{2}\cos[(Z_r \pm P_{es1})\theta \pm \theta_{es1} - Z_r\Omega_r t] \\ & + \frac{F_{es2}}{2}\cos[(Z_r \pm P_{es2})\theta \pm \theta_{es2} - Z_r\Omega_r t]\Bigg\} \end{aligned} \tag{5.42}$$

The resultant no-load back EMF generated in a single coil can be derived as follows:

$$\begin{aligned} e_c(t) = \frac{\mu_0}{g}\pi r_g l_{st}\Omega_r\lambda_{r1}\Bigg[& F_{es1}\frac{Z_r}{|Z_r \pm P_{es1}|}\sin|Z_r \pm P_{es1}|\frac{\pi}{Z_s} \times \sin(Z_r\Omega_r t \mp \theta_{es1}) \\ & + F_{es2}\frac{Z_r}{|Z_r \pm P_{es2}|}\sin|Z_r \pm P_{es2}|\frac{\pi}{Z_s} \times \sin(Z_r\Omega_r t \mp \theta_{es2})\Bigg] \end{aligned} \tag{5.43}$$

Analysis reveals that four rotating magnetic fields are present within the machine, all generating back-EMFs at identical frequencies. For these fields to function as working magnetic fields, the following three conditions must be met. Subsequent discussion will ascertain the validity of these conditions or establish the prerequisites for their fulfillment:

1. The four magnetic fields are tooth harmonics of one another.

As detailed in Sect. 5.5.5, the pole pair number of the magneto-motive force harmonic adheres to two distinct configurations as specified in Eq. (5.24). Assuming the first configuration is applicable, it can be expressed as follows:

$$P_{es1}=n_{e1}Z_s, \quad P_{es2}=n_{e2}Z_s \tag{5.44}$$

According to Eq. (5.44), the four magnetic fields after the modulation are

$$\begin{aligned} P_{1+} &= n_{e1}Z_s + Z_r, \quad P_{1-} = |n_{e1}Z_s - Z_r| \\ P_{2+} &= n_{e2}Z_s + Z_r, \quad P_{2-} = |n_{e2}Z_s - Z_r| \end{aligned} \tag{5.45}$$

Because of

$$P_{1+} - P_{2+}=(n_{e1} - n_{e2})Z_s \tag{5.46}$$

Therefore, these two magnetic fields are tooth harmonics of each other. When $n_{e1}Z_s > Z_r$, the following relationship is observed:

$$P_{1+} + P_{1-} = 2n_{e1}Z_s \tag{5.47}$$

Conversely, when $n_{e1}Z_s < Z_r$, the relationship can be represented as follows:

$$P_{1+} - P_{1-} = 2n_{e1}Z_s \tag{5.48}$$

It is evident that P_{1+} and P_{1-} are tooth harmonics of each other in both scenarios. Similarly, P_{2+} and P_{2-} are tooth harmonics of each other. Consequently, all four magnetic fields qualify as tooth harmonics. When the pole pair number of the magneto-motive force meets the second criterion specified in Eq. (5.24), this conclusion remains valid. Therefore, the derivation process will not be reiterated here.

2. The back-EMFs of the four mentioned magnetic fields exhibit the same phase sequence.

According to winding theory, for magnetic fields that are tooth harmonics, given their pole pair numbers are P_1 and P_2, the following conditions apply:

(1) If $P_1 + P_2 = nZ_s$, the induced back-EMFs display opposite phase sequences when the magnetic fields corresponding to these pole pair numbers rotate in the same direction. Conversely, when these fields rotate in opposite directions, the induced back-EMFs will have the same phase sequence.
(2) If $P_1 - P_2 = nZ_s$, the induced back-EMFs exhibit the same phase sequence when the magnetic fields of these pole pair numbers rotate in the same direction. If they rotate in opposite directions, the induced back-EMFs will have opposite phase sequences.

According to Eq. (5.42), P_{1+} and P_{2+} rotate in the counterclockwise direction and fulfill the second condition previously described. Consequently, their induced back-EMFs possess the same phase sequence. When $n_{e1}Z_s > Z_r$, P_{1+} and P_{1-} meet the first condition as per Eq. (5.47). Moreover, since P_{1-} rotates in the clockwise direction, as specified in Eq. (5.45), their induced back-EMFs maintain the same phase sequence. When $n_{e1}Z_s < Z_r$, P_{1+} and P_{1-} adhere to the second condition according to Eq. (5.48). Given that P_{1-} also rotates in the counterclockwise direction, their induced back-EMFs consistently exhibit the same phase sequence. Similarly, it is observed that the back-EMFs generated by P_{2+} and P_{2-} also share the same phase sequence. In summary, the back-EMFs induced by all operational harmonics display identical phase sequences.

3. The phases of the four magnetic fields are as coherent as possible.

These four magnetic fields are confirmed to be tooth harmonics of each other, sharing the same frequency and phase sequence for their back EMFs as noted above. Therefore, the phase relationship of the back-EMFs they induce in single coils aligns with that of the phase windings. According to Eq. (5.43), the phases of the back-EMFs induced by these four magnetic fields are delineated in Table 5.7.

According to Eq. (5.43), the pole ratios of these four magnetic fields differ. Typically, the two magnetic fields with pole pairs denoted by $|Z_r - P_{esh}|$ $(h = 1, 2)$ exhibit relatively high pole ratios, significantly influencing the magnitude of the back-EMF. Therefore, the initial objective is to align their phases:

$$\frac{\sin|Z_r - P_{es1}|\frac{\pi}{Z_s}}{\sin|Z_r - P_{es2}|\frac{\pi}{Z_s}} > 0 \tag{5.49}$$

If the judgment conditions for these two magnetic fields yield the same sign, it indicates that $\theta_{es1} = \theta_{es2}$, implying that the spatial phases of these two magnetic fields are identical. Conversely, if $\theta_{es1} + \theta_{es2} = 180°$, their spatial phases are opposite.

Further analysis of the detailed phase occurs under the following conditions:

$$\frac{\sin|Z_r + P_{esh}|\frac{\pi}{Z_s}}{\sin|Z_r - P_{esh}|\frac{\pi}{Z_s}} > 0 \tag{5.50}$$

In such scenarios, if only the back EMF generated by F_{esh} (where $h = 1$ or 2) is considered, the back-EMFs from the two no-load magnetic fields will be in the same

Table 5.7 The phase of back-EMF produced by different working magnetic fields

Pole pair number	Phase
$P_{esh} + Z_r$ $h = 1, 2$	$-\theta_{esh}$, if $\sin|Z_r + P_{esh}|\frac{\pi}{Z_s} > 0$
	$-\theta_{esh} + 180°$, if $\sin|Z_r + P_{esh}|\frac{\pi}{Z_s} < 0$
$|Z_r - P_{esh}|$ $h = 1, 2$	θ_{esh}, if $\sin|Z_r - P_{esh}|\frac{\pi}{Z_s} > 0$
	$\theta_{esh} + 180°$, if $\sin|Z_r - P_{esh}|\frac{\pi}{Z_s} < 0$

phase when $\theta_{esh} = 0$, according to Table 5.7. Conversely, when

$$\frac{\sin|Z_r + P_{esh}|\frac{\pi}{Z_s}}{\sin|Z_r - P_{esh}|\frac{\pi}{Z_s}} < 0 \tag{5.51}$$

In these circumstances, setting $\theta_{esh} = -90°$ aligns the back-EMFs in the same phase.

However, under many conditions, it is not feasible for all four components of the back EMF to have exactly the same phase. For instance, when Eq. (5.52) is satisfied:

$$\begin{aligned} &\sin|Z_r - P_{es1}|\frac{\pi}{Z_s} > 0 \quad \sin|Z_r - P_{es2}|\frac{\pi}{Z_s} > 0 \\ &\sin|Z_r + P_{es1}|\frac{\pi}{Z_s} > 0 \quad \sin|Z_r + P_{es2}|\frac{\pi}{Z_s} < 0 \end{aligned} \tag{5.52}$$

To ensure that the back EMFs contributed by the working magnetic fields of $|Z_r - P_{esh}|$ (h = 1, 2) pole pairs are in phase, $\theta_{es1} = \theta_{es2}$ is necessary. According to Eq. (5.50), setting $\theta_{es1} = 0$ leads to $\theta_{es2} = 0$. However, this condition results in the two components of the back-EMF generated by F_{es2} having opposite phases, thereby indicating a loss of back-EMF.

Continuing the analysis, only specific slot-pole combinations allow all back-EMF components to share the same phase. Detailed constraints and the selection of the phase of the magneto-motive force are listed in Table 5.8.

In Table 5.8, the phases of the two magneto-motive forces are either all 0 or all −90°. When the phases are all 0, the magneto-motive forces are symmetric along the tooth's center line. As a result, the corresponding permanent magnet arrays exhibit symmetry along the tooth's center line, a condition met by the two topologies presented in Fig. 5.36b, c. Conversely, when the phases are all −90°, the magneto-motive forces and the permanent magnet arrays are antisymmetric along the tooth's center line.

The selection of the ideal pole pair number for the working magneto-motive force in a multi-working magnetic motive force flux-reversal machine is crucial. Considering a bi-level permanent magnet array as an example, the stator's magneto-motive force can be expressed as follows, independent of the final distribution of the permanent magnets:

Table 5.8 Constraints and choices of magneto-motive force's phase for in-phase back-EMF components

$\sin\|Z_r - P_{es1}\|\frac{\pi}{Z_s}$	$\sin\|Z_r - P_{es2}\|\frac{\pi}{Z_s}$	$\sin\|Z_r + P_{es1}\|\frac{\pi}{Z_s}$	$\sin\|Z_r + P_{es2}\|\frac{\pi}{Z_s}$	θ_{es1}	θ_{es2}
+	+	+	+	0	0
+	+	−	−	−90°	−90°
−	−	−	−	0	0
−	−	+	+	−90°	−90°

$$F_{\mathrm{es}}(\theta)=\begin{cases}F_{\max} & \text{Region corresponding to N-pole}\\ -F_{\max} & \text{Region corresponding to S-pole}\end{cases} \tag{5.53}$$

Based on the Bessel inequality, it can be established that:

$$F_{\mathrm{es1}}^2+F_{\mathrm{es2}}^2\leq\frac{1}{\pi}\int_{-\pi}^{\pi}F_{\mathrm{es}}^2(\theta)\mathrm{d}\theta=2F_{\max}^2 \tag{5.54}$$

This reveals that although the multi-magnetic motive force permanent magnet array is used, there is still a mutual constraint relationship between the amplitudes of each magnetic motive force. This constraint implies that the amplitude of each force cannot be increased indefinitely. Therefore, introducing as many new magneto-motive forces as possible is recommended to generate the largest possible back-EMF and torque.

According to the principle of flux modulation, the slot-pole combination of the flux-reversal machine with uniformly distributed permanent magnets needs to satisfy the condition $|Z_{\mathrm{r}}-P_{\mathrm{es1}}|=P_{\mathrm{a}}$, and its pole ratio amplification factor is:

$$\mathrm{PR}_1=\frac{Z_{\mathrm{r}}}{|Z_{\mathrm{r}}-P_{\mathrm{es1}}|}=\frac{Z_{\mathrm{r}}}{P_{\mathrm{a}}} \tag{5.55}$$

Upon incorporating the multi-magnetic motive force permanent magnet array, the magneto-motive force of P_{es2} pole pairs must also be coordinated with a relatively high torque amplification factor. Consequently, the value of $|Z_{\mathrm{r}}-P_{\mathrm{es2}}|$ should be minimized to enhance the effectiveness of the torque amplification. In a machine with a fractional-slot concentrated-winding structure, $Z_{\mathrm{s}}-P_{\mathrm{a}}$ represents the pole pair number of the lowest order tooth harmonic. Thus, the appropriate pole pair number for the working magnetic field should be selected as follows:

$$|Z_{\mathrm{r}}-P_{\mathrm{es2}}|=Z_{\mathrm{s}}-P_{\mathrm{a}} \tag{5.56}$$

Under these conditions, the pole ratio amplification factor can be derived as:

$$\mathrm{PR}_2=\frac{Z_{\mathrm{r}}}{Z_{\mathrm{s}}-P_{\mathrm{a}}}\approx\mathrm{PR}_1 \tag{5.57}$$

The upper limit for improving the torque density of the multi-working magnetic motive force flux-reversal machine can be estimated using Eqs. (5.44) and (5.57). Following the principles of flux modulation and the Cauchy inequality, the torque generated by the multi-working magnetic motive force flux-reversal machine can be expressed as:

$$T_{\mathrm{pro}}\propto\mathrm{PR}_1F_{\mathrm{es1}}+\mathrm{PR}_2F_{\mathrm{es2}}\leq\sqrt{\left(\mathrm{PR}_1^2+\mathrm{PR}_2^2\right)\left(F_{\mathrm{es1}}^2+F_{\mathrm{es2}}^2\right)}$$

$$\leq \sqrt{2\left(\mathrm{PR}_1^2 + \mathrm{PR}_2^2\right)F_{\max}^2} \approx 2\mathrm{PR}_1 F_{\max} \tag{5.58}$$

In comparison, the torque generated by a flux-reversal machine with uniformly distributed permanent magnets is defined as:

$$T_{\mathrm{reg}} \propto \mathrm{PR}_1 F_{\mathrm{es}1} = \frac{4}{\pi}\mathrm{PR}_1 F_{\max} \approx 1.27\mathrm{PR}_1 F_{\max} \tag{5.59}$$

The ratio between these two torque values is 1.57, indicating significant potential for performance enhancement in the multi-working magnetic motive force flux-reversal machine. However, according to current research, the actual improvement in torque for this type of machine is still substantially below 57%, suggesting that the research into this technology is not yet fully developed and much remains to be explored through in-depth analysis.

5.7.3 *Performance Analysis of Multi-Working Magneto-Motive Force Flux Reversal Machine*

Further examination of the multi-working magnetic motive force flux-reversal machine is conducted with detailed examples. This analysis also includes a comparison with the conventional flux-reversal machine that utilizes uniformly distributed permanent magnets. The multi-working variant features a stator with 6 slots, an armature winding of 2 pole pairs, and a rotor tooth count of 14. Meanwhile, the classical design with uniformly distributed permanent magnets includes 12 pole pairs. This configuration necessitates the addition of a new working magneto-motive force harmonic following its modification to a multi-working magnetic-motive force structure. Given that the lowest order tooth harmonic of the armature winding is 4 pole pairs, the pole pair number for the new magneto-motive force is determined to be 14 + 4 = 18.

After evaluating the specified slot-pole combination against the criteria outlined in Table 5.8, it is determined that none of the conditions can be satisfied with this configuration. Consequently, this means that the components of the no-load back electromotive force (EMF) cannot maintain the same phase with such a slot-pole arrangement. To align the main working harmonics with the same phase, the spatial phases of the magneto-motive forces have been set to 0. Following further design iterations, the final topology is illustrated in Fig. 5.36b.

Comparative analysis of the no-load air gap magnetic flux density between a multi-working magnetic motive force flux-reversal machine and two flux-reversal machines with uniformly distributed permanent magnets is depicted in Fig. 5.51. It reveals that the adoption of the multi-working magnetic motive force permanent magnet array results in only a minor reduction in the magnetic motive force of 12 pole pairs, with the corresponding working magnetic field of 2 pole pairs being comparable to that of the flux-reversal machine with uniformly distributed permanent magnets. However, the magnetic motive force of 18 pole pairs is significantly lower, at 50% smaller than that found in the uniformly distributed permanent magnets flux-reversal machine.

The back-EMFs of the three machines are compared in Fig. 5.52. The flux-reversal machine with uniformly distributed permanent magnets of 18 pole pairs, having a relatively low pole ratio, exhibits the smallest fundamental wave of the back-EMF. In comparison, the fundamental back-EMF of the multi-working magnetic motive force flux-reversal machine shows only a 12% increase over that of the flux-reversal machine with uniformly distributed permanent magnets of 12 pole pairs. The electromagnetic torques of the three machines are analyzed in Fig. 5.53, where the torque improvement in the multi-working magnetic motive force flux-reversal machine is less than 10%. These results indicate that the advantages of utilizing the multi-working magnetic motive force flux-reversal machine are not yet pronounced, suggesting the need for continued research and optimization to fully utilize the potential benefits of this technology.

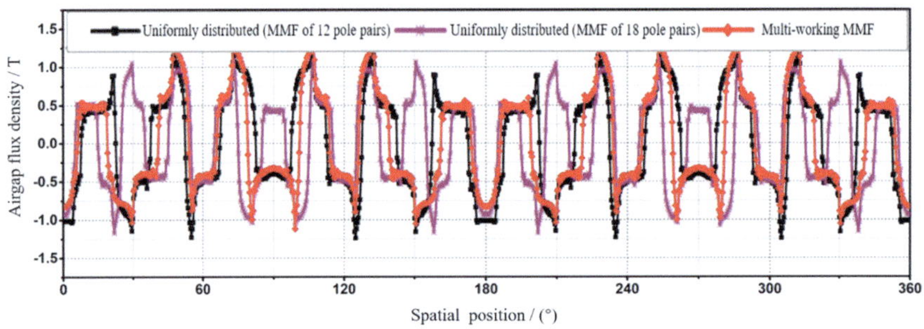

(a)

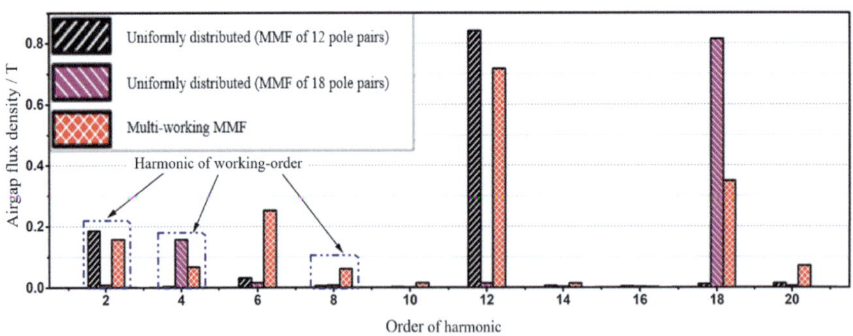

(b)

Fig. 5.51 Comparison of no-load air gap magnetic flux densities between a multi-working magnetic motive force flux-reversal machine and two uniformly distributed permanent magnet flux-reversal machines **a** waveform, **b** spectrum

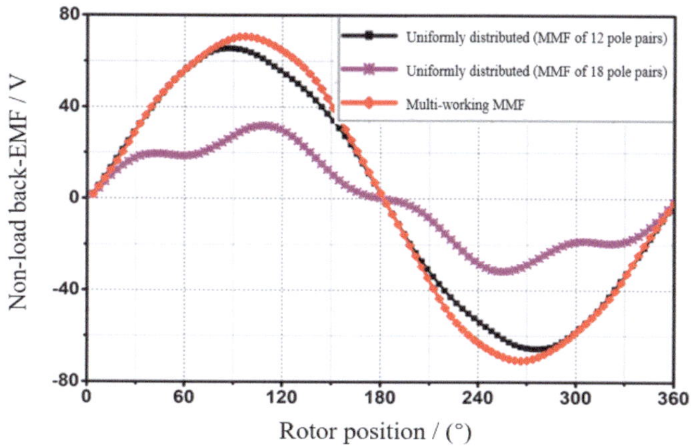

(a)

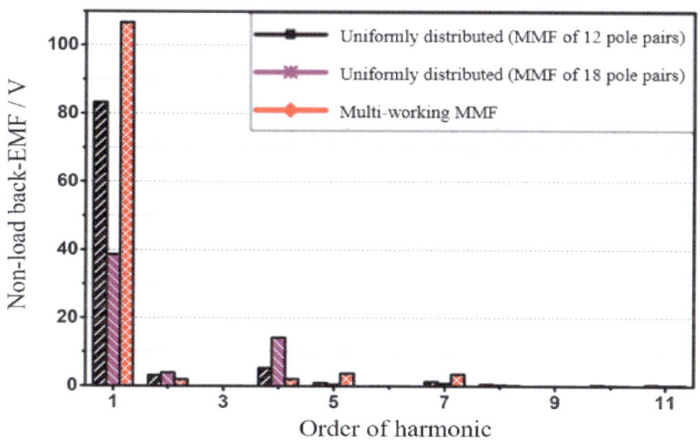

(b)

Fig. 5.52 Comparison of no-load back-EMFs between a multi-working magnetic motive force flux-reversal machine and two uniformly distributed permanent magnet flux-reversal machines **a** waveform, **b** spectrum

As detailed in Sect. 5.2.1, adopting the consequent-pole excitation structure in the FMM leads to notable improvements in the machine's back-EMF and torque. Transitioning to a consequent-pole structure from the multi-working magnetic motive force permanent magnet array, the resulting topology is depicted in Fig. 5.54. When compared to the consequent-pole topology with uniformly distributed permanent magnets, the adaptation of the three aforementioned topologies to the consequent-pole structure results in enhanced torque performance under identical line loads. Notably, the performance enhancement in the multi-working magnetic motive force topology is particularly significant—a 27% improvement over the consequent-pole

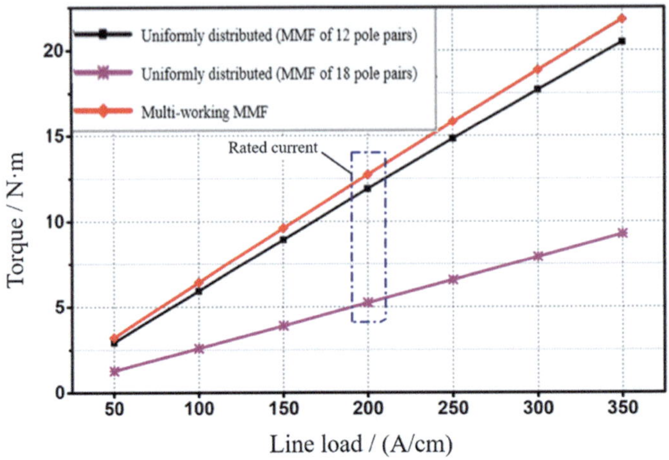

Fig. 5.53 Comparison of electromagnetic torques between a multi-working magnetic motive force flux-reversal machine and two uniformly distributed permanent magnet flux-reversal machines

flux-reversal machine with uniformly distributed permanent magnets of 12 pole pairs, as illustrated in Fig. 5.55. Moreover, the consequent-pole structure tends to reduce the equivalent air gap length, potentially leading to saturation more readily. This attribute explains why the three torque-current characteristic curves rapidly reach their inflection points, highlighting one of the drawbacks of the consequent-pole FMM.

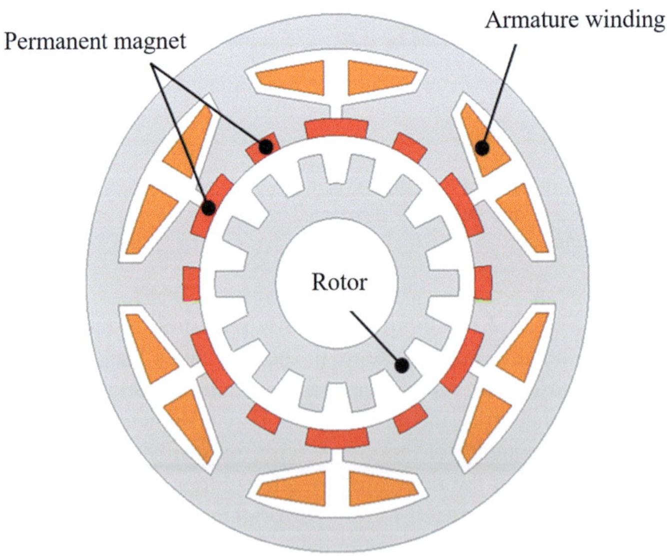

Fig. 5.54 Comparison of electromagnetic torques among three consequent-pole machines

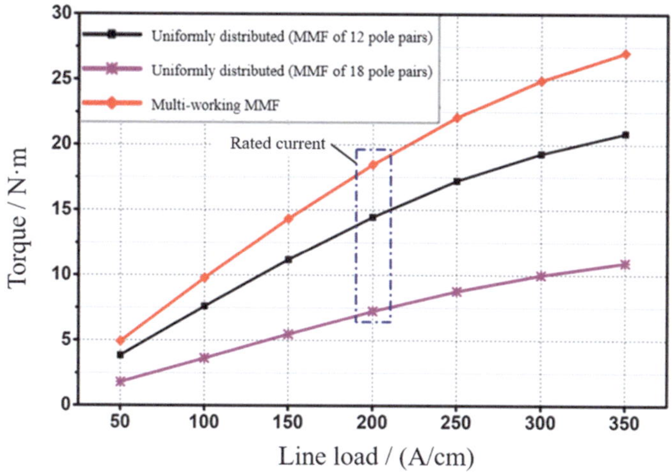

Fig. 5.55 Comparison of three consequent-pole machines' electromagnetic torques

5.8 Bilateral Excitation Integrated Flux Modulation Machine

The bilateral excitation integrated FMM, introduced earlier in Sect. 5.5.6, is recognized for its exceptional torque density. This promising feature warrants further analysis of its feasible topologies and electromagnetic characteristics. It is important to note that the topologies illustrated in Fig. 5.40a–c differ in their operational magnetic fields. The stator's specific permeances in Fig. 5.40a, b do not contribute to the machine's performance, whereas the topology in Fig. 5.40c is actively involved. Given that the earlier topologies have a comparatively larger equivalent air gap length, which may detract from optimal performance, only the topology in Fig. 5.40c will be further analyzed in this section.

5.8.1 Magnetic Field Analysis and Topology Constraint

In the development of the bilateral excitation integrated FMM, the analysis of the pole pair numbers for the working magneto-motive force and the specific permeance has been initially explored. To simplify further analysis, the following constraints have been established for its slot-pole combination:

(1) Equality of pole pair numbers on rotor side: The pole pair number of the working magneto-motive force and that of the specific permeance on the rotor side must be identical, denoted as P_r. This alignment is a critical constraint for such machines, ensuring synchronicity and effective interaction between the magneto-motive force and the specific permeance.

(2) Simplicity in harmonic construction on stator side: There should be only one main working harmonic for both the working magneto-motive force at the stator side and the specific permeance at the stator side, and their pole pair numbers must be identical, denoted as $P_{es1} = P_{er1}$. It is important to distinguish this from the armature winding's pole pair number, P_a to avoid confusion during design and analysis.

Under these constraints, both the stator's and rotor's permanent magnet arrays require a similar consequent-pole structure. At this point, the axes of the magneto-motive force and the specific permeance are offset by 180°. These configurations can be expressed as follows:

$$\begin{aligned} F_{es}(\theta) &= F_{es1}\cos(P_s\theta + \theta_{es}) \\ \lambda_s(\theta) &= \lambda_{s0} + \lambda_{s1}\cos(P_s\theta + \theta_{es} + \pi) \\ F_{er}(\theta, t) &= F_{er1}\cos[P_r(\theta - \Omega_r t) + \theta_{er}] \\ \lambda_r(\theta, t) &= \lambda_{r0} + \lambda_{r1}\cos[P_r(\theta - \Omega_r t) + \theta_{er} + \pi] \end{aligned} \tag{5.60}$$

It is important to note that since the magnetization direction of the permanent magnet can be either radially outward or inward, it is stipulated that outward magnetization (N-polarity) corresponds to a positive magneto-motive force in the relevant equation. Conversely, inward magnetization is considered negative. This notation helps in clearly defining the nature of magnetic interactions within the machine. The no-load air gap magnetic flux density, which results from these interactions, can be described as follows:

$$\begin{aligned} B_e(\theta, t) &= \frac{\mu_0}{g}[\lambda_{s0}F_{es}(\theta)\lambda_r(\theta, t) + \lambda_{r0}F_{er}(\theta, t)\lambda_s(\theta)] \\ &= \frac{\mu_0}{g}\left\{\begin{array}{l} F_{es1}\lambda_{s0}\lambda_{r0}\cos(P_s\theta + \theta_{es}) + F_{er1}\lambda_{s0}\lambda_{r0}\cos[P_r(\theta - \Omega_r t) + \theta_{er}] \\ +\left(\dfrac{F_{s1}\lambda_{s0}\lambda_{r1}}{2} + \dfrac{F_{r1}\lambda_{s1}\lambda_{r0}}{2}\right)\cos[(P_r \pm P_s)\theta - P_r\Omega_r t + \theta_{er} \pm \theta_{es} + \pi] \end{array}\right\} \end{aligned} \tag{5.61}$$

According to Eq. (5.61), the air gap magnetic fields within the bilateral excitation integrated FMM can be categorized into four distinct types. The first category consists of magnetic fields with P_{es} pole pairs that are stationary and do not contribute to electromechanical energy conversion; hence, they can be disregarded in further analyses. The remaining categories, which feature rotatory magnetic fields, all operate at the same electrical frequency. Specifically, the second category comprises the non-modulated magnetic field generated by the rotor. The third and fourth categories consist of modulated magnetic fields with $P_r + P_s$ and $|P_r - P_s|$ pole pairs, respectively.

Notably, the magnetic field of the fourth category, which has the lowest pole pair number and the highest torque amplification factor, is identified as the optimal working magnetic field for the machine. Furthermore, the magnetic fields of the second and third categories also function as working magnetic fields when they are tooth harmonics of the fourth category's magnetic field. It is established that the necessary and sufficient condition for these three categories of magnetic fields to be tooth harmonics of each other is $P_s = nZ_s$.

The working magnetic field with $|P_r - P_s|$ pole pairs merits further analysis. This magnetic field originates from both the stator and the rotor. To ensure that the magnetic fields produced by these two sources are in phase, the magneto-motive forces F_{es1} and F_{er1} should either be both positive or both negative. This alignment means that the permanent magnets in the stator and rotor should share the same magnetization direction to avoid cancelling each other's effects, which would otherwise reduce the output torque. For simplicity in subsequent analyses, it is assumed that the permanent magnets are of N-polarity, implying that both F_{es1} and F_{er1} are positive. When $P_s \neq nZ_s$, the scenario predominantly involves the modulated working magnetic field. After confirming that the magnetization direction of both the stator and the rotor are aligned, these two sources inherently share the same spatial phase, thus the back-EMFs generated also share the same phase. This synchronicity simplifies the design as no additional constraints are necessary for aligning the spatial phases within the stator structure.

Conversely, when $P_s = nZ_s$, a more complex situation arises with a relatively large number of interacting magnetic fields. The no-load back-EMF for a single coil under this condition can be derived using Eq. (5.61).

$$\begin{aligned} e_c(t) = \frac{\mu_0}{g}\pi r_g l_{st}\Omega_r\Big[&2F_{es1}\lambda_{s0}\lambda_{r0}k_{paP_r} \times \sin(P_r\Omega_r t - \theta_{er}) \\ &- (F_{es1}\lambda_{s0}\lambda_{r1} + F_{er1}\lambda_{s1}\lambda_{r0})\frac{P_r}{P_a}k_{paP_a} \times \sin(P_r\Omega_r t - \theta_{er} + \theta_{es}) \\ &- (F_{es1}\lambda_{s0}\lambda_{r1} + F_{er1}\lambda_{s1}\lambda_{r0})\frac{P_r}{(P_r + P_s)}k_{pa(P_r+P_s)} \times \sin(P_r\Omega_r t - \theta_{er} - \theta_{es})\Big] \end{aligned} \tag{5.62}$$

where $P_a = |P_r - P_s|$ represents the pole pair number of the armature winding. According to Eq. (5.62), there are four terms in the expression for the fundamental wave back-EMF of the bilateral excitation integrated FMM. Among these, the first two terms are particularly significant. To maximize the overall amplitude of the back-EMF, it is crucial to ensure that the phases of these two predominant terms are aligned. The short-pitch factor of the armature winding, represented by k_{pa}, is linked with the pole pair number and is typically close to 1, assuming a positive sign. Thus, the values of θ_{es} should be determined based on the sign of k_{paPr}.

$$\theta_{es} = \begin{cases} 0, & k_{paP_r} < 0 \\ \pi, & k_{paP_r} > 0 \end{cases} \tag{5.63}$$

When $k_{paPr} < 0$, the centerline of the magneto-motive force should coincide with the axis of the coil. When $k_{paPr} > 0$, the centerline of the specific permeance should align with the axis of the coil.

In an analysis approach similar to that used in Chap. 4 for vernier permanent magnet machines and differential modulation, the virtual coil method can be employed. By substituting the virtual coil's coil-span $\theta_{es} = \pi/Z_s$ into the formula for the short-pitch factor, the following can be deduced:

$$\theta_{es} = \begin{cases} 0, \ \sin P_r \frac{\pi}{Z_s} < 0 \\ \pi, \ \sin P_r \frac{\pi}{Z_s} > 0 \end{cases} \tag{5.64}$$

It is important to note that θ_{es} in Eq. (5.64) carries a different meaning compared to that in Eq. (5.63). In Eq. (5.64), θ_{es} represents the electrical angle between the axis of the stator's magneto-motive force (typically along the tooth center line) and the axis of the virtual coil. This angle helps to determine the relative positioning of the stator's magneto-motive force, specific permeance, and the tooth.

By applying the slot-pole combination of the bilateral excitation integrated FMM, where $nZ_s = P_r \pm P_a$, and substituting this into Eq. (5.64), it can be derived that:

$$\theta_{es} = \begin{cases} 0, \ n \text{ is even number with } + \text{ sign or } n \text{ is odd number with } - \text{ sign} \\ \pi, \ n \text{ is odd number with } + \text{ sign or } n \text{ is even number with } - \text{ sign} \end{cases} \tag{5.65}$$

In Eq. (5.65), a prevalent condition is $n = 1$, under which the configuration resembles a vernier permanent magnet machine. The slot-pole combination denoted by a "+" is classified as summation modulation, while the combination marked by a "−" is classified as differential modulation. In the context of summation modulation, the alignment of the tooth center line with the specific permeance center line is required. In contrast, for differential modulation, the center lines should exhibit an angular displacement of 180°. Regarding the vernier permanent magnet machine, the alignment of the stator tooth center line with the specific permeance center line is prescribed by its topology characterization. Therefore, the adoption of summation modulation is necessary to align the phases of the back EMF generated by the modulated and non-modulated magnetic fields; employing differential modulation would result in phase opposition.

5.8.2 *Feasible Stator Topology*

The rotor of the bilateral excitation integrated FMM features a consequent-pole structure. Although the stator's magneto-motive force and specific permeance share identical pole pair numbers, the topology of the stator varies. This discussion delineates feasible stator topologies. The stator's yoke, tooth, and inner slot structures are fixed; the variable component is the thin layer near the air gap, as illustrated in Fig. 5.56a. In terms of topology construction, three material options for this region are considered: permanent magnet, iron core, and air, each corresponding to the slot notch. These options result in differing distribution methodologies.

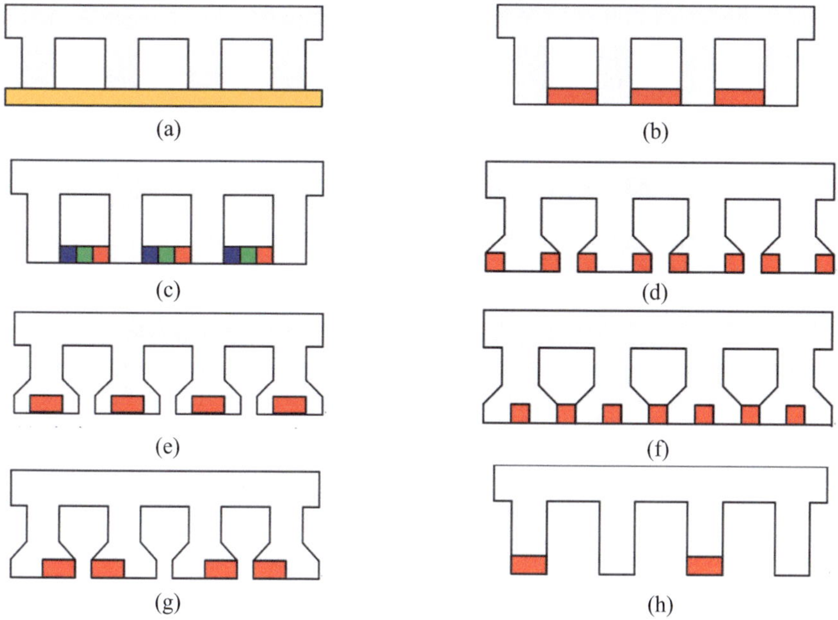

Fig. 5.56 Feasible stator topologies for the bilateral excitation integrated FMM: **a** region for topology construction, **b** placement of surface-mounted permanent magnet at the slot notch, **c** placement of Halbach permanent magnet at the slot notch, **d** placement of permanent magnet at the sides of the tooth. **e** Placement of permanent magnet at the middle of teeth, **f** auxiliary tooth structure, **g** asymmetric tooth structure **h** placement of permanent magnet at the tooth

Assuming that the air only occupies a minor fraction and is negligible, the lower region in Fig. 5.56 predominantly comprises the consequent-pole structure, consisting of both the permanent magnet and the iron core. When $P_s = Z_s$, a straightforward approach involves using the stator tooth as the iron core of the consequent-pole, with the permanent magnet positioned at the slot notch, as depicted in Fig. 5.56b. The primary issue with this topology is the absence of an iron yoke to form a complete magnetic circuit, which diminishes the magnetic load capability of the stator's magnetic field. To address this issue, the radial magnetization permanent magnet could be substituted with a Halbach array configuration, as illustrated in Fig. 5.56c, with the complete topology presented in Fig. 5.40c [26]. Although this arrangement offers optimized electromagnetic performance, the installation of the permanent magnet becomes considerably more complex, particularly with the Halbach array, due to the need to fit multiple magnetic steels at the slot notch, which complicates the manufacturing process.

An alternative solution involves incorporating an iron yoke alongside the permanent magnet in the setup shown in Fig. 5.56b, as represented in Fig. 5.56d. In this revised topology, the tooth tip at the bottom is broadened, resembling a semi-closed slot structure. Furthermore, this arrangement eliminates the need for a permanent magnet without an iron yoke at the slot notch. While the electromagnetic performance of this configuration may not reach the levels achieved with the Halbach array, it offers a more manufacturable solution.

For the three topologies described, the center line of the specific permeance aligns with the stator tooth, making them suitable for the slot-pole combination of summation modulation. In contrast, a topology conducive to differential modulation can be devised by modifying the arrangement shown in Fig. 5.56b. This involves relocating the permanent magnet to the tooth center and removing a small section in the middle of the consequent-pole iron core to create a slot notch, as shown in Fig. 5.56e. In this configuration, the center line of the stator's specific permeance is positioned at the slot notch, aligning it effectively for differential modulation applications.

When $P_s = nZ_s$, given that the number of consequent-pole iron cores is a multiple of the stator tooth number, an auxiliary tooth structure is created within the stator. In this configuration, permanent magnets are alternately positioned in the middle of the auxiliary teeth, as illustrated in Fig. 5.56f.

In the above topology, the stator pole pair number P_s is indeed a multiple of the stator tooth number. In contrast, a novel topology presented in Fig. 5.56g features P_s as half of the tooth number. This configuration can be envisioned as being created by affixing a consequent-pole excitation structure of the corresponding pole pair number onto the surface of the stator tooth. Specifically, for a single tooth, the lower half is occupied by the permanent magnet, while the upper half is filled with the iron core. To ensure the pole pair number is $Z_s/2$, the permanent magnet placement alternates from the left side in one tooth to the right side in the adjacent tooth. From another perspective, this topology serves as the stator in a consequent-pole flux-reversal machine.

Lastly, a topology involving a relatively large air gap at the slot opening is considered. In such a design, only a limited space is available to house both the permanent magnet and the iron core, resulting in the topology depicted in Fig. 5.56h. In this arrangement, permanent magnets are alternately placed at the tooth tips and feature a pole pair number of $P_s = Z_s/2$.

5.8.3 Electromagnetic Performance of Consequent-Pole Halbach Structure

The 7 topologies depicted in Fig. 5.56 are each characterized by high torque density, making them highly advantageous for applications requiring strong performance from compact designs. Among these, the Halbach array embedded at the slot notch, despite its manufacturing complexity, stands out for its exceptional torque density. This topology is therefore chosen for a detailed examination to further elucidate the electromagnetic performance of such machines, allowing for a deeper understanding of their capabilities.

For comparison purposes, a surface-mounted vernier permanent magnet machine is selected as a benchmark. Both machines boast a 12-slot stator, an 11-pole-pair rotor, and a 1-pole-pair winding, establishing a slot-pole ratio of 11:1. They also share the same main dimensions, as detailed in Table 5.9.

1. No-load performance

The magnetic field distributions of both the surface-mounted vernier permanent magnet machine and the bilateral excitation integrated FMM are showcased in Fig. 5.57. It is observed that both machines exhibit similar magnetic flux line distributions, attributable to the identical pole pair numbers of their working magnetic fields. However, the bilateral excitation integrated FMM enhances the internal magnetic field by positioning permanent magnets on both the stator and rotor sides.

To further analyze these characteristics, their no-load air gap magnetic fields are compared in Fig. 5.58. Initially, the rotor magneto-motive forces interact with the DC term of the air gap specific permeance, creating a magnetic field harmonic of 11 pole pairs. This magnetic field serves as the tooth harmonic of the 1-pole-pair magnetic field, thus capable of linking with the winding. Additionally, the presence of permanent magnets on both the stator and rotor contributes to generating working magnetic

Table 5.9 Main parameters of the models

Parameter	Value	Parameter	Value
Stator outer diameter	124 mm	Slot number / Pole number	12/22
Stator inner diameter	74.4 mm	Pole arc	0.9
Air gap length	0.5 mm	Slot depth	14.8 mm
PM thickness	3 mm	Slot opening width	13.6 mm
Stack length	70 mm		

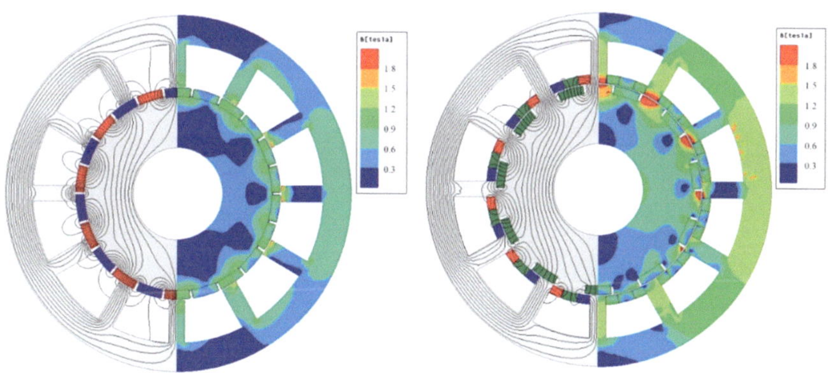

(a) (b)

Fig. 5.57 Comparison of the surface-mounted vernier permanent magnet machine's magnetic field distribution and bilateral excitation integrated FMM's magnetic field distribution: **a** surface-mounted permanent magnet vernier machine, **b** bilateral excitation integrated FMM

field fundamental waves of 1 pole pair due to the flux modulation effect. Given that both sides utilize consequent-pole structures, this configuration not only preserves but potentially enhances the working magnetic field compared to conventional surface-mounted vernier and flux-reversal machines. Consequently, the working magnetic field strength of the bilateral excitation integrated FMM can achieve twice that of the surface-mounted vernier permanent magnet machine, reaching a significant 0.4 T.

The no-load back-EMF waveforms of the bilateral excitation integrated FMM are displayed in Fig. 5.59. This machine configuration can be conceptualized as a combination of a consequent-pole vernier permanent magnet machine with rotor excitation and a consequent-pole flux-reversal machine with stator excitation. To analyze the contributions of each component, the permanent magnets on both the stator and rotor sides are independently removed in simulations, and the resulting

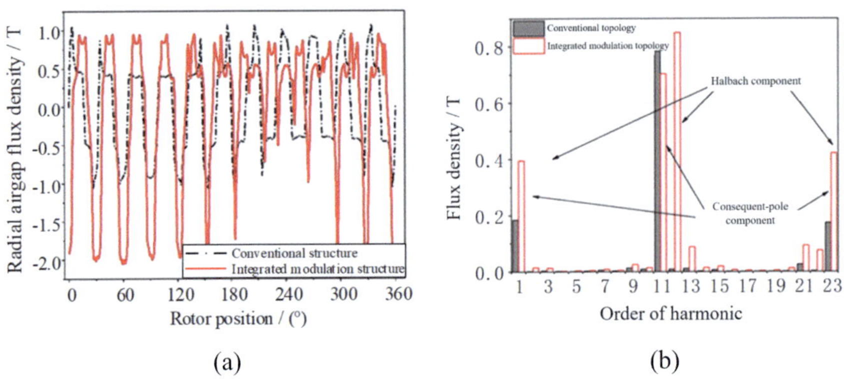

Fig. 5.58 Comparison of the surface-mounted vernier permanent magnet machine's no-load air gap magnetic field and bilateral excitation integrated FMM's no-load air gap magnetic field. **a** Waveform of air gap flux density. **b** Fourier decomposition of air gap flux density

Fig. 5.59 Waveforms of the bilateral excitation integrated FMM's no-load back-EMFs

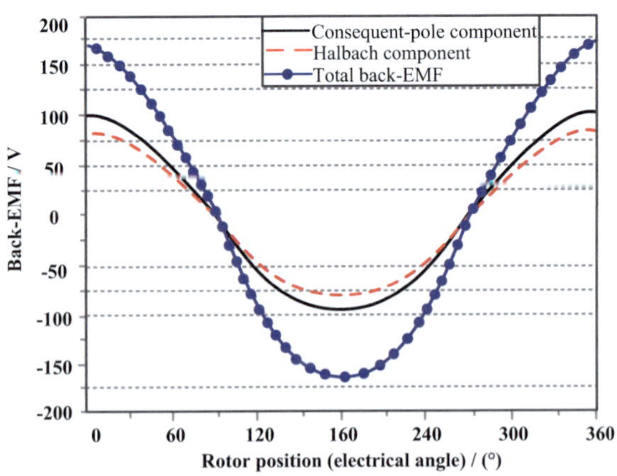

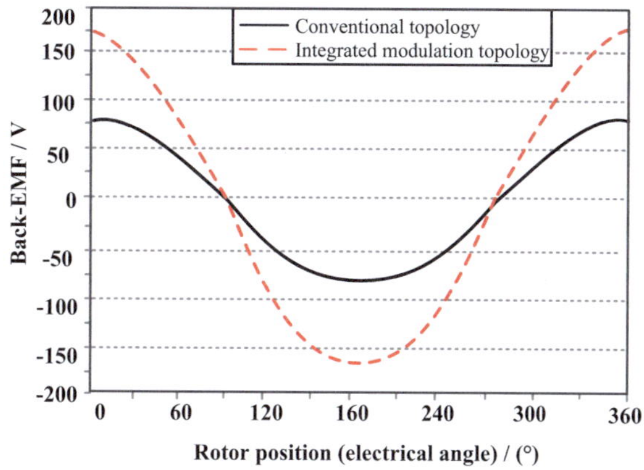

Fig. 5.60 Comparison of no-load back-EMFs' waveforms

no-load back-EMFs are then superimposed. As observed in the figure, the back-EMF generated by the flux-reversal machine component is marginally lower than that produced by the consequent-pole vernier permanent magnet machine component. The latter benefits from an additional working magnetic field of 11 pole pairs. Additionally, since the machine operates far from saturation under no-load conditions, the actual no-load back-EMF is effectively the sum of the contributions from both parts.

A comparative analysis of the no-load back-EMF between the conventional structure and the integrated modulation structure is depicted in Fig. 5.60. Here, the amplitude of the back-EMF from the bilateral excitation integrated FMM is double that of the vernier permanent magnet machine. This increased performance is attributed to the higher back-EMF of the consequent-pole vernier permanent magnet machine compared to the surface-mounted structure. While the Halbach component produces a relatively lower back-EMF (Fig. 5.59), the combined effect results in an overall back-EMF that is twice that of the conventional topology, underscoring the substantial performance advantage of the bilateral excitation integrated FMM.

Following this, the cogging torque of the machine is analyzed. Research indicates that both the machine's back-EMF and cogging torque are significantly influenced by the dimensions of the permanent magnets, specifically their width and height. Figure 5.61 and Eq. (5.66) detail the definitions of these dimension parameters, which are critical for understanding the machine's performance and optimizing its design.

$$k_{\mathrm{m}} = \frac{\theta_{\mathrm{radial}}}{\theta_{\mathrm{total}}} \tag{5.66}$$

According to Fig. 5.61, k_{m} represents the proportion of the radially magnetized component within the entire Halbach permanent magnet array. As depicted in Fig. 5.62, when $k_{\mathrm{m}} = 40\%$, the machine's back-EMF achieves its maximum

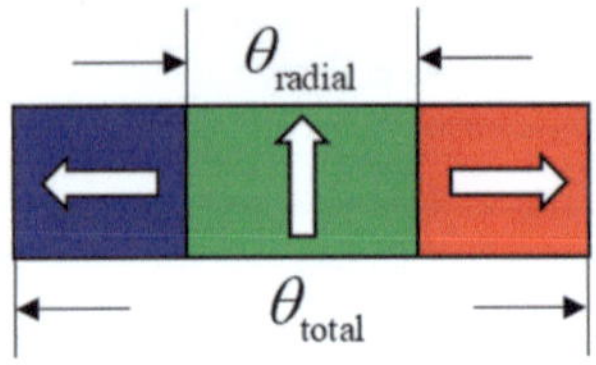

Fig. 5.61 Structural diagram of the Halbach permanent magnet

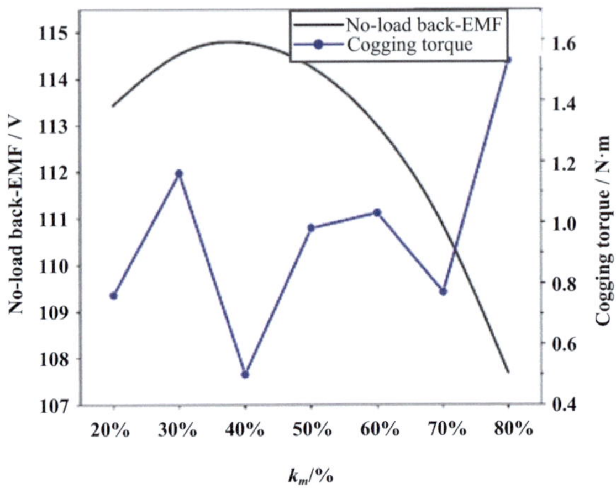

Fig. 5.62 Relationship between k_m and the no-load back-EMF and cogging torque

value, and the cogging torque reaches its minimum value. Consequently, this dimension is identified as the optimized value for this type of machine. With this specific dimension, the cogging torque of the surface-mounted vernier permanent magnet machine and that of the bilateral excitation integrated FMM are compared in Fig. 5.63. The analysis indicates that due to the unique slot-pole combination of the surface-mounted vernier permanent magnet machine, the least common multiple of the stator slot number and rotor pole number is relatively large, resulting in a comparatively low cogging torque. The bilateral excitation integrated FMM shares this slot-pole combination with the surface-mounted vernier permanent magnet machine, implying that these two machines exhibit the same cogging torque frequency. However, due to its consequent-pole structure, the cogging torque of the bilateral excitation integrated FMM is relatively higher. Additionally, beyond the cogging torque components generated by the stator's and rotor's permanent magnets, a new cogging torque component arises from the interaction between these two parts of permanent magnets, doubling the cogging torque compared to that of the surface-mounted vernier permanent magnet machine.

2. Load characteristics

Neither machine exhibits a significant salient-pole effect, whether it be the surface-mounted vernier permanent magnet machine or the bilateral excitation integrated FMM. The d-axis inductances of both machines can be considered equivalent, as

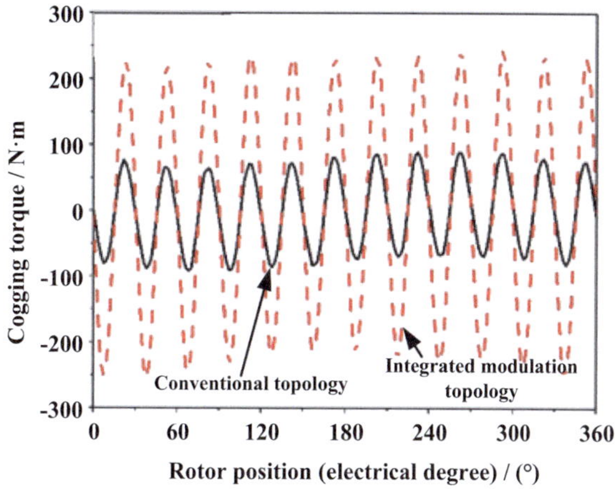

Fig. 5.63 Comparison of the surface-mounted vernier permanent magnet machine's cogging torque and the bilateral excitation integrated FMM's cogging torque

can the q-axis inductances. Therefore, implementing a control strategy of $I_d = 0$ maximizes the output torque. It is important to note that the Halbach permanent magnet occupies a certain slot area, thus to achieve an optimal output torque and relatively low torque ripple, the thickness of the magnet needs to be optimized while keeping the copper loss unchanged. As shown in Fig. 5.64, a magnet thickness of 4.4 mm allows the machine to reach its maximum output torque and minimum cogging torque, making this parameter the optimized value for further research.

The torque characteristics of the surface-mounted vernier permanent magnet machine and the bilateral excitation integrated FMM are analyzed in Fig. 5.65. Currently, the torque ripple of the bilateral excitation integrated FMM is relatively high, reaching 2.1%. To reduce this torque ripple, optimization of the machine can be achieved by implementing chamfering on the rotor structure, as shown in Fig. 5.66. Although this modification results in a 10% reduction in output torque, it successfully decreases the torque ripple to 1.8%. When compared to the surface-mounted vernier

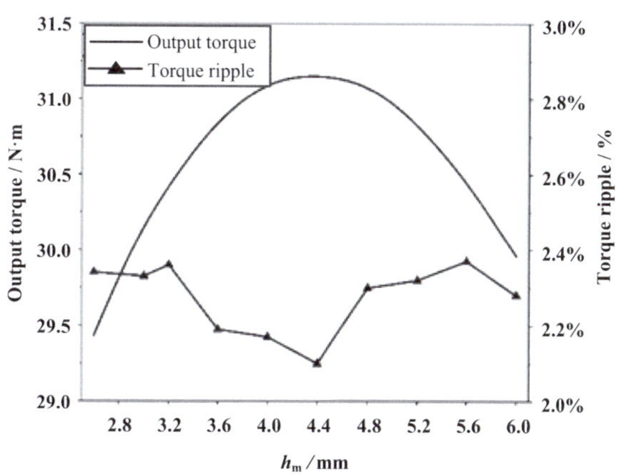

Fig. 5.64 Relationship between h_m and the output torque and the torque ripple

permanent magnet machine, the torque ripple of the bilateral excitation integrated FMM remains comparatively higher. This is primarily due to its higher cogging torque and the introduction of additional back-EMF harmonics by the bilateral consequent-pole structure. The torque ripples of the two machines under various input currents are further examined in Fig. 5.67, revealing that the torque ripple of the bilateral excitation integrated FMM consistently exceeds that of the surface-mounted vernier permanent magnet machine across all operating conditions.

The torque-current characteristic curves of the conventional surface-mounted vernier permanent magnet machine and the bilateral excitation integrated FMM are compared in Fig. 5.68. Based on the analysis in Fig. 5.60, the back-EMF of the bilateral excitation integrated FMM is twice that of the surface-mounted vernier permanent magnet machine. Consequently, at lower currents where the torque is largely proportional to the back-EMF, the bilateral excitation integrated FMM enjoys a torque advantage, effectively doubling its output under low-load conditions. However, as the current increases, the bilateral excitation integrated FMM is more prone to reaching saturation more quickly for two reasons: Firstly, it inherently exhibits a higher magnetic flux density under no-load conditions. Secondly, the use of a consequent-pole structure results in a relatively smaller equivalent air gap for the

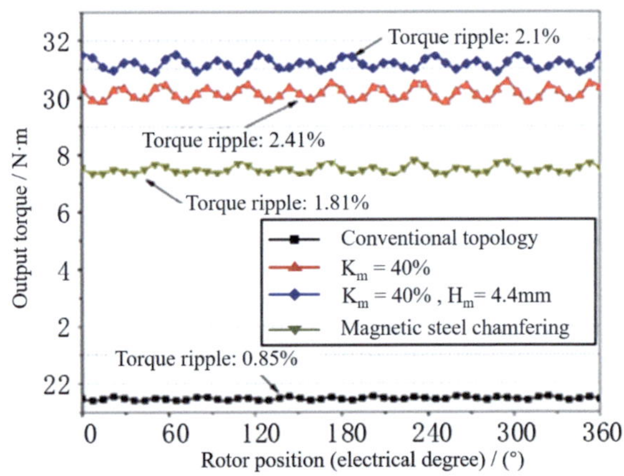

Fig. 5.65 Comparison of the surface-mounted vernier permanent magnet machine's output torque and the bilateral excitation integrated FMM's output torque

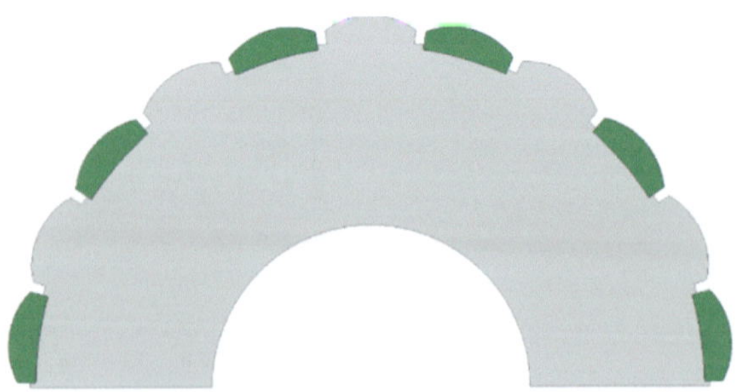

Fig. 5.66 Setting the chamfer on the rotor

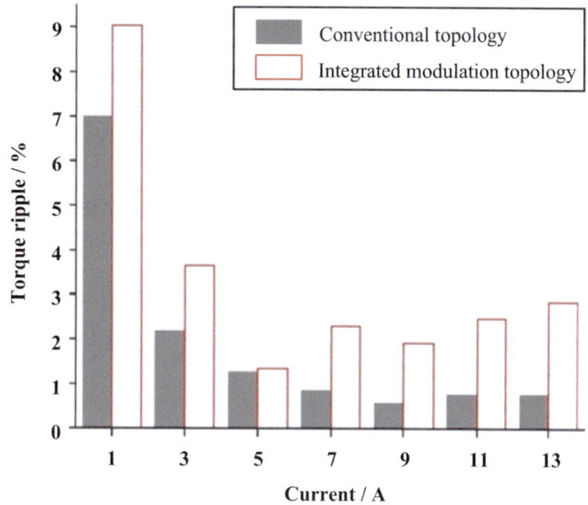

Fig. 5.67 Comparison of the surface-mounted vernier permanent magnet machine's torque ripple and the bilateral excitation integrated FMM's torque ripple

armature magnetic field, leading to a stronger armature reaction and higher inductance. When compared to the surface-mounted vernier permanent magnet machine, this propensity for easy saturation diminishes the torque advantage of the bilateral excitation integrated FMM under high-load conditions. For instance, at an input current of 3 A, the torque of the bilateral excitation integrated FMM is only 34% higher—significantly less than the 100% increase in back-EMF initially noted.

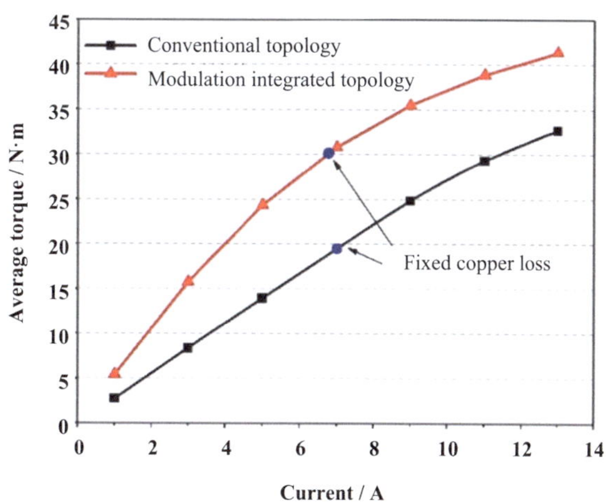

Fig. 5.68 Comparison of the torque-current characteristics of the surface-mounted vernier permanent magnet machine and the bilateral excitation integrated FMM

5.9 Summary of This Chapter

This chapter focuses on the methods and strategies for enhancing the electromagnetic performance of FMMs. Various types of high-performance FMMs are introduced, each characterized by multiple working harmonics. Consequently, performance improvement is approached in two primary ways: enhancing the amplitude of a single working magnetic field and increasing the number of working magnetic fields. For the first approach, the improvement is categorized into three areas: enhancing excitation performance, improving modulation performance, and boosting armature performance. Each area targeted specific aspects of machine function to optimize overall efficiency and output.

The second approach involves the establishment of flux modulation integration theory. Within this framework, eight groups of working magnetic fields within the FMM are identified. The integration methods are then divided into two main concepts: inner-group integration and intergroup integration. The former introduces two types of FMMs—multi-working permeance FMM and multi-working magnetic motive force FMM—while the latter discusses topologies constructed by integrating eight groups of magnetic fields in pairs. Using these methodologies, a comprehensive system of performance-improving techniques for the FMM is constructed in this chapter, laying the groundwork for future developments in this field. This chapter focuses on several key strategies to enhance the electromagnetic performance of FMMs:

1. *Changing the distribution method of permanent magnets*: This approach directly improves excitation. Unlike conventional permanent magnet machines, the consequent-pole FMM reduces the ripple of the magnetic potential, thereby decreasing the necessity for high levels of permanent magnet use. This adjustment results in increased back-EMF and output torque. Despite the magnetic barrier effect—which necessitates additional magnetic circuits for magnetic fields with low pole pair numbers—various topologies such as SAPMVMB, SACPPMVM, and others are explored.
2. *Improving modulation performance*: Enhancing the modulation performance of the FMM is challenging and could only be achieved by integrating superconducting materials into the slot notch to create a magnetic shielding effect. Currently, this method is not feasible for practical engineering applications. However, theoretical research on this model offers insights into the interactions between the FMM's external characteristics and internal parameters, such as torque density, power factor, and air gap specific permeance, providing a theoretical basis for further improvements in electromagnetic performance.
3. *Enhancing armature performance*: This is accomplished by addressing the trade-off between increasing torque density and reducing the end winding length. Various topologies of the circular winding vernier permanent magnet machine, including radial-flux and axial-flux topologies, are introduced. These designs not only minimize the end winding length but also potentially double the torque density.

4. *Implementing an auxiliary tooth structure in the multi-working permeance vernier permanent magnet machine*: This type of machine poses challenges for quantitative analysis. Using the analytical method based on the virtual air gap technique, it is discovered that the adoption of a nonuniform tooth pitch structure introduced numerous specific permeance harmonics. This alteration shifts phases of the original specific permeances by 180°. Consequently, the machine generates significant additional torque components, and some originally negative torque components turn positive due to this phase shift. Ultimately, these changes result in an overall torque improvement of more than 20%.
5. *Constructing the multi-working magnetic motive force in the flux-reversal machine*: The key is the superposition of working magneto-motive forces. To ensure that each working magneto-motive force harmonic contributes positively to torque, a specific selection of pole pair numbers and phases is necessary, with detailed principles outlined within this chapter. Although such machines have the potential to enhance torque by approximately 60%, the surface-mounted multi-working magnetic motive force flux-reversal machine exhibits only a 10% improvement in both torque and back-EMF compared to the surface-mounted uniformly distributed permanent magnet machine, indicating that its full torque potential remains untapped. Conversely, the consequent-pole multi-magnetic motive force flux-reversal machine significantly improves back-EMF by over 20%.
6. *Utilizing the bilateral excitation integrated FMM*: This strategy involves integrating FMMs with permanent magnets on both the stator and rotor, where both are configured in a consequent-pole structure. The chapter explores the working magnetic fields under various slot-pole combinations. When the stator excitation's pole pair number is not an integral multiple of the tooth number, only certain modulated magnetic fields act as working harmonics. Ensuring that the permanent magnets on the stator and rotor share the same magnetization direction is crucial for phase alignment of the working magnetic fields. When the stator excitation's pole pair number is an integral multiple of the tooth number, the rotor's non-modulated magnetic field also serves as a working harmonic. The choice between summation or differential modulation impacts the selection of the stator's magnetic motive force and specific permeance. The chapter also introduces potential topologies for such machines, highlighting an optimal topology that incorporates a Halbach permanent magnet embedded at the slot notch. This configuration significantly outperforms the surface-mounted vernier permanent magnet machine, with no-load back-EMF improved by 100% and torque increased by over 60% under natural cooling conditions.

These comprehensive discussions and analyses have established a robust framework for advancing the design and application of high-performance FMMs, setting the stage for future innovations in machine design and efficiency optimization.

References

1. Li DW, Qu RH, Li J (2015) Topologies and analysis of flux-modulation machines. In: IEEE Energy conversion congress and exposition, Montreal. IEEE, pp 2153–2160
2. Li D (2015) Research on the permanent magnet flux modulation machine (in Chinese). Huazhong University of Science and Technology, Wuhan
3. Cheng M, Han P, Hua W (2017) IEEE Trans Ind Electron 64(8):6063
4. Zhou Y, Shi CJ, Li DW et al (2021) IEEE Trans Ind Appl 57(6):5841
5. Li DW, Qu RH, Li J et al (2015) IEEE Trans Ind Appl 51(6):4470
6. Gao YT, Qu RH, Li DW et al (2016) IEEE Trans Appl Supercond 26(4):1
7. Zou TJ, Li DW, Qu RH et al (2017) IEEE Trans Magn 53(6):1
8. Kim K, Lipo T (2016) IEEE Trans Ind Appl 52(1):217
9. Ren X, Li DW, Qu RH et al (2018) IEEE Trans Energy Convers 33(4):2112
10. Ren X (2019) Research on the multi-mechanical-port multi-electrical-port machine (in Chinese). Huazhong University of Science and Technology, Wuhan
11. Gao YT, Li DW, Qu RH et al (2018) IEEE Trans Energy Convers 33(4):2153
12. Zhang YZ, Li DW, Yan P et al (2022) IEEE J Emerg Sel Top Power Electron 10(2):1756
13. Li DW, Qu RH, Lipo T (2014) IEEE Trans Ind Appl 50(6):3664
14. Li DW, Qu RH, Xu W et al (2015) IEEE Trans Ind Appl 51(4):2972
15. Zhao Y, Li DW, Lin MX et al (2022) IEEE J Emerg Sel Top Power Electron 10(3):2859
16. Cheng Y (2021) Research on the critical technique of the large-scale double-stator superconducting flux modulation machine for wind generation (in Chinese). Huazhong University of Science and Technology, Wuhan
17. Zou TJ, Li DW, Qu RH et al (2017) IEEE Trans Ind Appl 55(3):1920
18. Xie K (2020) The research based on the electromagnetic integration and permanent magnet machine's theory and topology (in Chinese). Huazhong University of Science and Technology, Wuhan
19. Zou TJ, Li DW, Qu RH et al (2017) IEEE Trans Ind Appl 53(6):5295
20. Zhu ZQ, Chen JT, Pang Y et al (2008) IEEE Trans Magn 44(11):4313
21. Min W, Chen JT, Zhu ZQ et al (2011) IEEE Trans Magn 47(8):2134
22. Chen CR, Ren X, Li DW et al (2021) IEEE Trans Transp Electrif 7(4):2623
23. Gao YT, Qu RH, Li DW et al (2017) IEEE Trans Magn 53(11):1
24. Fang L, Li DW, Ren X et al (2022) IEEE Trans Ind Electron 69(6):6058
25. Zou TJ, Li DW, Chen CR et al (2018) IEEE Trans Magn 54(11):1
26. Xie KF, Li DW, Qu RH et al (2017) IEEE Trans Magn 53(6):1

Chapter 6
Summary and Future Outlook of Flux Modulation Machines

The preceding chapters provide a comprehensive introduction to the fundamental theory, performance characteristics, and advanced configurations of FMMs. Over the years, FMMs have emerged as a distinct research area in the domain of electrical machines, encompassing a range of novel machines that differ significantly in performance from traditional machines. This chapter will revisit the historical development and overall research of FMMs, setting the stage for a discussion on future research priorities in this field.

6.1 Historical Development

The historical evolution of FMMs can be divided into three phases. Initially, the phenomenon of flux modulation was observed in certain electrical machines. Subsequently, academic research began to explore electrical machines that operate on the principles of flux modulation. Lastly, this study introduced the topological unification theory, which established the research field of FMMs.

(1) **Flux modulation phenomena in electrical machines**

As noted in Chap. 1, historically, the term "modulation" was not associated with electrical machines. In power electronics, "modulation" pertains to the process of imposing an original signal onto a carrier according to specific rules, ensuring the modulated signal retains the information of the original signal. However, researchers discovered similar phenomena when analyzing electrical machines. For instance, Heller and Hamata investigated the slotting effect on the air gap magnetic field of an induction machine. They observed that the air gap's specific permeance harmonics introduced additional harmonics to the MMF [1]. From a physical perspective, this process can be regarded as a form of "modulation", where the MMF acts as the original signal and the specific permeance as the carrier wave. These components

R. Qu et al., *Flux Modulation Machine*, https://doi.org/10.1007/978-981-95-4746-3_6

interact through multiplication, ultimately generating the air gap flux density signal. Additionally, Professor Tang in China observed that in a 48-slots 8-pole permanent magnet machine, the 11th harmonic in the no-load back EMF was significantly higher than theoretical predictions. This discrepancy was attributed to the modulation of the 11th MMF harmonic of permanent magnets (i.e., 44-pole-pair) by the fundamental specific permeance (i.e., 48-pole-pair), resulting in a 4-pole-pair air gap magnetic field. This 4-pole-pair air gap magnetic field significantly amplified the 11th back-EMF harmonic in the 4-pole-pair armature winding [2]. These observations have led to the intriguing possibility that electrical machines might operate on the principle of "modulation".

(2) **Early academic research on FMMs**

Prior to the twenty-first century, academic research into flux modulation was still in its nascent stages, with certain electrical machines already exhibiting principles of "modulation". For instance, the vernier reluctance machine, originally classified as a type of synchronous reluctance machine, exhibited interaction between the stator and rotor reluctances, producing a new reluctance component that aligned with the stator pole pair number [3]. Similarly, the flux switching machine, also known as a switched flux machine, derived its name from the role of rotor reluctance in modulating the direction of the main flux [4]. Prof. Lipo discovered that in permanent magnet machines, the modulation effect of the air gap specific permeance on the MMF led to a marked increase in torque [5]. These studies represented the embryonic stages of "flux modulation". However, each machine maintained distinct "original signals" and "carrier waves", indicating that a unified principle had not yet been established.

(3) **The formation of flux modulation theory and FMMs**

By 2014, scholars had determined that the operational principle common to all electrical machines based on "modulation" involves the excitation MMF (original signal) being modulated by the specific permeance (carrier wave) produced by the non-uniform air gap, ultimately generating the air gap flux density (modulated signal) [6–8]. Consequently, these machines are collectively classified as FMMs, thereby standardizing the flux modulation principle. Structurally, for the flux modulation process to occur, the presence of an excitation MMF and a non-uniform specific permeance within the machine is essential, alongside an armature winding that can interlink with the modulated air gap flux density. Therefore, FMM must incorporate three functional units: excitation unit, modulation unit, and armature unit. This "three-unit" structure has now been adopted as the basic model for FMMs. Additionally, FMMs exhibit similar performance characteristics. Firstly, the non-uniform specific permeance modulates the excitation MMF, producing no-load air gap flux density, resulting in rich air gap magnetic field harmonics and significant iron loss. Second, the torque formula for FMMs includes a pole ratio coefficient that enhances torque density. Lastly, a strong armature reaction predisposes FMMs to rapid saturation and a low power factor. The consolidation of flux modulation theory, FMM structures, and performance characteristics underscores the prominence of FMMs as a distinct research area within the field of electrical machines.

(4) **Development and promotion of flux modulation theory for electrical machinery**

The development of flux modulation theory and the FMM family has profoundly impacted traditional electrical machine theory, particularly in terms of structure and torque capability. Traditionally, electrical machine theory mandates that the pole pair number of the excitation unit must align with that of the armature unit, and their rotational speeds must also match to facilitate electromechanical energy conversion. However, FMMs challenge these traditional constraints. In FMMs, the pole pair number of the excitation unit often significantly exceeds that of the armature unit, and the rotational speed of the excitation unit is considerably lower than that of the armature field.

From another perspective, if the excitation and modulation units are regarded as a single entity, the pole pair number and the speed of the modulated magnetic field correspond to those of the armature field, thereby aligning with traditional electrical machine theory. Thus, flux modulation theory reconciles with and extends traditional principles by introducing the concept of indirect excitation. Here, the MMF generated by the excitation unit undergoes transformation—specifically, modulation—and the resultant modulated magnetic field serves as the actual excitation field. Consequently, flux modulation theory opens new avenues in the design and topology of electrical machines, offering enhanced design flexibility and innovative structural configurations.

In terms of torque capability, traditional electrical machine theory posits that electromagnetic torque is fundamentally constrained by electric and magnetic loadings. This notion is deeply rooted in electromagnetic principles, where torque is essentially perceived as the force exerted on an energized conductor within a magnetic field. Consequently, the magnitude of electromagnetic torque is inherently limited by the current (i.e., electric loading) and the magnetic field strength (i.e., magnetic loading). These constraints have traditionally been regarded as unalterable.

FMMs challenge this conventional understanding. The established "iron law" typically pertains to the electromagnetic torque of the armature unit. Most electrical machines traditionally comprise only an excitation unit and an armature unit. According to Newton's third law, the torques on the armature and excitation units are equal. However, FMMs incorporate an additional modulation unit, complicating this dynamic. Regardless of whether the excitation or modulation unit is considered as the output, the resultant output torque equals the combined torques of the other two units, potentially allowing the rotor torque to substantially exceed that of the armature unit.

However, this does not imply that rotor torque is unlimited and can increase indefinitely. The torque formula for FMMs indicates that it is still restricted by factors such as the amplification factor (i.e., pole ratio) as well as electric and magnetic loadings, which ultimately govern the maximum achievable torque.

6.2 Research Status Summary

The emergence of flux modulation theory marks a new phase in research for FMMs, establishing it as a distinguished field and guiding recent scholarly inquiry. As previous chapters have thoroughly introduced related FMM research, this section will provide only a concise summary.

(1) **Theoretical study**

Following the unification of theory and structure in FMMs, researchers have explored numerous possibilities for each unit of the basic model, aiming to develop additional FMM topologies. These possibilities are summarized in Table 3.1. It has been observed that many electrical machines previously deemed "special" now fall within the FMM category. Building on this, several scholars have applied flux modulation theory to re-analyze machines such as the flux reversal machine [9] and electrically excited vernier reluctance machine [10], deriving analytical expressions for their performance and elucidating output characteristics such as cogging torque and average torque.

(2) **Topology research**

Given FMMs' capability for torque amplification, investigating how to further leverage this advantage to develop topologies with higher torque density has emerged as a key research focus. Predominantly, there are two approaches based on the three-unit basic structure. One approach involves enhancing the capacity of one or more units, leading to three enhanced topology types: excitation-enhanced, modulation-enhanced, and armature-enhanced. The other approach entails incorporating multiple three-unit structures within a single FMM, resulting in a variety of integrated FMM topologies. These topologies have been extensively discussed in Chap. 5 and will not be reiterated here. All these high torque density topologies represent evolutionary developments under the guidance of flux modulation theory.

(3) **Defect compensation**

FMMs are known to exhibit inherent shortcomings, such as rich magnetic field harmonics, high iron loss, strong armature reaction, susceptibility to saturation, low power factor, small armature pole pair number, and lengthy end windings. To address these issues, researchers have focused on three main areas: advanced topology, optimal design, and control strategies. In the realm of advanced topology, innovations such as the dual-stator spoke-array vernier machine [11] and low-pole-ratio structures [12] have been proposed. The former, an excitation-enhanced topology, not only increases the power factor but also enhances the output torque. The latter, while offering a slight reduction in torque capability, boasts several benefits including a high power factor, robust overload capacity, and shortened end winding length. In the area of optimal design, an optimization method has been developed to simultaneously maximize torque density and power factor [13]. Regarding control strategies, significant research has been conducted on optimizing the distribution of d-axis and q-axis currents to improve the power factor and mitigate saturation [14], as well as on accommodating the nonlinear variations of inductances due to saturation and cross-saturation effects [15].

6.3 Future Research Directions

Since the invention of the first FMM in the 1960s, the field has demonstrated remarkable resilience and innovation over more than five decades. The introduction of flux modulation theory has spurred a wealth of research, leading to the development of cutting-edge topologies, new principles, and methodologies. However, it is crucial for continued progress in any research field to challenge and transcend established paradigms. The goal of such endeavors is not destruction but rather to build upon existing knowledge and advance further. Herein lies the potential for transcending current research in the field of FMMs.

(1) **Further development of modulation theory**

Originally, the term "modulation" was associated with signal processing. In this book, we observe that the process of generating a magnetic field mirrors that of signal generation, prompting us to apply the term "modulation" to magnetic fields and thereby broadening its application. This expansion prompts further reflection on whether similar phenomena might exist in other fields beyond magnetism. For example, "EMF" in the electric field corresponds to "MMF" in the magnetic field. In electric fields, when a voltage signal is applied across a non-uniform air gap, the process of generating electric field intensity resembles that of producing magnetic flux density, suggesting the existence of electric field modulation. Moreover, in the analysis of machine vibration, when radial force is applied to a non-uniform air gap, the resulting deformation of the iron core does not directly correspond to the waveform of the radial force, indicating modulation within the force field as well.

This observation raises the question of whether modulation could be a universal phenomenon across other physical fields, such as thermal and fluid dynamics. The broader application of modulation theory could lead to significant breakthroughs in physical research, potentially extending far beyond the scope of flux modulation theory and FMMs.

(2) **Further development of FMM topologies**

Chapter 5 presents a variety of high-performance FMM topologies developed based on the three-unit basic model. These advanced FMMs are categorized by the number and movement of MMF and specific permeance, as illustrated in Table 6.1. This categorization encompasses not only traditional electrical machines but also introduces new combinations of machine designs. Currently, specific machine structures corresponding to these novel combinations remain unidentified. However, Table 6.1 serves as a catalyst for scholars to explore these new directions and address these gaps.

Table 6.1 Classification of FMMs by number and motion of MMF and specific permeance

MMF and permeance classification	One-stationary MMF	One-rotary MMF	Multi-stationary MMF	Multi-rotary MMF	Stationary and rotary MMF
One-stationary permeance	DC Machine	Induction machine; Synchronous machine	Unknown	Stator-modulation magnetic geared machine	Unknown
One-rotary permeance	Evenly distributed flux reversal permanent magnet machine; Unipolar machine	Magnetic geared machine	Stator multi-working-MMF FMM	Magnetic geared continuously variable transmission; BLDDM	Magnetic geared machine; Stator and rotor permanent magnet BLDDM
Multi-stationary permeance	Unknown	Vernier permanent magnet machine	Unknown	Magnetic geared vernier machine	Unknown
Multi-rotary permeance	Rotor multi-working-permeance flux reversal machine	Rotor multi-working-permeance BLDDM	Unknown	Bilateral modulation permanent magnet brushless dual-rotor machine	Unknown
Stationary and rotary permeance	Unknown	Stator and rotor multi-working-permeance BLDDM	Unknown	Unknown	Bilateral modulation permanent magnet machine; Bilateral modulation magnetic geared machine

Tables 3.1 and 6.1, along with Fig. 5.1, are all based on the three-unit basic model of FMMs. This model has effectively addressed issues of topological induction and prediction. However, the question remains: why should the development of topologies continue to be a focus for future FMM research? The answer lies in the nature of the three-unit model itself, which is a fundamental structure derived and abstracted from existing topologies. Although this model has facilitated the creation of new topologies, these are essentially permutations and combinations of known unit structures. For instance, to enhance excitation capabilities, a spoke-magnetization permanent magnet structure is incorporated within the excitation unit. This choice is based on the known ability of such structures to enhance excitation. Consequently, the ultimate capability of the three-unit model is to enumerate all possible combinations of existing unit structures.

This realization indicates that the three-unit model may not be conducive to the discovery of new and varied FMM topologies, as its guidance becomes less effective with the need to explore beyond known configurations. Therefore, it is imperative to critically examine the current three-unit model, considering how the specific structure of each unit evolved and identifying the constraints each unit imposes. By challenging and ultimately removing these constraints, we can develop a new theory of topology construction and invent more advanced FMM topologies. The creation of these new topologies does not depend on the combination of existing frameworks. It represents a novel theoretical advancement and aligns more closely with the fundamental principles of flux modulation.

(3) **Power factor improvement**

FMMs include a pole ratio amplification coefficient in their torque expression, endowing them with high torque density capabilities. However, FMMs are also plagued by several defects, such as high iron loss, large rotor pole pair numbers, and a low power factor. The issues of high iron loss and large rotor pole pair numbers can be alleviated by restricting the use of FMMs to low and medium speeds. However, this method does not address the low power factor issue. According to the analysis in Sect. 4.5, a low power factor not only necessitates an increase in inverter capacity, system weight, volume, and cost but also results in low output torque under heavy load conditions, thereby undermining the theoretical advantage of high torque density. Furthermore, in power generation, a low power factor contributes to increased grid reactive power. In summary, the low power factor is a significant problem that limits the broader application of FMMs and thus represents the most urgent research area in this field.

Currently, there are two technical approaches to improving the power factor of FMMs. The first involves enhancing the control systems of FMMs through methods such as capacitor reactive power compensation or drive circuit optimization, which can reduce the voltage on power switches. However, these measures only improve the power factor at the electrical terminal and do not address the issue of low output torque under heavy load conditions. The second approach is to modify the structure of FMMs. By optimizing the design, such as choosing a lower pole ratio and increasing the width of the flux path in the iron cores, the power factor can be improved and saturation effects under heavy loads can be reduced. However, these methods represent compromises, improving the power factor at the expense of torque density, and therefore, they do not provide definitive solutions to the low power factor problem.

From the perspective of the armature magnetic field, the FMM's torque amplification effect arises from its low armature pole pair and high MMF amplitude, which facilitate the generation of a substantial modulated magnetic field following the flux modulation effect. However, this also results in a significant non-modulated magnetic field, which is primarily responsible for the low power factor. Therefore, high torque density and low power factor are inherently interconnected. To address the low power factor issue without compromising torque density, the development of a new principle is necessary.

6.4 Application Prospects

As a relatively new type of electrical machines, FMMs are predominantly in the theoretical research phase, with practical applications still under development. This section will explore the future application prospects of FMMs based on their performance characteristics. Analysis from Chaps. 3 and 4 indicates that FMMs can be categorized into three distinct groups.

First category: This group includes machines that possess characteristics similar to traditional electrical machines but with the added benefit of higher torque density. Examples include the vernier permanent magnet machine, transverse-flux permanent magnet machine, and vernier reluctance machine. Second category: Machines in this category may not offer a significant advantage in torque density, but they feature simple and robust rotor structures. This group includes stator excitation FMMs. Third category: These machines are characterized by unique features and functions, such as BLDDM and brushless doubly-fed reluctance machine. Their potential applications and performance are separately analyzed below.

(1) **Vernier permanent magnet machine and transverse-flux permanent magnet machine**

Vernier machine and transverse-flux machines are types of permanent magnet FMMs characterized by high torque density and large rotor pole pair numbers, making them well-suited for low- and medium-speed applications. Moreover, due to strong armature reactions, both the vernier and transverse-flux machines exhibit limited overload capacity, rendering them more suitable for environments with constrained cooling conditions. Here are some examples of application areas.

First, as the electrification of transportation deepens, the demand for electric drive machines in vehicles is increasing. At present, electric vehicles represent one of the most dynamic areas in this field. However, the vernier machine is not ideal for the main drive systems of electric vehicles due to their requirements for high rotational speeds and heavy loads, typically managed with water or oil cooling. These conditions negate the performance benefits of vernier machines, making them unsuitable as primary drive machines. In contrast, in-wheel drive systems, which operate at lower rotational speeds because they directly drive the wheels, are well-suited for vernier machines due to their high torque density needs. For electric trucks and armored vehicles, in-wheel systems are often water- or oil-cooled. In such instances, it is advantageous for the vernier machine to have a pole ratio lower than 5 to enhance torque performance under heavy load conditions. Conversely, electric bicycles, which often use naturally cooled in-wheel drive systems due to cost constraints and simpler designs, can accommodate vernier machines with pole ratios higher than 5.

The transverse-flux machine, while also boasting high torque density, has a complex mechanical structure that generally precludes its use in in-wheel applications. However, it offers significant advantages in terms of weight and volume, making it particularly advantageous for large ship propulsion machines.

Servo machines are crucial components in machine tools and other processing equipment. The torque density of a servo machine influences the system response speed, while cogging torque and torque ripple affect processing accuracy. Therefore, servo machines should have high torque density, along with low cogging torque and torque ripple. Vernier machines meet these criteria well, making them highly suitable for low- and medium-speed servo machine applications. Moreover, many servo machines utilize natural cooling, circumventing the issue of reduced output torque under heavy load conditions that vernier machines often face. Beyond processing equipment, robot joints and actuators also benefit from the attributes of servo machines, presenting additional potential applications for vernier machines.

In applications that require low speed and high torque, such as wind power generators and mining machinery, the vernier machine is advantageous for significantly reducing the weight and volume of the drive system while maintaining the same output torque. This reduction can lead to cost savings in processing, assembly, and transportation of these machines. However, it is important to note that the electric loading of larger machines is often high, which may diminish the power factor and torque density of vernier machines. Thus, maintaining the power factor and torque density advantage in large vernier machines warrants further investigation.

(2) **Vernier reluctance machine**

Compared to the synchronous reluctance machine, the vernier reluctance machine offers higher torque density and does not utilize permanent magnets, making it a cost-effective alternative for applications with stringent budget constraints. However, the vernier reluctance machine typically has a lower power factor, requiring a larger inverter capacity. Consequently, it is generally preferred for smaller power applications. Before the twenty-first century, vernier reluctance machines were employed in compact devices such as cash counting machines and fax machines.

(3) **Stator electrically excited FMM**

Based on flux modulation theory, the excitation unit in synchronous machines can be moved to the stator, resulting in the development of the stator electrically excited FMM. This configuration eliminates issues associated with rotating excitation, marking a significant technical advancement. However, the torque density of stator electrically excited FMMs is lower than that of traditional synchronous machines. Therefore, the feasibility of stator electrically excited FMMs as replacements for synchronous machines depends on whether the power density of the former exceeds that of the whole system comprising the latter and its excitation units. When dealing with very high machine power, the weight and volume of the exciters are considerably less significant compared to the machine itself. In such scenarios, the power density of the machine and exciter system is predominantly influenced by the machine. Given that stator electrically excited FMMs have a lower torque density than synchronous machines, they are not advantageous for applications requiring high power. However, in the aircraft starter-generator system, where power demands are moderate, exciters typically constitute a substantial portion of the total weight and volume of the system—up to 50%. Therefore, stator electrically excited FMMs exhibit promising application potential in this area, offering a more efficient and compact solution.

(4) **Stator permanent magnet FMM**

While the torque density of stator permanent magnet FMMs does not substantially exceed that of conventional permanent magnet machines, their design, which places both windings and permanent magnets within the stator, simplifies cooling and reduces the risk of irreversible demagnetization. Additionally, the rotor structure is straightforward and robust, making it suitable for both low-speed and high-speed operations. However, it is important to note that at high speeds, the rotor pole pair number should not be excessively high to avoid increased frequency, iron loss, and AC loss. Furthermore, transforming a stator permanent magnet FMM into a straight configuration allows for the realization of primary permanent magnet linear FMMs. These are particularly advantageous for long-stroke linear applications, such as railway transportation, where traditional permanent magnet linear machines require extensive and costly secondary magnet structures. In contrast, primary permanent magnet linear FMMs utilize magnets in the shorter primary, significantly reducing magnet usage and overall cost in long-stroke applications.

(5) **Functional FMM**

FM-BLDDM and the doubly-fed reluctance machine differ from traditional FMMs in terms of external characteristics and functionalities, leading to specific application areas. The FM BLDDM, serving as a type of continuously variable transmission device, is suitable for hybrid power systems in vehicles and vessels. Meanwhile, the doubly-fed reluctance machine is suitable for variable-speed constant-frequency applications, such as wind power generation. However, both machines exhibit inherent drawbacks. In hybrid power systems, the mismatch between the high internal combustion engine speed at the input mechanical port and the low wheel speed at the output mechanical port necessitates a mechanical gearbox for FM-BLDDM. This requirement can increase the overall weight and volume of the system. In wind power generations, the doubly-fed reluctance machine shares the common limitation of low power factor seen in FMMs, resulting in substantial reactive power consumption and increased transmission losses in the power grid. These issues should be carefully considered in the deployment of these machines.

References

1. Heller B, Hamata V (1980) Harmonic field effects in induction machines (trans: Zhang MT) (in Chinese). Machinery Industry Press
2. Tang YQ (2014) Electrical machinery (in Chinese). Machinery Industry Press, Beijing
3. Li HM, Li QF (1982) Electromagnetic reduction machine (in Chinese). Machinery Industry Press, Beijing
4. Rauch S, Johnson L (1955) Trans Am Inst Electr Eng Part III: Power Appar Syst 74(3):1261
5. Toba A, Lipo T (1999) Novel dual-excitation permanent magnet vernier machine. In: IEEE Thirty-fourth IAS annual meeting, Phoenix. IEEE, pp 2539–2544
6. Zhu ZQ, Evans D (2014) Overview of recent advances in innovative electrical machines—with particular reference to magnetically geared switched flux machines. In: 2014 17th International conference on electrical machines and systems (ICEMS), Hangzhou, pp 1–10
7. Li DW, Qu RH, Li J (2015) Topologies and analysis of flux-modulation machines. In: IEEE Energy conversion congress and exposition (ECCE), Montreal. IEEE, pp 2153–2160
8. Cheng M, Han P, Hua W (2017) IEEE Trans Ind Electron 64(8):6063
9. Gao YT, Qu RH, Li DW et al (2017) IEEE Trans Ind Appl 53(3):2110
10. Jia SF, Qu RH, Li J et al (2016) IEEE Trans Energy Convers 31(3):935
11. Li DW, Qu RH, Lipo T (2014) IEEE Trans Ind Appl 50(6):3664
12. Li DW, Zou TJ, Qu RH et al (2018) IEEE Trans Ind Appl 54(2):1320
13. Wu DY, Xiang ZX, Zhu XY et al (2021) IEEE Trans Ind Electron 68(10):9265
14. Yu ZX, Kong WB, Li DW et al (2019) IEEE Trans Ind Appl 55(5):4643
15. Zou TJ, Han X, Jiang D et al (2018) IEEE Trans Ind Appl 54(3):2175

MIX
Papier aus verantwortungsvollen Quellen
Paper from responsible sources
FSC® C105338

If you have any concerns about our products,
you can contact us on
ProductSafety@springernature.com

In case Publisher is established outside the EU,
the EU authorized representative is:
Springer Nature Customer Service Center GmbH
Europaplatz 3, 69115 Heidelberg, Germany

Printed by Libri Plureos GmbH
in Hamburg, Germany